RECENT DEVELOPMENTS IN
CONDENSED MATTER PHYSICS

Volume 2 • Metals, Disordered Systems,
Surfaces, and Interfaces

RECENT DEVELOPMENTS IN
CONDENSED MATTER PHYSICS

RECENT DEVELOPMENTS IN CONDENSED MATTER PHYSICS

Volume 2 • Metals, Disordered Systems, Surfaces, and Interfaces

Edited by
J. T. DEVREESE
Chairman of the Conference
University of Antwerpen (RUCA and UIA)

L. F. LEMMENS
University of Antwerpen (RUCA)

V. E. VAN DOREN
University of Antwerpen (RUCA)

and

J. VAN ROYEN
University of Antwerpen (UIA)

PLENUM PRESS • NEW YORK AND LONDON

Library of Congress Cataloging in Publication Data

Main entry under title:

Recent developments in condensed matter physics.

"Invited papers presented at the first general conference of the Condensed Matter Division of the European Physical Society, held April 9-11, 1980, at the University of Antwerp (RUCA and UIA), Antwerp, Belgium."
Includes indexes.
Contents: v. 1. Invited papers—v. 2. Metals, disordered systems, surfaces, and interfaces— v. 3. Impurities, excitons, polarons, and polaritons—[etc.]
1. Condensed matter—Congresses. I. Devreese, J. T. (Jozef T.) II. European physical Society. Condensed Matter Division.

| QC176.AIR42 | 530.4'1 | 80-280-67 |
| ISBN 0-306-40647-0 | | AACR2 |

Contributed papers presented at the first General Conference of the Condensed Matter Division of the European Physical Society, held April 9-11, 1980, at the University of Antwerp (RUCA and UIA), Antwerp, Belgium

© 1981 Plenum Press, New York
A Division of Plenum Publishing Corporation
233 Spring Street, New York, N. Y. 10013

CONFERENCE CHAIRMAN

J. T. DEVREESE, RUCA & UIA, Antwerpen

LOCAL COMMITTEE

L. F. LEMMENS, RUCA, Antwerpen
V. E. VAN DOREN, RUCA, Antwerpen
J. VAN ROYEN, UIA, Antwerpen

INTERNATIONAL ADVISORY COMMITTEE

M. BALKANSKI, Paris, France
A. ABRIKOSOV, Moscow, USSR
V. M. AGRANOVITCH, Moscow, USSR
P. AVERBUCH, Grenoble, France
G. BENEDEK, Milan, Italy
H. B. CASIMIR, Heeze, The Netherlands
B. R. COLES, London, UK
S. R. DE GROOT, Amsterdam, The Netherlands
A. J. FORTY, Warwick, UK
D. FRÖHLICH, Dortmund, FRG
A. FROVA, Rome, Italy
H. GRIMMEISS, Lund, Sweden
H. HAKEN, Stuttgart, FRG
C. HILSUM, Malvern, UK

W. J. HUISKAMP, Leiden, The Netherlands
G. M. KALVIUS, Munich, FRG
A. R. MACKINTOSH, Copenhagen, Denmark
N. H. MARCH, Oxford, UK
E. MOOSER, Lausanne, Switzerland
N. F. MOTT, Cambridge, UK
K. A. MÜLLER, Zürich, Switzerland
S. NIKITINE, Strasbourg, France
C. J. TODD, Ipswich, UK
M. VOOS, Paris, France
E. P. WOHLFARTH, London, UK
E. WOLF, Stuttgart, FRG
P. WYDER, Nijmegen, The Netherlands
H. R. ZELLER, Baden, Switzerland

INTERNATIONAL PROGRAM COMMITTEE

N. N. BOGOLUBOV, Moscow, USSR
J. BOK, Paris, France
M. CARDONA, Stuttgart, FRG
E. COURTENS, Rüschlikon, Switzerland
S. F. EDWARDS, Cambridge, UK
R. ELLIOTT, Oxford, UK
J. FRIEDEL, Orsay, France
H. FRÖHLICH, Liverpool, UK
F. GARCIA MOLINER, Madrid, Spain
G. HARBEKE, Zürich, Switzerland
H. R. KIRCHMAYR, Vienna, Austria
S. LUNDQVIST, Göteborg, Sweden
S. METHFESSEL, Bochum, FRG

E. MITCHELL, Oxford, UK
F. MUELLER, Nijmegen, The Netherlands
R. PEIERLS, Oxford, UK
H. J. QUEISSER, Stuttgart, FRG
D. SETTE, Rome, Italy
H. THOMAS, Basel, Switzerland
M. TOSI, Trieste, Italy
F. VAN DER MAESEN, Eindhoven, The Netherlands
J. ZAK, Haifa, Israel
A. ZAWADOWSKI, Budapest, Hungary
W. ZAWADZKI, Warsaw, Poland
W. ZINN, Jülich, FRG
A. ZYLBERSZTEJN, Orsay, France

NATIONAL ADVISORY COMMITTEE

S. AMELINCKX, SCK/CEN, Mol
F. CARDON, RUG, Gent
R. EVRARD, ULg, Liège
R. GEVERS, RUCA, Antwerpen
J. P. ISSI, UCL, Louvain-la-Neuve
L. LAUDE, UEM, Mons
A. LUCAS, FUN, Namur

K. H. MICHEL, UIA, Antwerpen
J. NIHOUL, SCK/CEN, Mol
J. PIRENNE, ULg, Liège
D. SCHOEMAKER, UIA, Antwerpen
L. STALS, LUC, Diepenbeek
R. VAN GEEN, VUB, Brussels
L. VAN GERVEN, KUL, Leuven

INTRODUCTION

These volumes contain the invited and contributed talks of the first general Conference of the Condensed Matter Division of the European Physical Society, which took place at the campus of the University of Antwerpen (Universitaire Instelling Antwerpen) from April 9 till 11, 1980.

The invited talks give a broad perspective of the current state in Europe of research in condensed matter physics. New developments and advances in experiments as well as theory are reported for 28 topics. Some of these developments, such as the recent stabilization of mono-atomic hydrogen, with the challenging prospect of Bose condensation, can be considered as major break-throughs in condensed matter physics.

Of the 65 invited lecturers, 54 have submitted a manuscript. The remaining talks are published as abstracts.

The contents of this first volume consists of 9 plenary papers. Among the topics treated in these papers are:
- electronic structure computations of iron
- the density functional theory
- hydrogen in amorphous Si
- topologically disordered materials
- nuclear antiferromagnetism
- stabilization of mono-atomic hydrogen gas
- covalent and metallic glasses
- nonlinear excitations in ferroelectrics.

The other 56 papers of the first volume are divided among 17 symposia and 12 sessions. The different topics treated are:
- localization and disorder
- metals and alloys
- fluids
- excitons and electron-hole droplets
- surface physics
- dielectric properties of metals
- experimental techniques

  - electronic properties of semiconductors
  - low-dimensional systems
  - defects and impurities
  - spin waves and magnetism
  - phase changes
  - superionic conductors
  - dielectric properties
  - polarons
  - molecular crystals
  - superconductivity
  - spin glasses
  - photo-emission
  - polaritons and electron-phonon interaction

A review of these invited talks is given in the Closing Address.

Volumes 2, 3 and 4 contain the contributed papers of the participants to this Conference. All the 170 contributions deal with a wide range of topics including the same subjects reviewed in the invited papers.

The conference itself was organized in collaboration and with the financial support of the University of Antwerpen (Rijksuniversitair Centrum Antwerpen and Universitaire Instelling Antwerpen), the Belgian National Science Foundation and Control Data Corporation, Belgium. Co-sponsors were: Agfa-Gevaert N.V., Bell Telephone Mfg. Co., Coherent, Esso Belgium, I.B.M., Interlaboratoire N.V., I. Komkommer, Labofina, Metallurgie Hoboken-Overpelt, Spectra Physics.

Previous meetings of the Condensed Matter Division of the European Physical Society emphasized topical conferences treating "Metals and Phase Transitions" (Firenze, 1971), "Dielectrics and Phonons" (Budapest, 1974), "Molecular Solids and Electronic Transport" (Leeds, 1977). The promising start of 1971, however, did not evolve into a continuous success, and only 80 participants attended the 1977 meeting.

When the Board of the Condensed Matter Division invited me to act as Chairman for this Conference, several reasons could have been invoked for feeling reluctant about accepting this invitation. It was nevertheless decided to undertake this task, because we were convinced that there exists a genuine need in the European Physical Community for an international forum similar to the successful "March Meeting" of the American Physical Society. This March Meeting is not organized as a meeting for the physical societies of the different states but for all physicists and laboratories individually.

At the early stages of the organization of this Conference, about 800 solid state physicists were members of the Condensed Matter Division. However, as a result of an analysis which was made in Antwerpen, it soon became clear that in Europe not less than 8000 solid state physicists are active in several fields of research.

Therefore as a first step announcements were sent to all these physicists individually. The large response to these announcements, as reflected in the presence of about 600 solid state physicists at this Conference, proves that the need for such a forum indeed is present in the European physics community and that the time is ripe for the organization of an annual conference.

Subsequently, the International Advisory and Program Committees of this Conference were formed with great care. A large number of distinguished physicists were invited to serve as members of these committees. They represented not only the different fields and sections of condensed matter physics but also the different member states of the European Physical Society.

It was also the task of both the International Advisory and Program Committees to guide in establishing the conference program. As a result of a first consultation a large number of topics and speakers were suggested to be incorporated in the program. In order to make a fair and balanced choice, a ballot with all the suggestions was sent out to the Committee members. The result led to a first selection of speakers and topics. This number was gradually supplemented with names and topics resulting from private consultations to a total of 28 topics and 65 invited speakers. The final program consisted of 9 plenary sessions, 11 symposia each with 3 or 4 invited speakers and a maximum of 8 parallel sessions, in which 320 contributed papers were presented. Among the invited and contributed papers, there were several contributions from the U.S.A. Also scientists from Japan and China gave invited and contributed talks.

I should like to thank the invited speakers for their collaboration in preparing the manuscripts of their talks. This volume together with the volumes of the proceedings of the contributed papers will give the scientific community a review of the state of affairs of condensed matter physics in Europe together with several new developments from the U.S.A., Japan and China. It gives a report on the most recent advances in theory as well as experiment. Publication of these proceedings will serve as a guide in the years to come not only to the participants of this conference but to the scientific community at large.

It gives me great pleasure to thank the President of the Universitaire Instelling Antwerpen, Dr. jur. P. Van Remoortere and the Rector, Prof. Dr. R. Clara, for their continuous interest and

support for this Conference and for making available so promptly
and effectively the whole infrastructure of the university conference
center.

Finally, I wish to thank Miss R.M. Vandekerkhove, the administrative secretary of the Conference. My thanks are also due to Miss
H. Evans and Mr. M. De Moor for their administrative and technical
assistance in the organization, and to all those who helped to ensure a smooth operation of the Conference.

J.T. Devreese
Professor of Theoretical Physics
Chairman of the Conference

CONTENTS

1. METALS AND ALLOYS

2. DIELECTRIC PROPERTIES OF METALS

5.   GLASSES

6.   CHALCOGENIDES

8.   MOLECULAR CRYSTALS

μSR AND MUON KNIGHT SHIFT MEASUREMENTS ON $LaNi_5H_x$

F.N. Gygax, A. Hintermann, W. Ruegg, A. Schenck,
and W. Studer
Laboratorium für Hochenergiephysik, ETHZ, c/o SIN
CH-5234 Villigen

L. Schlapbach and F. Stucki
Laboratorium für Festkörperphysik, ETHZ
CH-8093 Zürich, Switzerland

Using the positive muon as a proton substitute, we have made spin
relaxation and Knight shift measurements for three series of
samples, which had undergone up to 40 hydrogenation cycles :
$LaNi_5H_6$, $LaNi_5$, and $LaNi_5$ annealed. This enables us to separate
the influence of hydrogen absorption on the muon Knight shift
(+80 ppm) from the effect of Ni precipitations formed at the
surface and from possible muon trapping at lattice defects.

$LaNi_5$ is a promising hydrogen storage material. It
absorbs[1] at room temperature below 2 bar hydrogen up to $LaNi_5H_6$.
Thermodynamic[2] and magnetic[3] properties, structure[4], H-diffusion
[5][6][7] and catalytic properties of the surface[8] have been
investigated, but so far only very little is known about the
electronic structure of $LaNi_5H_6$.

Hydrogenation expands the $LaNi_5$ lattice by 25 vol %
and ends in the disintegration of the material into a fine
powder (20 μm) with a large specific surface area and probably
a fairly high lattice defect concentration. $LaNi_5$ is an exchange
enhanced Pauli paramagnet. It decomposes at the surface into La
oxide/hydroxide and superparamagnetic Ni particles of 6000 Ni
atoms per particle[3]. The decomposition and thus the formation
of fresh Ni particles continues with each cycle of hydrogenation
and upon air exposure of the samples. The Ni particles dominate
the magnetic properties of several times hydrogenated and
dehydrogenated $LaNi_5$.

1

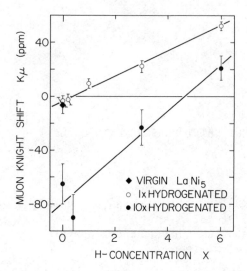

Fig. 1: Knight shift at a positive muon in LaNi₅Hₓ as a function
         of the hydrogen content x for samples submitted to
         various numbers of hydrogenation cycles.

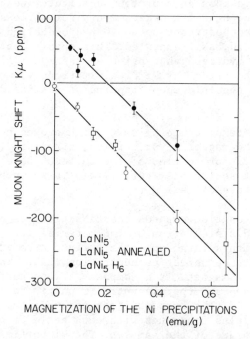

Fig. 2: Knight shift in LaNi₅H₆, LaNi₅ and LaNi₅ annealed after
         various hydrogenation cycles, plotted versus the
         magnetization of the Ni precipitations.

Using the positive muon μ^+ as a proton substitute we have made spin relaxation and Knight shift Kμ measurements. Our preliminary μSR experiments[9] on a several times cycled sample indicated a positive Kμ with increasing H concentration. However, we could not distinguish between the variation of Kμ due to H absorption, increasing amount of Ni precipitations and possible μ^+ trapping at lattice defects. We now have investigated three series of samples (LaNi$_5$H$_6$, LaNi$_5$ and LaNi$_5$ annealed) which had undergone different numbers of hydrogenation cycles (up to 40). The μSR results were completed by magnetization measurements.

The preparation and characterization of the LaNi$_5$ samples is described in [8]. About 100 g LaNi$_5$ was hydrogenated and dehydrogenated in a stainless steel reactor vessel using 99.999 % pure H$_2$. The reactor vessel was then opened in an Ar dry box and twice 15g of the sample was filled into quartz vessels of spherical shape. The remaining material was further cycled. The one of the 15g samples was sealed under vacuum. The other one was hydrogenated at 10 bar and 20°C, then cooled to liquid nitrogen temperature and sealed. The magnetization of the sealed samples was measured using a Faraday type balance.

The muon Knight shift measurements have been performed at the main superconducting muon channel of the Swiss Institute for Nuclear Research (SIN) in Villigen. A stroboscopic measurement method, taking advantage of the periodicity of the muon beam bursts and of the fully available intensity was applied. Apparatus and experimental procedure are described in ref.[10]. Depending upon the intensity produced by the accelerator, 250000 to 600000 muons/s could be stopped in the target, and 10000 to 24000 decay positrons/s (registrated event rate) could be observed in the positron telescope. A typical run took about three hours, yielding a Kμ value with a precision of the order of 3 to 6 ppm. From the observed line broadening, the relaxation rate of the μ^+ polarization in the target could be determined.

Fig. 1 shows the Knight shift Kμ at the μ^+ in LaNi$_5$H$_x$ as a function of hydrogen content. For the virgin LaNi$_5$ sample, Kμ is slightly negative. For the samples which had been hydrogenated once, the measured shift is zero or slightly negative for x = 0, and rises apparently linearly with x to a value of about 50 ppm for x = 6. For samples hydrogenated several times, Kμ is well negative for x = 0, and rises also, but even faster than for the probes only hydrogenated once, reaching a positive value for x = 6. The decrease of Kμ observed as a function of hydrogenation cycles can be explained by the shielding of the magnetic field through the Ni precipitations formed at the powder grain surface. At larger H concentration, i.e. higher H$_2$ pressure, the surface of the Ni precipitations becomes demagnetized due to H chemisorption[11]. This weakens the shielding and appears as a faster increase of Kμ.

In Fig. 2 are shown the Kμ values measured for $LaNi_5$ and $LaNi_5H_6$ after different numbers of hydrogenation cycles as a function of the magnetization of the Ni precipitations. For both types of samples Kμ decreases approximately in parallel, linearly with the Ni magnetization. Some of the $LaNi_5$ samples were annealed 4 days at 400°C in the sealed quartz vessel and then remeasured. Their Kμ fits the values of the not annealed samples indicating that $μ^+$ trapping on lattice defects is of minor importance.

The muon spin relaxation has also been determined for these samples. It increases linearly with the Ni magnetization starting around zero for the extrapolated value of zero Ni magnetization. For $LaNi_5H_6$ the increase is weaker whereas for annealed $LaNi_5$ it is stronger compared to the relaxation increase for unannealed $LaNi_5$. The general trend shows clearly a decrease of the field homogeneity due to increasing amount of Ni segregation.

We conclude that hydrogenation causes a positive Knight shift of about 80 ppm. In virgin $LaNi_5$, the muon is in a hydrogen α-phase environment. The measured Kμ, zero or slightly negative corresponds to the typical values observed for transition metals, where d-electrons can influence the impurity spin. Hydrogenation causes an increase of β-phase environment. In that case the electrons showing magnetic response are by large only the conduction electrons, as in simple metals.

REFERENCES

1. F. A. Kuijpers, Philips Res. Rep. Suppl. 2 (1973).

2. S. Tanaka and T. B. Flanagan, J. Less Common Metals 51, 79 (1977).

3. L. Schlapbach, F. Stucki, A. Seiler, H. C. Siegmann, to appear in J. of Magnetism and Magnetic Materials (1980).

4. P. Fischer, A. Furrer, G. Busch, and L. Schlapbach, Helv. Phys. Acta 50, 421 (1977).

5. E. Lebsanft, D. Richter, and J. M. Töpler, to appear in Z. physikal. Chemie (1979).

6. R. F. Karlicek, and I. J. Lowe, Solid State Commun. 31, 163 (1979).

7. R. C. Bowman, D. M. Gruen, and H. M. Mendelsohn, Solid State Comm. 32, 501 (1979).

8. H. C. Siegmann, L. Schlapbach, and C. R. Brundle, Phys. Rev. Lett. 40, 972 (1978).

9. M. Camani, F. N. Gygax, W. Ruegg, A. Schenck, H. Schilling, F. Stucki, and L. Schlapbach, Hyperfine Interactions <u>6</u>, 305 (1979).

10. M. Camani, F. N. Gygax, E. Klempt, W. Ruegg, A. Schenck, H. Schilling, R. Schulze, and H. Wolf, Phys. Lett. <u>77B</u>, 326 (1978).

11. P. W. Selwood, Chemisorption and Magnetization, Academic Press, N.Y. 1975.

TRAPPING OF POSITIVE MUONS IN DILUTE ALUMINIUM ALLOYS

O. Hartmann and L.O. Norlin
Institute of Physics, Box 530, 2-751 21 Uppsala, Sweden

and

K.W. Kehr, D. Richter, and J.M. Welter
Institut für Festkörperforschung, KFA, Postfach 1913,
D-517 Jülich, BRD

and

E. Karlsson and T.O. Niinikoski
EP Division, CERN, CH-1211 Genève 23, Switzerland

and

A. Yaouanc
CENG, 85 X 38041 Grenoble Cedex, France

The precession of positive muons in $\underline{Al}Mn_x$ samples with x ranging
from 5 to 1300 ppm was studied. All the samples show a maximum
in the μSR linewidth around 17 K, with a linewidth depending
on the Mn concentration. In samples of $\underline{Al}Mg_x$, $\underline{Al}Li_x$ and $\underline{Al}Ag_x$
with x = 40 - 120 ppm the maximum occurs at different temperatures
17 - 45 K. This suggests that the muons are trapped in a strain
field characteristic for each kind of impurity atom.

A number of measurements of muon diffusion in aluminium
have shown that positive muons are highly mobile in very pure Al
byt can become localized by impurities[1,2,3,4,5] or other defects[6,7].
The present report deals with an extension of our $\underline{Al}Mn_x$ studies to
several concentrations in the region x = 5 - 70 ppm, and
measurements on systems with other impurity atoms, namely Ag, Li
and Mg.

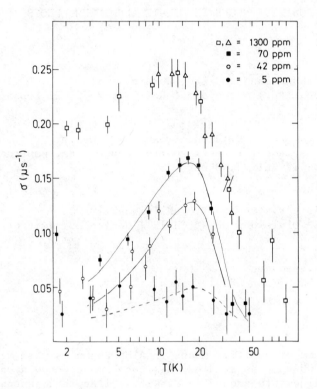

<u>Fig.1</u> : Gaussian linewidth (σ) for muon precession in aluminium
        doped with manganese as function of temperature and
        concentration.

        Full curves show fits to the two-state model.
        The broken curve for the 5 ppm sample is obtained from
        an extrapolation of the 42, 57 and 70 ppm Mn alloy results.

The measurements were performed at the 600 MeV synchrocyclotron at CERN, Geneva, using conventional μSR techniques and transverse field arrangement. The studies were concentrated to the temperature region 2-100 K. Data on two AlMn alloys measured at lower temperatures have been published elsewhere[5].

The samples were prepared at KFA, Jülich, from a starting material of 6N aluminium. The concentration of the alloys was determined by optical absorption spectroscopy.

The μSR spectra were fitted to the usual expression

$$N = N_o \exp(-t/\tau\mu)[1+a_o P(t)\cos(\omega t+\phi)] + Bg \qquad (1)$$

where the polarization decay can be written as

$$P(t) = P(0)\exp(-\sigma^2 t^2) \quad \text{or} \quad P(t) = P(0)\exp(-\lambda t).$$

Figures 1-2 give the linewidth parameter σ obtained with the $\exp(-\sigma^2 t^2)$ dependence (Gaussian damping) as function of temperature (about the same goodness of fit could be obtained with the $\exp(-\lambda t)$ damping). The data were corrected for the fraction of muons stopping in the cryostat walls etc. This fraction was determined by use of dummy samples[5].

As seen in Fig. 1 all the AlMn samples exhibit a pronounced maximum linewidth around 17 K, with the maximum width dependent on the concentration. The position of this peak is the same for the different Mn concentrations. Another sample with 57 ppm Mn (not included in the figure) gives linewidths intermediate between the 42 and 70 ppm samples.

Measurements on Al samples with other kinds of impurity atoms show linewidth maxima at different temperatures (Fig. 2). The AlMg measurement should be compared with that of Kossler et al[4] on a sample with much higher Mg concentration (1000 ppm). Also here the steep decrease of the linewidth starts around 40 K. Taken in conjunction with the AlMn data it appears that the muons are caught in a trap, and that the escape process from the traps takes place at a temperature characteristic for each kind of impurity atom.

We analyze the linewidth data quantitatively using a two-state model introduced earlier to describe the μ^+ diffusion in doped Nb samples[8]. In this model we consider the random walk through a crystal containing trapping regions as well as more or less undisturbed parts of the lattice. A muon diffusing in the undisturbed lattice (characterized by a polarization decay $P_1(t)$ is caught by a trap after an average time τ_1; a muon is a trap

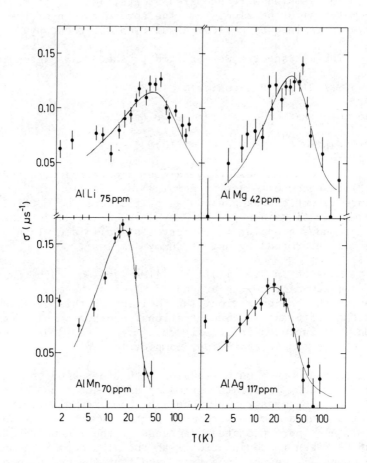

Fig.2 : Gaussian linewidth (σ) for muon precession in aluminium doped with lithium, magnesium, manganese and silver.

Full curves are fitted functions from the two-state model.

(polarization decay $P_o(t)$ escapes with t arate $1/\tau_o$. The total polarization decay $P(t) = P_o(t) + P_1(t)$ can be described approximately by the system of equations[9] :

$$\frac{\partial}{\partial t} P_1 = - \frac{1}{\tau_1} P_1 - 2 \sigma^2 \tau_c (\exp(-t/\tau_c) - 1) P_1 + \frac{1}{\tau_o} P_o \qquad (2a)$$

$$\frac{\partial}{\partial t} P_o = - \frac{1}{\tau_o} P_o - 2 \sigma^2 \tau_o (\exp(-t/\tau_o) - 1) P_o + \frac{1}{\tau_o} P \qquad (2b)$$

The fit was done assuming fast diffusion in the undisturbed lattice, i.e. τ_c is very small and the corresponding term in Eq.(2a) can be neglected.

The data have been fitted to Eq.(2) and the following results appear : i) the trapping rate $1/\tau_1$ is very weakly temperature dependent; $\tau_1^{-1} = \Gamma_1 T^\alpha$ with $\alpha \stackrel{<}{\sim} 1$. ii) The activation energies for escape from the traps are of the order of 100 K ($\tau_o^{-1} = \Gamma_o \exp(-E_a/kT)$. The fitted parameter values are summarized in the Table.

As an example we give the results found for the AlMn 70 ppm sample, for the temperature 17 K at the linewidth maximum :

$$\tau_1 = 3.3 \; \mu s \qquad\qquad \tau_o = 40 \; \mu s$$

This means that i) only a fraction of the muons finds a trap ($\tau_\mu = 2,2 \; \mu s$, observation time $\sim 7 \; \mu s$). ii) If however a muon is trapped, it has a large probability to stay there.

Similar results are obtained for all the samples. Thus, the actual maximum linewidth reached is a mesure of the fraction of muons trapped.

The trapping rates for $AlMn_x$ which have been measured for 3 different concentrations follow quite accurately a linear concentration dependence confirming thereby the consistency of the applied model.

The actual nature of the traps causing the strong linewidth maxima can be discussed in several ways. We find it plausible that the traps consist of extended regions in the lattice around each impurity. The evidence for the large sizes of the traps are :
a) The measurements on the AlMn 42 ppm sample : This sample was measured down to 0.1 K[5] yielding about the same nearly linear T dependence of linewidth between 0.1 - 2 K as observed for τ_1^{-1} in the region 5 - 15 K. This suggests muon diffusion at low

TABLE: Fitted parameters for the different Al samples, with
$\tau_1^{-1} = \Gamma_1 T^\alpha$ and $\tau_0^{-1} = \Gamma_0 \exp(-Ea/T)$.

Sample	Conc. (ppm)	Γ_1 (s^{-1})	α	Γ_0 (s^{-1})	E_a (K)
AlMn	42	$1.62(12) \cdot 10^4$			
AlMn	57	$2.02(14) \cdot 10^4$	0.89(3)	$2.86(35) \cdot 10^7$	120(6)
AlMn	70	$2.43(14) \cdot 10^4$			
AlLi	75	$4.0 \ (10) \cdot 10^4$	0.36(7)	$0.85(15) \cdot 10^6$	145(27)
AlMg	42	$3.3 \ (4) \ \cdot 10^4$	0.54(4)	$2.03(44) \cdot 10^6$	142(7)
AlAg	117	$3.9 \ (5) \ \cdot 10^4$	0.45(5)	$2.45(32) \cdot 10^6$	87(7)

temperatures and trapping at higher temperature to be controlled
by the same diffusion mechanism ($\tau_1^{-1} \propto \tau_c^{-1}$). Applying reaction
theory of point defects ($\tau_1^{-1} = 4\pi R_t D_c$) we can estimate the mean
radius of a trap to be in the order of some 10 Å depending on the
extrapolation of the low T data.
b) The population of the traps is higher than expected from
thermodynamical arguments when the relevant activation energies
and impurity concentrations are considered.
c) In the earlier AlMn measurements[2] it was concluded from the
value of the electric field gradient at the Al nuclei that the
muon is not next neighbor to the Mn atom. This could easily be
understood in the picture of an extended trap surrounding the
impurity atom.

With this model of extended trap regions of considerable
size we find that the 1300 ppm AlMn sample is dominated by trapping
regions. The space for the free muon diffusion is very limited,
and fitting of data with the two-state model has not been
attempted for this sample.

It is tempting to correlate the effect of various
impurity elements on muon diffusion in Al to the volume dilatation
$\Delta V/V$ of the lattice. For the impurities investigated here there is
no such correlation. Neutron scattering experiments on Al show in
fact that large atomic displacements can be present close to
impurity atoms even when the net volume dilatation is small[11].

REFERENCES

1. O. Hartmann, E. Karlsson, K. Pernestål, M. Borghini, T.O. Niinikoski and L.O. Norlin, Physics Letters 61A, 141 (1977).

2. O. Hartmann, E. Karlsson, L.O. Norlin, D. Richter and T.O. Niinikoski, Physical Review Letters 41, 1055 (1978).

3. W.J. Kossler, A.T. Fiory, W.F. Lankford, J. Lindemuth, K.G. Lynn, S. Mahajan, R.P. Minnich, K.G. Petzinger and C.E. Stronach, Physical Review Letters 41, 1558 (1978).

4. W.J. Kossler, A.T. Fiory, W.F. Lankford, K.G. Lynn, R.P. Minnich and C.E. Stronach, Hyperfine Interactions 6, 295 (1979).

5. O. Hartmann, E. Karlsson, L.O. Norlin, T.O. Niinikoski, K.W. Kehr, D. Richter, J.M. Welter, A. Yaouanc, J. Le Hericy, Physical Review Letters 44, 337 (1980).

6. K. Dorenburg, M. Gladisch, D. Herlach, W. Mansel, H. Metz, H. Orth, G. zu Putlitz, A. Seeger, W. Wahl and M. Wigand, Zeitschrift für Physik B31, 165 (1978).

7. J.A. Brown, R.H. Heffner, M. Leon, M.E. Schillaci, D.W. Cooke, W.B. Gauster, Physical Review Letters 43, 1513 (1979).

8. M. Borghini, T.O. Niinikoski, J.C. Soulié, O. Hartmann, E. Karlsson, L.O. Norlin, K. Pernestål, K.W. Kehr, D. Richter and E. Walker, Physical Review Letters 40, 1723 (1978).

9. K.W. Kehr, G. Honing and D. Richter, Zeitschrift für Physik B32, 49 (1978).

10. see for instance K. Schroeder in Point Defects in Metals II; Springer Tracts in Modern Physics vol 87, Berlin, 1980.

11. K. Werner, KFA Jülich, private communication.

A NEW THEORETICAL APPROACH TO CALCULATE TOTAL ENERGY DIFFERENCES IN SOLIDS. APPLICATION FOR THE TETRAGONAL SHEAR MODULI OF FCC TRANSITION METALS

J. Ashkenazi, M. Dacorogna, and M. Peter

Département de Physique de la Matière Condensée
Université de Genève, 24, Quai Ernest-Ansermet

1211 Genève 4, Switzerland

A new approach for ground state energy calculations, based on the density functional formalism, is proposed. It uses the extremum property that first order variations of the electronic density around the ground state, result in second order total energy variations, and also an analogous property obtained for the ground state kinetic energy of non-interacting electrons. In addition to this, a new method, the LECA, is introduced, which enables us to gain a further order accuracy by error cancellation. So, having the possibility to use approximate densities, we can calculate the total energy differences, including the leading non-local exchange-correlation terms which are minimized to zero. The method is applied for the tetragonal shear moduli of the seven fcc transition metals. The results agree with experiment and include a prediction for Rhodium.

Using the density functional formalism [1], the total ground state energy of a system of electrons, under the influence of an external potential $v(\underline{r})$, is the minimum, among the admissible electronic densities $n(\underline{r})$, of a functional of v and n denoted here by $\mathcal{E}\{v,n\}$. Let us consider an external potential which has a smooth dependence on a parameter β : $v(\beta,\underline{r})$ and denote by $n(\beta,\underline{r})$ and $\mathcal{E}(\beta)$ the corresponding ground state density and energy respectively, i.e. :

$$\mathcal{E}(\beta) = \mathcal{E}\{v(\beta), n(\beta)\} \tag{1}$$

A "β variation" is defined as the one associated with the variation of β. By derivation of Eq.(1) we get :

$$\frac{d\mathcal{E}(\beta)}{d\beta} = \int \frac{\delta\mathcal{E}\{v,n\}}{\delta v(\beta,\underline{r})} \frac{dv(\beta,\underline{r})}{d\beta} d^3r$$

$$+ \int \frac{\delta\mathcal{E}\{v,n\}}{\delta n(\beta,\underline{r})} \frac{dn(\beta,\underline{r})}{d\beta} d^3r \tag{2}$$

However the functional derivative $\delta\mathcal{E}/\delta n(\underline{r})$ of a system at a ground state is equal, at any point $\underline{r}$, to the Fermi energy .
So, taking into account that the number of electrons is conserved, the last term in Eq.(2) falls, and we get that variations of $n(\beta,\underline{r})$ (alone) do not vary $\mathcal{E}$ to first order. So for two points β_o, β_1 :

$$\mathcal{E}(\beta_o) = \mathcal{E}\{v(\beta_o), n(\beta_1)\} + 0[(\beta_1-\beta_o)^2] \tag{3}$$

(the error includes also a term from non-linear dependence of n on β).

The external potential v influencing the electrons in a crystal is the Coulomb potential of the nuclei. By adding to it a constant (in $\underline{r}$), $\mathcal{E}\{v,n\}$ can be renormalized to include nucleus-nucleus interactions.

It has been shown [2] that the ground state problem can be solved on the basis of an equivalent (i.e. having the same ground state density n) system of non-interacting electrons subject to a potential $v_n(\underline{r})$ given by :

$$v_n(\underline{r}) = v(\underline{r}) + v_c(\underline{r}) + v_{xc}(\underline{r}) \tag{4}$$

where $v_c(\underline{r})$ and $v_{xc}(\underline{r})$ are the so called "Coulomb" and "exchange-correlation" potentials [3] . $\mathcal{E}\{v,n\}$ can be expressed as a sum of three functionals :

$$\mathcal{E}\{v,n\} = T_n\{n\} + U_c\{v,n\} + \mathcal{E}_{xc}\{n\} \tag{5}$$

where T_n is the kinetic energy of the equivalent non-interacting electrons system, U_c is the total classical Coulomb energy of the crystal, and $\mathcal{E}_{xc}$ is the effective exchange-correlation energy. The equivalent system is determined by the effective potential v_n which yields energies E_k in bands in the Brillouin zone (BZ), wave functions $\phi_k(\underline{r})$, Fermi energy E_F and density of states $N(E)$. Its ground state energy, which is naturally a functional of v_n, is given by :

$$\mathcal{E}_n\{v_n\} = \int_{BZ} d^3k\,\theta(E_F-E_{\underline{k}})E_{\underline{k}} = \int_{-\infty}^{E_F} E\,N(E)dE \tag{6}$$

The BZ integration includes also a band summation which is
omitted for simplicity. Let us define for any $v_n(\underline{r})$ and $n(\underline{r})$
a functional $T_n\{v_n,n\}$ by the identity :

$$T_n\{v_n,n\} = \mathcal{E}_n\{v_n\} - \int v_n(\underline{r}) \, n(\underline{r}) \, d^3r \tag{7}$$

It coincides with $T_n\{n\}$, defined in (5), only for densities
given by :

$$n(\underline{r}) = \int_{BZ} d^3k \, \theta(E_F-E_{\underline{k}}) \, |\phi_{\underline{k}}(\underline{r})|^2 \tag{8}$$

Let us consider, in this system, an effective potential which
has a smooth dependence on a parameter α : $v_n(\alpha,\underline{r})$, and let us
denote by $n(\alpha,\underline{r})$ the density connected to it by (8).
Consequently :

$$T_n\{n(\alpha)\} = T_n\{v_n(\alpha), \, n(\alpha)\} \tag{9}$$

An "α variation" is defined as the one associated with the
variation of . Eq.(4) yields that any β variation is related
to an α variation through density equality : $n(\beta,\underline{r}) = n(\alpha,\underline{r})$.
By derivation of Eq.(9), and use of (6)-(8), we get :

$$\frac{d}{d\alpha} T_n\{n(\alpha)\} = \int_{BZ} d^3k \, \theta(E_F-E_{\underline{k}}) \left[\frac{dE_{\underline{k}}}{d\alpha} - \langle\phi_{\underline{k}}| \frac{dv_n}{d\alpha} |\phi_{\underline{k}}\rangle \right]$$

$$+ E_F \int_{FS} \frac{dS_{\underline{k}}}{|\nabla_{\underline{k}}E_{\underline{k}}|} \frac{d}{d\alpha} (E_F-E_{\underline{k}})$$

$$- \int v_n(\alpha,\underline{r}) \frac{dn(\alpha,\underline{r})}{d\alpha} d^3r \tag{10}$$

where the second integral is over the Fermi surface (FS). It
turns out that the BZ integral falls by perturbation theory,
and that the FS integral falls by the definition of $dE_F(\alpha)/d\alpha$.
So we have proven that variations of $v_n(\alpha,\underline{r})$ (alone) do not vary
T_n to first order, and in analogy to (3), one has the relation :

$$T_n\{n(\alpha_1)\} = T_n\{v_n(\alpha_0), n(\alpha_1)\} + 0[(\alpha_1-\alpha_0)^2] \tag{11}$$

By combination of Eqs.(3), (5) and (11), one can get an
approximation to the ground state energy. Let us do it in terms
of a parameter θ, such that $\alpha = \alpha(\theta)$, $\beta = \beta(\theta)$, the points θ_0,θ_1
satisfy :

$$\alpha_o = \alpha(\theta_o), \ \beta_o = \beta(\theta_o), \ \alpha_1 = \alpha(\theta_1), \ \beta_1 = \beta(\theta_1) \qquad (12)$$

and for any $\underline{r}$:

$$n(\alpha_1, \underline{r}) = n(\beta_1, \underline{r}) \qquad (13)$$

The point θ_1 is chosen to give a simple density, which yields (by (3), (5) and (11)) an approximation to $\mathcal{E}(\beta_o)$ in terms of a simple expression $\mathcal{E}(\theta_1) = \mathcal{E}(\beta_o) = \tilde{\mathcal{E}}(\theta_1) + 0[(\theta_1 - \theta_o)^2]$. Since in general, a smooth function $F(\theta)$ can be expanded as $F(\theta_1) + 0(\theta - \theta_1)$, the following expression is valid :

$$\tilde{\mathcal{E}}(\theta) = T_n\{v_n(\alpha_o), n[\alpha(\theta)]\} + U_c\{v(\beta_o), n[\beta(\theta)]\}$$
$$+ \mathcal{E}_{xc}\{n[\beta(\theta)]\} + 0(\theta - \theta_i) \qquad (14)$$

Using the same type of Taylor expansion on $0[(\theta_1 - \theta_o)^2]$, we get that there exists a smooth function $f(\theta)$, such that :

$$\mathcal{E}(\beta_o) = \tilde{\mathcal{E}}(\theta) + f(\theta)(\theta - \theta_1) + 0[(\theta - \theta_o)^2] \qquad (15)$$

For cases where $\mathcal{E}(\theta_1)$ is not expected to be a good approximation to $\mathcal{E}(\beta_o)$, we introduce the "linear error cancellation approximation" (LECA), which consists of choosing θ and $\tilde{\mathcal{E}}(\theta)$ such that $f(\theta)$ is well approximated, in the interval $[\theta_o, \theta_1]$, by a constant b. This means that the dominant non-linear terms in $(\theta - \theta_1)$ in (15) are included in $\mathcal{E}(\theta)$. (One should be aware of the fact that non-linear dependence of the densities n on θ also contributes to $f(\theta)$). So for $\theta \in [\theta_o, \theta_1]$:

$$\mathcal{E}(\beta_o) \simeq \tilde{\mathcal{E}}(\theta) + b(\theta - \theta_1) + 0[(\theta - \theta_o)^2] \qquad (16)$$

By derivation at $\theta = \theta_o$, we get : $b = -d\tilde{\mathcal{E}}(\theta_o)/d\theta$, and by substitution in (16), for $\theta = \theta_o$:

$$\mathcal{E}(\beta_o) \simeq \tilde{\mathcal{E}}(\theta_o) + \frac{d\tilde{\mathcal{E}}(\theta_o)}{d\theta} (\theta_1 - \theta_o) \qquad (17)$$

So, when the problem can be approached by the above procedure, one can get by the LECA a more accurate value for $\mathcal{E}(\beta_o)$ than the one based directly on a linear expansion.

Within such an expansion, density variations are related to potential variations through the expression (8) :

$$\frac{dn(\alpha,\underline{r})}{d\alpha} = \int_{FS} \frac{dS_{\underline{k}}}{|\nabla_{\underline{k}} E_{\underline{k}}|} \frac{d}{d\alpha} (E_F - E_{\underline{k}}) |\phi_{\underline{k}}(\underline{r})|^2$$

$$+ \int_{BZ} d^3k \; \theta(E_F - E_{\underline{k}}) \frac{d}{d\alpha} |\phi_{\underline{k}}(\underline{r})|^2 \qquad (18)$$

It contains two terms : a FS term, which exists only for metals, and is due to the occupation variation of $\underline{k}$ states close to the FS, and a BZ term due to the variation of the occupied $\phi_{\underline{k}}(\underline{r})$. The BZ term is obtained from non-diagonal matrix elements of $dv_n/d\alpha$, which mix states of the same $\underline{k}$ (for periodicity conserving variations) in different bands. It can be shown that contributions to $dn/d\alpha$ from mixing states below the Fermi level cancel each other. So for regular variations, this term is dominated by contributions of mixed states of the same $\underline{k}$ close to the Fermi level on both sides (due to the energy difference denominator). Let us consider a one-atom crystal under a symmetry removing shear expressed by a strain parameter γ [4]. The original and the sheared crystals have different Wigner-Seitz (WS) cells, but with the same volume Ω and radius s. Denoting the spherically averaged density, around the WS cell center, by n(r), we define :

$$Z \equiv \Omega n(s) \qquad (19)$$

Let us introduce symmetry conserving β and β' variations for the original and the sheared crystals respectively, and denote the actual ground state densities (and similarly other quantities) by $n(\underline{r}) \equiv n(\beta_o, \underline{r})$ and $n_\gamma(\underline{r}) \equiv n_\gamma(\beta_o', \underline{r})$ respectively. Following the spirit of this work, we want to vary these densities to $n(\beta_1, \underline{r})$ and $n(\beta_1', \underline{r})$, which should be as close as possible to model densities $\tilde{n}(\underline{r})$ and $\tilde{n}(\underline{r})$ determined by the following conditions :

(a) They are identical within the WS cell (in an average manner close to its boundaries).
(b) They are spherically symmetric around the cell center.
(c) They are flat around the cell boundaries, starting at the smallest muffin-tin radius.

Similarly, let us introduce to these crystals symmetry conserving α and α' variations, such that Eq.(13) is satisfied for both of them, and that the "varied back" effective potentials $v_n(\alpha_o, \underline{r})$ and $v_{n\gamma}(\alpha_o', \underline{r})$ are as close as possible to $v_n(\underline{r})$ and $v_{n\gamma}(\underline{r})$ defined by the conditions :

(d) $v_n(\underline{r})$ is spherically symmetric within a WS cell around the cell center, and related to $v(\underline{r})$ ($\equiv v(\beta_o,\underline{r})$) by the spherical average of (4).

(e) $v_n(\underline{r})$ is identical to $v_n(\underline{r})$ within a cell (in an average manner close to its boundaries).

Let ε be a typical relative density variation applied by conditions (a)-(e). In order to calculate the energy variation under shear to order γ^2, our approximation to it should be correct to order ε^2 (since (a) and (e) impose that ε is partly determined by γ). This can be achieved by the LECA. It remains to discuss how close can conditions (a)-(e) be approaches within our scheme. The density variation under shear is due to FS and BZ terms (of the type given to first order in Eq.(18)), associated with the potential variation [4]. It contains symmetry removing and conserving (e.g. radial) components of orders γ and γ^2 respectively.

Conditions (a) can be approached by choosing $v_\gamma(\beta',\underline{r})$ such that counter terms are introduced to n_γ. The remaining discrepancies between $n_\gamma(\beta_1',\underline{r})$ and $\tilde{n}_\gamma(\underline{r})$ have symmetry removing and conserving components of orders γ^2 and γ^3 respectively, which cause errors of order γ^3 in the components of the energy functional in Eq.(5). Condition (b) can be approached by appropriately chosen angular components of $dv/d\beta$. The angular discrepancies between $n(\beta_1,\underline{r})$ and $\tilde{n}(\underline{r})$ are expected to be of order ε^2, but this causes higher order errors in the components of the energy functional (since, locally, the non-spherical effects in a crystal can be considered as symmetry removing perturbations of the atomic spherical symmetry).

In order to approach condition (c), there remains to minimize to zero as many derivatives as possible, of the spherically averaged density $n(r)$ at $r = s$. It can be approximated as :

$$n(r) = \frac{1}{4\pi} \sum_\ell \int_{-\infty}^{E_F} N_\ell(E)\ \phi_\ell(E,r)^2\ dE \qquad (20)$$

where $N_\ell(E)$ is the ℓth partial density of states, and $\phi_\ell(E,r)$ is the solution (normalized in the WS cell) of the radial Schrödinger (or Dirac) equation for ℓ, the energy E, and the appropriate spherically averaged potential $v_n(r)$. One can show by Eq.(18), and the discussion following it, that a first order variation of $n(r)$ can be approximated in terms if an effective variation of $N_\ell(E_F)$ in Eq.(20). This is clear concerning the FS term in (18). For the BZ term, this is obtained when approximating the products $\phi_\ell(E,r)\phi_\ell(E',r)$ by $\phi_\ell(E_F,r)^2$, where E and E' are the

energies of the mixed states close to E_F from its two sides. This
effective variation of $N_\ell(E_F)$ is subject to two limitations,
first the conservation of $\Sigma_\ell N_\ell(E_F)$ linked with the conservation
of $\int_{WS} n(\underline{r})d^3r$, and second, by crystal symmetry, the value of
the first r derivative $n^{(1)}(s)$ is to be kept close to zero. If
the states ϕ_k close to E_F (or more precisely the variations
between them) are combinations of ϕ_ℓ with essentially three
ℓ-values, then there is (approximately) only one degree of freedom
for first order β variations of $n(\beta_o,r)$, and condition (c) can be
approached by minimizing to zero also the second derivative $n^{(2)}(s)$.
This is the case for transition metals. Conditions (d) and (e) can
be approached by making a self-consistent band calculation on the
original crystal with a "good" choice for v_{xc}, and leaving the
obtained potential "frozen" under shear.

The variation of the ground state energy under shear can
be expressed as (5) :

$$\delta_\gamma \mathcal{E}\{v,n\} = \delta_\gamma T_n\{n\} + \delta_\gamma U_c\{v,n\} + \delta_\gamma \mathcal{E}_{xc}\{n\} \qquad (21)$$

where we have introduced the notation $\delta_\gamma T_n\{n\} \equiv T_n\{n_\gamma\} - T_n\{n\}$,
and similarly for other quantities. Conditions (a)-(e) yield :
$\delta_\gamma T_n\{v_n,\tilde{n}\}$ is equal by (a), (e) and Eq.(7) to $\delta_\gamma \mathcal{E}_n\{v_n\}$; $\delta_\gamma U_c\{v,\tilde{n}\}$
can be expressed, by (a), (b) and (c), as $\delta_\gamma \mathcal{E}_M[Z(\alpha_1)] = Z(\alpha_1)^2 \delta_\gamma \mathcal{E}_M(1)$,
where Z has been defined in Eq.(19), and $\mathcal{E}_M(1)$ is the classical
Madelung energy, for the given crystal structure, of unit charge
point ions in a uniform neutralizing background [5] ; $\delta_\gamma \mathcal{E}_{xc}\{\tilde{n}\}$
vanishes by (a) and (c). This is because $\mathcal{E}_{xc}\{\tilde{n}\}$ can be expressed
as a sum of a local term (function of $\tilde{n}$), and a non-local term,
expanded in gradients of $n(\underline{r})$ [1], which are zero in the region
where the WS cells match. The densities which can be achieved
by the β and β' variations, introduce to the above expressions
errors of order ε^3 and of high order terms in the density gradient
expansion around r = s.

So if we could find an appropriate parameter θ, and
correlate α, β, α' and β' by Eq.(12), then the expression (see
Eqs.(14)-(17)) :

$$\delta_\gamma \tilde{\mathcal{E}}(\theta) = \delta_\gamma \mathcal{E}_n\{v_n\} + Z[\alpha(\theta)]^2 \delta_\gamma \mathcal{E}_M(1) \qquad (22)$$

would have been useful to calculate $\delta_\gamma \mathcal{E}\{v,n\}$ within the LECA.
Under condition (d), such a parameter is given by :

$\theta = n^{(2)}[\alpha(\theta),s]$, $(\theta_1 = 0)$. The variations of the radial components of n and v_n are carried out in two steps. In the first step the spherically averaged densities are slightly varied such that they have all the same radial dependence $n(\alpha_o,r)$, and this introduces by Eqs. (3) and (11) only energy errors of orders ε^3 and γ^3. In the second step, all the spherically averaged densities are varied with the same radial dependence towards $\tilde{n}(r)$. The LECA condition (16) then means that one can use a linear approximation for the energy effect of the deviation from condition (c) (expressed by θ) on Eq. (22), which is reasonable within the required accuracy. As for the variations of the angular components of n and v_n, since in the sheared crystal there exists an approximative symmetry of the original one, and in both crystals there exists, locally, an approximate spherical symmetry, such variations are symmetry removing, and their energy effect is decreased by an order of ε or γ. So the linear approximation of Eqs. (3) and (11) gives for them the energy correctly to orders ε^2 and γ^2. By using Eq. (17), it turns out that Eq. (22) represents the correct $\delta_\gamma \tilde{\varepsilon}$ when $Z[\alpha(\theta)]^2$ is replaced by :

$$Z_{eff}^2 = Z_o^2 - 2Z_o\, n^{(2)}(\alpha_o,s)\, \frac{dZ(\alpha_o)}{d\alpha}\, \Big/\, \frac{dn^{(2)}(\alpha_o,s)}{d\alpha} \qquad (23)$$

where $Z_o \equiv Z(\alpha_o)$ is the band structure value for Z (defined in (19)). As was mentioned following Eq. (20), these α-derivatives are sharply determined for transition metals. It is important to note that a spherical approximation of $v_n(r)$, and slight radial modifications (of order ε^2), cause here an error of order ε^3 in $\delta_\gamma \tilde{\varepsilon}$, while the effect on the band structure is of order ε^2.

The shear modulus, defined as [4] : $C = 2/3 \lim \delta_\gamma \tilde{\varepsilon}/\gamma^2$, can then be expressed as a sum of two terms : $C = C_b + C_M$, where C_b corresponds to $\delta_\gamma \tilde{\varepsilon}_n$ and C_M to $\delta_\gamma \tilde{\varepsilon}_M(Z_{eff})$.

The idea of "freezing" the potential, when calculating an expression similar to $\delta_\gamma \tilde{\varepsilon}_n$, was pointed out by Pettifor [6], for bulk modulus. This idea was generalized in ref.3, and O.K. Andersen (private communication) has shown that in a shear calculation, it should be valid to order γ^2 within a spherical potential approximation.

Expressions for $\delta_\gamma \tilde{\varepsilon}$, of a form similar to Eq. (22), have been suggested in the past [6-9]. However the effective values of Z proposed there are close to Z_o of this work, (which might be true here for simple metals and insulators).

TABLE I

Results for the tetragonal shear moduli of fcc transition metals in comparison with experiment (C'_{exp}) (quantities defined in the text)

units	Cu	Rh	Pd	Ag	Ir	Pt	Au
e							
Z_0	2.6	3.8	3.2	2.8	4.5	4.0	3.4
Z_{eff}	4.2	5.7	5.8	4.6	6.4	6.6	5.8
Ryd/atm							
C'_b	-0.25 ± 0.05	$+0.02\pm0.05$	-0.38 ± 0.05	-0.20 ± 0.05	$+0.12\pm0.05$	-0.51 ± 0.05	-0.50 ± 0.05
C'_M	$+0.39\pm0.04$	$+0.68\pm0.07$	$+0.68\pm0.07$	$+0.41\pm0.04$	$+0.85\pm0.09$	$+0.88\pm0.09$	$+0.65\pm0.06$
C'_{calc}	$+0.14\pm0.09$	$+0.70\pm0.12$	$+0.30\pm0.12$	$+0.21\pm0.09$	$+0.97\pm0.14$	$+0.37\pm0.14$	$+0.15\pm0.11$
10^{10} N/m^2							
C'_{calc}	2.6 ± 1.6	11.0 ± 2.0	4.4 ± 1.8	2.7 ± 1.1	14.7 ± 2.1	5.3 ± 2.0	1.9 ± 1.4
C'_{exp}	2.56		2.9	1.71	17.2	5.22	1.6

Using self-consistently the LMTO-ASA band method [10] with $\ell = 0,1,2,3$, we have calculated by the present method the tetragonal shear moduli $C' = \frac{1}{2}(C_{11}-C_{12})$ of the seven non magnetic fcc transition metals. The Madelung sums have been calculated by the method of Harris and Monkhorst [11], and a polynomial decomposition has been applied to $\delta_\gamma \mathcal{E}$ values for $\gamma = \pm 0.02$, ± 0.04. The relative error in Z^2_{eff} and in C'_M is estimated to be of 10 %, which is larger by more than a factor of two than the inaccuracy introduced by the method. The error in C'_b is estimated to be about 50 mRyd/atm. The results, including the error bars, are represented in table I. Clearly there is an agreement between theory and experiment considering the error bars. They become relatively large when C'_b and C'_M have opposite signs and close magnitudes. This is not be case for Rh where C' has not been measured yet, and is given here a prediction with a quite small error bar.

Acknowledgement - The authors benefited from collaboration with O.K. Andersen in the initial stages of this work, and from the fruitful criticism of W. Kohn.

REFERENCES

1. P. Hohenberg and W. Kohn, Phys.Rev. 136, B864 (1964).

2. W. Kohn and L.J. Sham, Phys.Rev. 140, A1133 (1965).

3. A.R. Mackintosh and O.K. Andersen in : "Electrons at the Fermi Surface", edited by M. Springford (Cambridge University Press, 1980), p.149.

4. J. Ashkenazi, M. Dacorogna, M. Peter, Y. Talmor, E. Walker and S. Steinemann, Phys.Rev. B18, 4120 (1978).

5. M.P. Tosi, Solid State Physics, 16, 1-120 (1964).

6. D.G. Pettifor, Commun. Phys. 1, 141 (1976); J. Phys. F8, 219 (1978).

7. W.A. Harrison : "Pseudo-Potentials in the Theory of Metals" Ed. W.A. Benjamin Inc. New-York (1966); R.W. Shaw Jr., J. Phys. C2, 2335 (1969).

8. V. Heine and D. Weaire, Solid State Physics 24, 363 (1970).

9. S.H. Vosko, R. Taylor and G.H. Koech, Can. J. Phys. 43, 1187 (1965).

10. O.K. Andersen, Phys. Rev. B12, 3060 (1975); Y. Glötzel, D. Glötzel and O.K. Andersen, to be published.

11. F.E. Harris and H.J. Monkhorst, Phys. Rev. B2, 4400 (1970) and Errata, Phys. Rev. B9, 3946 (1974).

EVIDENCE FOR ELECTRON-ELECTRON SCATTERING IN COPPER AND SILVER FROM RADIO FREQUENCY SIZE EFFECT MEASUREMENTS

W.M. MacInnes, P.-A. Probst, B. Collet, and R. Huguenin
Institut de Physique Expérimentale de l'Université de
Lausanne
CH-1015 Lausanne, Switzerland

V.A. Gasparov
Institute of Solid State Physics
Academy of Sciences of the USSR
142432 Chernogolovka, USSR

New high precision RFSE measurements down to 0.5 K on thick Cu
and Ag samples reveal deviations from the usual T^3 temperature
dependence due to electron-phonon scattering. If the observed
less rapid temperature dependence is associated with a T^2
contribution to the scattering frequency, then the coefficient
of this term is of the order of magnitude predicted by Lawrence
for electron-electron scattering in Cu and Ag.

1. INTRODUCTION

The search for the electron-electron scattering
contribution to the temperature dependence of the electrical
resistivity of metals has become feasable in recent years with
the availability of high purity samples and precise measurements
techniques. While the experimental observation of a resistivity
$\rho_{ee} \propto T^2$ in transition metals has been relatively easy, its
measurement in non-transition metals is more difficult for several
reasons. The coefficient of this contribution is indicated to be[1-3]
of the order of $1-10 \times 10^{-15}$ ΩmK^{-2}, nearly two orders of magnitude
less than predicted by a simple tight binding model calculation[4].
Thus measurements with a precision of about 1 ppm on very pure
samples are necessary. Furthermore, the reported sample and

impurity dependence of a contribution proportional to T^2 in K
but not in Al[5] makes the unambiguous identification of ρ_{ee}
difficult. Additionally, one often finds a size effect
contribution[6] to $\rho(T)$ also proportional to T^2 that may be
confused with ρ_{ee}.

The temperature dependence of the scattering frequency,
$\nu(T)$, of groups of electrons much more localized on the Fermi
surface (FS) can be measured using several techniques[7] : high
frequency cyclotron resonance[8], surface Landau levels, time of
flight effect[9], magneto-acoustic attenuation (MA)[10], and, the
radio frequency size effect[7,11] (RFSE). In these effects the
ratio of the signals at 1 and 4 K may range say, from 1.1 to
greater than 10. Thus with the usual precision of these
measurements (2-10 %), it has been possible to observe a T^2
contribution to $\nu(T)$ in some transition metals. T^2 temperature
dependencies have been observed in Nb[10], Mo[12], Re[13] and in
W[12,14].

The search for the electron-electron scattering
contribution to $\nu(T)$ in simple metals using the RFSE has
previously been unsuccessful[15] due to a lack of precision. The
smallness of ν_{ee} necessitates a precision of at least 0.5 % and
this permits fruitful measurements to 1 K and below, where the
remaining variation in the signal strength amounts to only a
few percent. We present here our preliminary RFSE measurements
on copper and silver that indicate the presence of electron-
electron scattering in these two metals that is of the same
order of magnitude as we have previously found in cadmium[16] and
as is predicted theoretically for these metals[1,3].

2. EXPERIMENTAL DETAILS

We will outline here briefly the experimental conditions
and leave a more elaborate discussion to a later publication. The
measurements were made on thick (d = 1.88 mm for Cu and d = 1.02
mm for the Ag) high purity monocrystalline samples prepared by
Dr. Gasparov, and oriented such that the sample normals were
parallel to <110>. Strict parallelism of the two sample faces
and of the external magnetic field B is essential for obtaining
narrow, well characterized RFSE signals. In our measurements we
readily obtained the condition $\delta/d = \Delta B/6B \leqslant 0.003$, at RF frequencies
of ~7 MHz. The marginal oscillator is optimized for RFSE
experiments[17] and additionally the oscillator and demodulation
transistors are held at constant temperature. The modulation
magnetic field B_m is stabilized, the demodulated RFSEamplitude
in phase and in quadrature with B_m is measured using phase

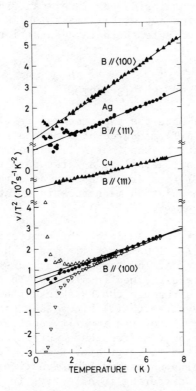

Fig.1 : $\nu(T)/T^2$ for Cu and Ag belly orbits.

sensitive detection at discreet dc magnetic field values, the
lock-in outputs are digitized, averaged on line, and, plotted
as a function of B on X-Y recorders. The overall stability of the
measurements of A(T) is presently about 0.5 % and further
improvements are being implimented to obtain a precision of 0.1 %.
In the cryostat, the sample rests on a copper plate containing
two Ge resistance thermometers calibrated from 0.3 K to 40 K.
This tiltable sample holder is located in a can filled with
exchange gas, and is isolated from the ^{4}He bath by a second can
exterior to it. Cooling to temperatures below 1 K is accomplished
by pumping on a ^{3}He pot inside the inner can. The sample
temperature is measured and stabilized to $\pm$ 1mK using the same
micro-computer that effects the data acquisition.

3. EXPERIMENTAL RESULTS AND DISCUSSION

We present here the experimental results deduced from
the RFSE amplitudes A(T) measured at a modulation frequency of
9 Hz and a phase setting such that A(T) was a maximum of 4 K.
We will take up in more detail at a later date the (small)
corrections[18] to A(T) that are necessary as a result of the poor
penetration of the modulation magnetic field into these high
purity samples.

Measurements were made on the central extremal orbits
in Cu and Ag with $\vec{B}$ // <100> and <111> where the electron-phonon
scattering is weakest. At low temperatures, a significant deviation
from linearity can be discerned in plots of ln A(T) versus T^3.
While multiple passes through the RF skin depth are possible, it
is unlikely that they affect A(T) since these samples are very
thick. The product, $\nu_0 t$, of the impurity scattering frequency
and the sample crossing time for the electrons on the orbits
considered is estimated to be greater than 2. Furthermore the
RFSE line shape remains unchanged as a function of temperature.
The electron-phonon scattering is effective ($\nu_{ep} \propto T^3$) under our
experimental conditions in the temperature range of these
measurements. We have analysed the T dependence of A(T) using

$$A(T) = A(0)\exp(-\nu t)$$
$$\nu = \alpha T^2 + \beta T^3 \tag{1}$$

The data and the fitted temperature dependencies are shown in
Figure 1, plotted as ν/T^2 versus T. The measured points are
defined as $\nu(T) = t^{-1} \ln(A(0)/A(T))$. The parameters α and β of
the fits together with their errors are listed in Table I below.

Table I : Parameters of $\nu(T) = \alpha T^2 + \beta T^3$

$$\alpha(\times 10^7 s^{-1} K^{-2}) \qquad \beta(\times 10^7 s^{-1} K^{-3})$$

		$\alpha(\times 10^7 s^{-1} K^{-2})$	$\beta(\times 10^7 s^{-1} K^{-3})$
Cu :	$\vec{B}$ // <100>	0.39 ± 0.25	0.3 ± 0.04
	$\vec{B}$ // <111>	0.14 ± 0.08	0.19 ± 0.01
Ag :	$\vec{B}$ // <100>	0.4 ± 0.4	0.6 ± 0.06
	$\vec{B}$ // <111>	1.0 ± 0.5	0.35 ± 0.08

The error given for a parameter corresponds to an increase of
100 % in the standard deviation of the fit, with all the other
parameters free. We show in the lower part of the Figure two
additional fits for the Cu $\vec{B}$ // <100> "belly" orbit. The lowest
line and open triangular points are for α fixed equal to zero
(outside the abovementioned uncertainty limits on α). The upper
line and open triangular points are for α increased by the error
listed in Table I. It is clear that the large uncertainty in β is
due to the uncertainty in α; if α is fixed then β is determined
with about an order of magnitude greater precision.

We obtain values for the coefficients of the T^3
electron–phonon scattering terms that are generally smaller
than those previously measured by Gantmakher and Gasparov[19,20]
due partly to the presence of the T^2 term in Eq.(1) for $\nu(T)$.
Nevertheless our values for β are larger than those obtained by
Johnson et al.[21,22] from their RFSE and MA experiments on these
metals. We have listed in Table II the total temperature dependent
scattering frequency at 7 K as this is rather independent of the
choice of A(0) and the decomposition of $\nu(T)$.

Table II : $\nu(7\ K)(\times 10^9 s^{-1})$

	Cu		Ag	
Ref.	B // <100>	B // <111>	B // <100>	B // <111>
RFSE (this work)	1.22	0.72	2.25	1.69
RFSE[19]	1.06	0.69		
RFSE[20]			2.02	1.65
MA[22]	0.93		1.44	
CR[23]	1.0			
RFSE[21]			1.45	

In our RFSE measurements on these thick samples we find $\nu(7\ K)$ to be comparable or larger than what was previously observed by Gantmakher and Gasparov[19,20] and considerably larger than $\nu(7\ K)$ obtained from other measurements. This may indicate that the electron-phonon scattering is not entirely effective in the MA experiments[22] or in RFSE measurements[21] on thin samples of Ag.

We turn now to the T^2 term. It is clear that the precision of the data is not sufficient to determine unambiguously the exponent of the low T contribution to $\nu(T)$. With the supposition of a T^2 temperature dependence we find that the coefficients of this contribution are of the order of magnitude predicted for electron-electron scattering in Cu and Ag. Our high precision, tilted field RFSE measurements[16] of the scattering frequency of electrons near <0001> in cadmium have also shown a T^2 contribution to $\nu(T)$ of the right order of magnitude for electron-electron scattering in this metal. Lawrence[1,2] estimates the basic (isotropic) electron-electron scattering frequency for Cu and Ag to be about the same : $\alpha = 0.46 \times 10^7\ s^{-1} K^{-2}$. Recently MacDonald[3] has shown that α will be enhanced in metals such as Al and Pb where an attractive phonon-mediated electron-electron interaction is stronger than the repulsive Coulomb interaction. In noble metals where the latter interaction is stronger, the predicted values for α are largely unchanged by the former interaction[2] :

$$\alpha(Cu,\ Ag) = (0.53,\ 0.56) \times 10^7\ s^{-1}\ K^{-2}.$$

It would seem that in RFSE measurements with the necessary precision one could expect to observe electron-electron scattering in other simple metals. Such measurements would be most useful to elucidate the strong dependence of ν_{ee} on the nature of the screening of the Coulomb potential in metals. However, Cu, Ag and Cd are not expected to differ greatly in this respect. The question of the validity of the Born approximation, and of the anisotropy of the electron-electron scattering are also important. The point scattering frequency $\nu_{ee}(\vec{k},T)$ is determined by a triple integration over the FS sheet(s), and, for metals with complicated Fermi surfaces, the conclusion from 2PW calculations[1] that ν_{ee} is only slightly anisotropic may have to be reexamined.

Acknowledgement - The authors would like to thank the Fonds National Suisse pour la Recherche Scientifique for financial support as well as the Office Suisse de la Science et de la Recherche and the Academy of Sciences of the USSR for making possible this collaborative research effort.

REFERENCES

1. WE Lawrence, Phys.Rev. B13, 5316 (1976).

2. AH MacDonald, Phys.Rev.Lett. 44, 489 (1980) and private communication.

3. JE Black, Can.J.Phys. 56, 708 (1978).

4. GT Morgan and C. Potter, J.Phys.F : Metal Phys. 9, 493 (1979).

5. H van Kempen, JS Lass, JHJM Ribot and P Wyder, Phys.Rev.Lett. 37, 1574 (1976); JHJM Ribot, J.Bass, H van Kempen and P Wyder, J.Phys.F : Metal Phys. 9, L117 (1979); B Levy, M Sinvani and AJ Greenfield, Phys.Rev.Lett.43, 1822 (1979).

6. J van der Maas, C Rizzuto and R Huguenin, this conference.

7. VF Gantmakher, Rep Prog.Phys. 37, 317 (1974).

8. R Carin, P Goy and WM MacInnes, J.Phys.F : Metal Phys.8, 2335 (1978) and references therein.

9. VA Gasparov, J Lebeck and K Saermark, Solid State Commun. (to be published).

10. JA Rayne and JR Liebowitz, 15th Int.Conf. on Low Temp.Phys., Grenoble, J. de Physique C6, 1006 (1978).

11. DK Wagner and R Bowers, Advances in Physics 27, 651 (1978).

12. VV Boiko, VF Gantmakher and VA Gasparov, Sov.Phys. - JETP 38, 604 (1974).

13. TL Ruthruff, CG Grenier and RG Goodrich, Phys.Rev. B17, 3070 (1978).

14. BP Nilaratna and A Myers, 15th Int.Conf. on Low Temp.Phys., Grenoble, J. de Physique C6, 1064 (1978).

15. VA Gasparov and MH Harutunian, Solid State Comm. 19, 189 (1976).

16. WM MacInnes, PA Probst and R Huguenin, 15th Int.Conf. on Low Temp.Phys., Grenoble, J. de Physique C6, 1062 (1978) (α values are a factor of 2 too large).

17. PA Probst, B Collet and WM MacInnes, Rev.Sci.Instrum. 47, 1522 (1976).

18. WM MacInnes, PA Probst, B Collet and R Huguenin, J.Phys.F : Metal Phys. 7, 655 (1977).

19. VF Gantmakher and VA Gasparov, Sov.Phys. - JETP 37, 864 (1973).

20. VA Gasparov, Sov.Phys. - JETP 41, 1129 (1976).

21. PB Johnson and RG Goodrich, Phys.Rev. B14, 3286 (1976).

22. PB Johnson and JA Rayne, Proc.3rd Int.Conf. on Phonon
    Scattering (to be published).

23. P Haussler and SJ Welles, Phys.Rev. 152, 675 (1966).

NEW EVIDENCE FOR INTERSHEET ELECTRON-PHONON SCATTERING FROM MEASUREMENTS ON THE DOPPLERON AND GANTMAKHER-KANER OSCILLATIONS IN CADMIUM

M. Berset, W.M. MacInnes, and R. Huguenin

Institut de Physique Expérimentale de
l'Université de Lausanne
CH-1015 Lausanne, Switzerland

The temperature dependence of the amplitudes of the 3rd band electron doppleron and Gantmakher-Kaner surface impedance oscillations has been measured in cadmium between 1.2 and 4.2 K in a magnetic field directed normal to the sample surface and along the <0001> direction. The temperature dependencies of these oscillations show deviations from a simple T^5 law, and we interpret them in terms of intersheet electron-phonon scattering. The magnetic field dependence of the scattering frequency is due to a shift of the belt of resonant electrons from near the rim of the lenticular 3rd band Fermi surface sheet towards the <0001> limiting point as the field is increased. The value of the scattering frequency is in good agreement with previous measurements (at 4 K).

1. INTRODUCTION

The properties of the electromagnetic waves that propagate in cadmium in strong magnetic fields directed perpendicular to the surface of pure monocrystalline samples have been elucidated both experimentally and theoretically by Fisher and co-workers[1]. Their published results concern only the situation where the magnetic field $\vec{B}$ is directed along (or near) the <0001> direction. Our interest has been mainly centered on the temperature dependence of the surface impedance oscillations due to these propagating waves since the attenuation length of the wave is essentially the mean free path, ℓ, of the resonant carriers when the magnetic field is strictly parallel to the sample normal (a crystallographic direction of high symmetry) and there is no collisionless Landau damping. We have made measurements of $\ell(T)$ of electrons on the 3rd band lenticular Fermi surface (FS) sheet of cadmium using this effect with $\vec{B}$ parallel to the sample normals : <0001>, <11$\bar{2}$0>, and <10$\bar{1}$0>[2]. These measurements of the temperature dependence of the electron mean free path (mfp) appear to give quite localized values of ℓ rather than orbitally average ones as is the case in the more usual parallel field radio-frequency size effect (RFSE) measurements[3]. This is because the waves are produced by the Doppler-shifted cyclotron resonance of a narrow belt of carriers around the line of intersection of a plane k_z = const. (perpendicular to $\vec{B}$) and the Fermi surface. The area of this section is $S(k_z)$. The resonant carriers are those which have an extremal value of $\partial S/\partial k_z$ at the particular value of the magnetic field of the surface impedance oscillation. The period of these oscillations in high magnetic fields, where the doppleron essentially becomes what has been termed a Gantmakher-Kaner oscillation, is determined by the maximum (extremal) value of $(\partial S/\partial k_z)$ on the Fermi surface. On the 3rd band FS sheet this means the resonant electrons are near the <0001> spherical limiting point ($\vec{B}$ // <0001> or near the rim – a limiting point with two fold symmetry ($\vec{B}$ // <11$\bar{2}$0> or <10$\bar{1}$0>). At lower magnetic fields this resonant belt of carriers moves to smaller values of k_z, the period of the oscillations with $\vec{B}$ // <0001> decreases, and the temperature dependence determines $\ell(T,k_z)$. In this paper we will present our new measurements with $\vec{B}$ // <0001> for comparison with those of Voloshin, Fisher and Yudin[1] in the same orientation and leave to a later date the more precise results obtained with $\vec{B}$ in the basal plane of cadmium.

2. EXPERIMENTAL DETAILS

        Voloshin et al.[1] have pointed out that the amplitude of the oscillations depends markedly on the nature of the electron reflection at the sample surface. In the case of diffuse reflection of the electrons it is necessary to measure the temperature dependence of both the oscillatory and the smoothly varying parts of the surface impedance of the sample plate. This was not done in our previous measurements with $\vec{B}$ // <0001>[2]. The new data presented here were obtained using a marginal oscillator optimized for RFSE experiments[4]. Its sensitivity is maintained constant by regulating the rf level by means of a feedback circuit acting on thermistors in parallel with the tank circuit in which the sample is placed. The measurements were carried out on a cadmium sample 0.455 mm thick in fields up to 30 kG and at a rf frequency of

2.5 MHz (linearly polarized). The field derivative of the real
part of the surface impedance dR/dB was measured using phase
sensitive detection and the output of the lock-in was integrated
to obtain R(B) at temperatures between 1.3 and 4.2 K.

3. EXPERIMENTAL RESULTS AND DISCUSSION

A typical recording of dR/dB is shown in Figure 1.
The amplitudes A(B,T) and the periods ΔB of the oscillations were
measured at the fields of ~10, 14 and 20 kG. The values of
R(B,T)−R(B=0,T) were also taken from simultaneous recordings of
the output of the integrator. The temperature dependence was
analysed using the following expressions[1] :

$$A(B,T) = A(B)\exp(-d/\ell(T)) \tag{1}$$

assuming specular reflection from the sample surfaces or

$$A(B,T) = A(B)\, R^2(B,T)\exp(-d/\ell(T)) \tag{2}$$

for the case where the surface reflections are taken to be
diffuse. A(B) is the line shape function of the Doppleron
Gantmakher-Kaner surface impedance oscillation. The T-dependence
of the reciprocal mfp can be modelled by a simple power law
$\ell^{-1}(T) \propto T^n$ as was done by Voloshin et al.[1]. They found n = 5
for the T dependence of the lens electrons at high fields, which
they attribute to ineffective electron—phonon scattering. On the
contrary we believe the electron—phonon scattering to be
effective. If it were ineffective one would expect $\ell^{-1}(T) \propto T^5$
at low temperatures becoming $\ell^{-1}(T) \propto T^3$ at higher temperatures
yet our data indicate a less rapid T dependence at low T.

Furthermore, the significant deviations from a T^5 law
we have observed in these and other RFSE measurements 3) can be
well described using the 1 plane wave expression for intra- and
intersheet electron—phonon scattering. This expression is
discussed extensively in refs. 3 and 5. The first term in it
is the normal (effective) intra-sheet scattering contribution
($\propto T^3$). Intersheet scattering via transversely polarized phonons
(sound velocity s_t) "turns on" exponentially for $T > T_t/10$ where
$T_t = \hbar s_t q_o/k_B$ is the "gap" temperature corresponding to the
minimum distance q_o in reciprocal space between the FS sheets.
The contribution due to intersheet scattering via longitudinally
polarized phonons sound velocity s_ℓ) turns on for $10T > T_\ell =$
$(s_\ell/s_t)T_t = 1.5 - 2.5)T_t$ in cadmium. In the temperature range of
our measurements the transversely polarized phonons effectively

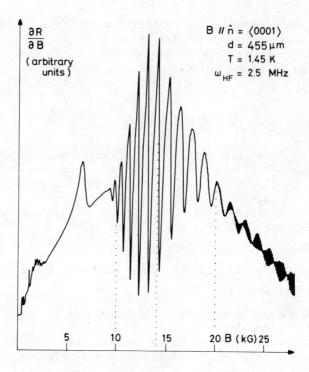

Fig.1 : Typical record of dR/dB versus B.

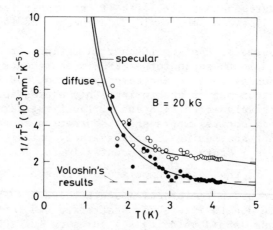

Fig.2 : $\ell^{-1}T^{-5}$ versus T obtained assuming specular (upper points) or diffuse (lower points) reflection of the electrons at the sample surface.

dominate ℓ^{-1}(T) above ~2K (T$_t$ ~ 20K). Below 2K the temperature
dependence is given by the T^3 term. In Figs.2,3 the T-dependence
of the mfp is plotted as ℓ^{-1}T^{-5} versus T to emphasize deviations
from a pure T^5 law. The scatter in the data is correspondingly
magnified at low T as is well known for this type of plot.

In Figure 2 we present the observed values of ℓ^{-1}T^{-5}(T)
together with the fitted ℓ^{-1}(T) (Eq.16, ref.3) and Eq.(1) (upper
points) or (2) (lower points). It can be seen that agreement is
obtained with Voloshin's result at 4K if diffuse electron
reflection at the sample surface is assumed. It is most important
to note that taking into account the T dependence of R^2(B,T) does
not suppress the observed deviations from a simple T^5 law at this
or any other magnetic field.

The field dependence of the mfp is shown in Figure 3.
The experimental points (filled circles) are obtained by fitting
Eq.(2) to the data whence $\ell^{-1} = d^{-1}(\ln(A(0)/A(T))-2\ln(R(0)/R(T)))$.
Parallel field and tilted field RFSE measurements are shown for
comparison. These orbits are shown on a cross-section of the 3rd
band FS sheet of cadmium (Fig.4). The data at the particular
values of θ were selected from a series (see ref.3) to correspond
approximately with the perpendicular field results at 4K for the
indicated field values.

As was mentioned in the introduction the mfp of the
resonant electrons (for example the shaded belt in Fig.4) is a
local value since the $\vec{k}$-states through which the magnetic field
drives these electrons can be supposed to have the same ℓ^{-1}(T,$\vec{k}$)
because of the high symmetry of the magnetic field direction.
For the other orbits shown in Figure 4, the observed mfp is an
orbitally averaged value. There remains the question of determining
the value k$_z$(B) of the belt of resonant electrons. In principle
it would be possible to obtain k$_z$(B) from the theoretical
expression for the field dependence of the surface impedance.
Here we shall only attempt to qualitatively locate the belt of
resonant electrons by comparison with our other RFSE measurements,
and by using the field dependence of the period of the
oscillations.

The ℓ^{-1}(T) measured at 20kG is slightly larger than
that observed on the tilted field orbit. As the electron-phonon
scattering is weakest at the <0001> point on the FS (most free
electron like point) we tentatively locate the belt of resonant
electrons at 20kG as shown by the shaded region in Figure 4.
At 14 and 10 kG, the observed ℓ^{-1}(T) are comparable with the
RFSE orbits for θ = 6° and 2° respectively and the belt of
resonant electrons is located closer to the rim of the FS sheet.

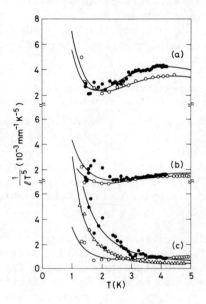

<u>Fig.3</u> : $\ell^{-1}T^{-5}$ measured at different perpendicular magnetic field
(o) : a) 10 kG, b) 14 kG, c) 20 kG. Parallel field RFSE
values of $\ell^{-1}T^{-5}(0)$ for the extremal orbits with $\vec{B}$ at
an angle θ = a) 2°, b) 6°, c) 20° from <0001>.
Tilted field values of $\ell^{-1}T^{-5}(\Delta)$.

At high fields the resonant electrons will be concentrated around the <0001> spherical limiting point and the period of the Gantmakher–Kaner oscillations is determined by the Gaussian radius of curvature $K^{-1/2}$ of the limiting point.

$$\Delta B_{GK} = \frac{2\pi}{ed} \hbar \, K^{-1/2}$$

At lower fields the Fermi velocity of the resonant electrons is at an angle α (see Figure 4) with respect to $\vec{B} \,/\!/\, <0001>$. If this belt is taken to be on a sphere whose radius of curvature is $K^{-1/2}(\alpha)$ then the ratio of the period at B to the Gantmakher–Kaner period is

$$\frac{\Delta B(\alpha)}{\Delta B_{GK}} = \cos \alpha \, \frac{K^{-1/2}(\alpha)}{K^{-1/2}(\alpha=0)} \tag{5}$$

If one assumes the free electron value $K^{-1/2}(0) = 1.42 \; \overset{\circ}{A}{}^{-1}$ then the period of the oscillation at 20 kG is exactly ΔB_{GK} and the belt of resonant electrons must be assumed to be much closer to the limiting point than shown in Figure 4 in contradiction with the arguments given above concerning its position. In fact, we have observed previously[2] that the period of these oscillations tends to saturate at the value such that $K^{-1/2}(0) \gtrsim 1.65 \; \overset{\circ}{A}{}^{-1}$, larger still than the value of $1.49 \; \overset{\circ}{A}{}^{-1}$ obtained by Fisher et al. Assuming $K^{-1/2}(0) = 1.65 \; \overset{\circ}{A}{}^{-1}$ we obtain $\Delta B/\Delta B_{GK} = 0.4, \; 0.75$ and

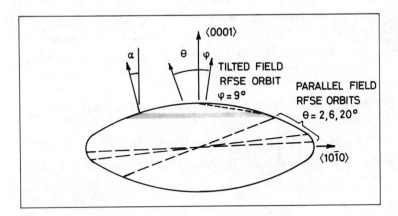

Fig.4 : Section of the 3rd band Fermi surface sheet of cadmium in the [11$\bar{2}$0] plane. The orbits shown by the dashed lines are referred to in Figure 3. The shaded band is the approximate position of the belt of resonant electrons in a perpendicular magnetic field at 20 kG.

0.9 at 10, 14 and 20 kG respectively. Using the observed variation of $K^{-1/2}(\alpha)$ (Fig.3, ref.6) one obtains the corresponding values $\alpha \cong 35°$, 25° and 14°. In Figure 4, $\alpha = 14°$ and this is seen to agree with our previous tentative assignment of the position of the belt of resonant electrons at 20 kG.

In conclusion we have shown that the electron-phonon scattering on the 3rd band FS sheet is effective and cannot be described by a T^5 law. The observed $\ell^{-1}(T)$ are well fit by a simple model which predicts that above ~2K the T dependence is dominated by intersheet scattering. The measurement of $\ell^{-1}(T)$ using dopplerons and Gantmakher-Kaner oscillations as a function of magnetic field can give a determination of the variation of $\ell^{-1}(T)$ on the Fermi surface in high symmetry directions where collisionless Landau damping is not present.

Acknowledgement — The authors would like to thank the Fonds National Suisse pour la Recherche Scientifique for financial support.

REFERENCES

1. IF Voloshin, LM Fisher and VA Yudin, Sov.Phys. Solid State 20, 405 (1978) and references therein.

2. M Berset, WM MacInnes and R Huguenin,, Helv.Phys.Acta 50, 381 (1977) and M Berset, Travail de diplôme (1976) (unpublished).

3. PA Probst, WM MacInnes and R Huguenin,, J.Low Temp.Physics 41 (to be published 1980) and PA Probst, Thèse de doctorat 1978 (unpublished).

4. PA Probst, B Collet and WM MacInnes, Rev.Sci.Instrum. 47, 1522 (1976).

5. DK Wagner and R Bowers, Advances in Physics 27, 651 (1978).

6. A Myers, SG Porter and RS Thompson, J.Phys.F : Metal Phys. 2 (1972).

ON THE HOMOMORPHIC CLUSTER CPA

J. van der Rest, P. Lambin, and F. Brouers

Institut de Physique
B5, Université de Liège
4000 Sart Tilman, Belgium

Among the various extentions of the CPA, the homomorphic cluster
CPA (HCPA) was presented recently as the only approximation (i)
having the correct analytical properties, (ii) taking multiple
scattering into account and (iii) preserving the symmetry of the
original lattice. Though very appealing mathematically, we show
in this paper that it does not give the right position for the
impurity states in the dilute limit with non zero diagonal disorder.
Consequently, even if HCPA satisfies points (i) to (iii) and even
if it gives the correct dilute limit for purely off-diagonal disorder, it is to be rejected as a method to study systems with non-
negligible diagonal disorder.

1. INTRODUCTION

The study of the electronic properties of disordered
systems made a decisive step forward in the late sixties with
Soven's proposal of the coherent potential approximation (CPA)[1].
This theory has been extensively studied and applied since them,
and its possibilities and limitations are now well established[2].
One of these limitations arises from the fact that, being a single
site approximation, all multiple scattering effects are neglected;
the general features of the density of states are thus correctly
given by CPA, but the details, which arise from multiple scattering
effects, are completely absent. Several methods have then been
proposed in order to overcome this shortcoming, but the first
attempts[34] did not satisfy the condition that the density of
states be continuous and single valued in the whole energy range[5].
The only extension of the CPA which takes multiple scattering into

account and which satisfies the analyticity requirements is the molecular CPA (MCPA)[6], but its numerical difficulties have limited its use up to now.

The homomorphic cluster CPA (HCPA) put forward by Yonezawa and Odagaki[7] seemed very promissing because it joints the advantages of the MCPA with the simplicity of the ordinary CPA in that it takes multiple scattering into account but avoids lengthly integrations in reciprocal space. Unfortunately, we show in this paper that HCPA leads to unphysical results because an electron, upon arrival on a given atom, diffuses only on a fraction (1/Z) of the potential on that atom, Z being the coordination number of the lattice. Consequently, HCPA does not give the correct position for the impurity states in the dilute limit with non zero diagonal disorder. Furthermore, we have also found numerically that the theory does not give the correct weight to the impurity band in the split band regime.

The outline of the paper is as follows : in a first part (§ 2) we analyse the mathematical and the physical aspects of the HCPA, in a second part (§ 3) we present some numerical results and in a last part (§ 4) we present our conclusions.

2. THEORETICAL ASPECTS OF THE HCPA

For simplicity, we will keep the same notations as [7] and discuss only the case when the lattice is divided into all possible pairs of nearest neighbours. The generalization to larger clusters is straightforward.

The essence of the HCPA is to divide the lattice into pairs of nearest neighbours atoms and to divide the potential of each atom between all the pairs each atom belongs to, the number of these pairs being equal to Z. All the pairs, but the one under study, are then replaced by effective pairs which are chosen in such a way that the average scattering on all the pairs under study vanishes.

The partitionning of the lattice into pairs is irreprochable, but the partitionning of the potentials into several parts is physically meaningless : when on electron arrives on a specific site, it encounters the whole potential of that site and not a fraction (A/Z) of that potential plus a fraction ((Z-1)/Z) of the effective potential.

This van be illustrated most clearly for the case of a pair of impurity atoms (A) in a pure host (B) with diagonal disorder only ($v_{AA} = v_{BB} = v_{AB} = v$, $\delta = \varepsilon_A - \varepsilon_B$). With Yonezawa's notation, the Green's function for the impurity pair is

$$G = \frac{\begin{pmatrix} G^B_{11} & G^B_{12} \\ G^B_{12} & G^B_{11} \end{pmatrix}}{\begin{pmatrix} 1 - \alpha\delta G^B_{11} & -\alpha\delta G^B_{12} \\ -\alpha\delta G^B_{12} & 1 - \alpha\delta G^B_{11} \end{pmatrix}} \tag{1}$$

with $\alpha = 1/Z$ in the HCPA and $\alpha = 1$ in the CCPA[3]. In the strong scattering limit, the states localized on the impurities arise at the energies corresponding to the zero's of the denominator, i.e. at the energies for which

$$(1 - \alpha\delta G^B_{11})^2 - (\alpha\delta G^B_{12})^2 = 0 \tag{2}$$

Now, it is well-known that a localized state arises on a single impurity when

$$1 - \delta G^B_{11} = 0 \tag{3}$$

and this condition should coincide with the condition for two non-interacting impurities (i.e. $G^B_{12} = 0$); this implies $\alpha = 1$ in equation (2). Consequently, the HCPA as proposed by Yonezawa does not give the correct dilute limit for non zero diagonal disorder.

3. NUMERICAL RESULTS

        The numerical results presented in 7 correspond to the very particular case of purely off-diagonal disorder where the deficiencies of the HCPA do not show up because the HCPA treats the off-diagonal disorder correctly but the diagonal disorder incorrectly. We present therefore on figure 1 the results obtained by the HCPA in the case of strong diagonal disorder and compare them with CPA[1], CCPA[3] and with the exact dilute limit in the case of a linear chain. We see immediately that the position and weight of the impurity band as given by HCPA differ completely from the results obtained by the other methods; in particular, the single (S) and pair (P) impurity states, which do not depend on any effective medium because they correspond to one or two impurities in a pure host, are completely misplaces by HCPA. In fact, the HCPA single and pair impurity states correspond to

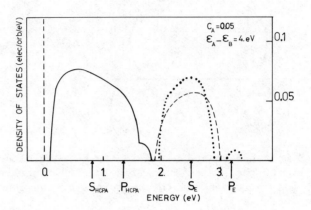

Fig.1 : Minority band of the density of states of a linear chain
with diagonal disorder only. Full line : HCPA; dashed line :
CPA; dotted line : CCPA. $S_E(S_{HCPA})$ and $P_E(P_{HCPA})$ indicate
the exact (HCPA) single and pair impurity states.

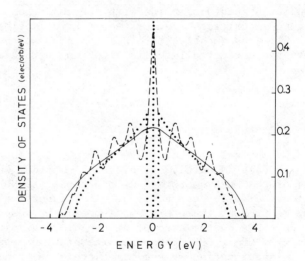

Fig.2 : Density of states of a simple cubic lattice with 60 % of
broken bonds. Full line : HCPA; dotted line : CCPA; dashed
line : reconstruction via a moment technique.

a scattering potential Z times too small; in a three dimensional lattice, the discrepancy between HCPA and the other methods is therefore still worst.

The inadequacy of HCPA to handle diagonal disorder being thus well established, let us comment on the numerical results presented in 7. On figure 2 we present the HCPA results (full line) obtained for a simple cubic lattice (Z = 6) where 60 % of the bonds between nearest neighbours have been broken : the comparison with the CCPA results (dotted line) and with a reconstruction via a moment technique (dashed line) shows that the overall agreement between the three methods is good. Taking a closer look at the figure, one can nevertheless point out that HCPA over-estimates the density of states (DOS) at the band's edges and under-estimates the DOS at the center of the band,while CCPA has the opposite tendency. This can be put more quantitatively if we look directly at the moments of the DOS : all HCPA's even moments (except the zeroth one) are larger than the corresponding moments obtained by simulation which in turn are larger than the CCPA's corresponding moments. HCPA seems therefore a reliable method to handle purely off-diagonal disorders, though the improvement with respect to CCPA is difficult to appreciate.

4. CONCLUSIONS

Among all the attempts to go beyond the CPA, the HCPA has really a unique position : it is the only method we know which reaches the three main goals all other attempts tryed, but failed, to reach, namely (i) to have the correct analytical properties, (ii) to take multiple scattering into account and (ii) to preserve the symmetry of the original lattice, but it is also the only method we know which does not give the correct dilute limit for non zero diagonal disorder. The use of the HCPA should therefore be limited to the case of very small diagonal disorder with any off-diagonal disorder; it is indeed easy to show that it gives the correct dilute limit for the case of purely off-diagonal disorder.

REFERENCES

1. Soven P., Phys.Rev. 156, 809 (1967).

2. Elliott R., Krumhansl J. and Leath P., Rev.Mod.Phys. 46, 465 (1974).

3. Nickel B. and Krumhansl J., Phys.Rev. B4, 4357 (1971); Ducastelle F., J.Phys. F2, 468 (1972).

4. Brouers F., Cyrot M. and Cyrot-Lackmann F., Phys.Rev. B7, 4370
   (1973); Brouers F., Ducastelle F., Gautier F. and van der
   Rest J., J.Phys. F3, 2120 (1973).

5. Nickel B. and Butler W., Phys.Rev.Letters 30, 373 (1973).

6. Tsukada M., J.Phys.Soc.Japan 27, 684 (1969); Ducastelle F.,
   J.Phys. C7, 1795 (1974).

7. Yonezawa F. and Odagaki T., Sol.Stat.Comm.27, 1199 (1978);
   Odagaki T. and Yonezawa F., Sol.Stat.Comm. 27, 1203 (1978).

PSEUDOPOTENTIALS AND ALLOYING BEHAVIOUR

A. de Rooy, E.W. van Royen, and J.T.W. de Hosson

Department of Applied Physics, Materials Science Centre
Rijksuniversiteit Groningen
Nijenborgh 18, 9747 AG Groningen, The Netherlands

The effective interatomic potentials of the transition metals are
reported which are calculated within the framework of the pseudopotential approximation based on a model potential of the Heine-
Abarenkov type. The results obtained are combined with the cluster
variation method in order to make predictions on the phase diagram
of the ternary system CuNiZn. The predictions on the phase diagram
are in reasonable agreement with experiments taking into account
more-body forces and are topologically correct.

1. INTRODUCTION

The study of phase diagrams of metal systems is important,
not just for the theory of metals but also for many practical issues
of physical metallurgy and metals engineering. Progress in the
study of the electronic structure of alloys will lead to a better
understanding of their metallurgical properties and even to the
possibility of making predictions on their phase diagram.

The problem of understanding the stability of phases
and predicting phase diagrams in terms of interactions between
the constituent atoms has a long history. However, there are still
many unanswered questions left. The theoretical interpretations of
order - disorder phase transformations have been based mostly on
various approximations to the classical Ising model. The various
models involve presuppositions regarding the extent of multisite
correlations and the manner in which these correlations influence
the alloy free energy. However, since the principal aim of these
theories is to provide a realistic description of the temperature

47

dependence of phase stability, it is important that the assumptions are based on first principle calculations whenever possible.

    In the present contribution the results of the calculation of the phase diagram of the ternary system CuNiZn are reported by a method which is as far as we know novel in this field. In this work we combine effective interatomic potentials which are calculated within the framework of the pseudopotential approximation, with the so called cluster variation method (CVM) [1], i.e. a combination between a pseudopotential theory for transition metals and a combinatorial theory for the configurational contribution to the alloy free energy. This approach is chosen because one of the more appealing features of the pseudopotential theory is that it permits calculations of the total energy as an analytic function of the configuration of the different kinds of individual ions. It greatly facilitates the study of the many physical properties which depend on the change in the total energy when the ions are arranged at constant volume. As a matter of course, the pseudopotential theory provides only effective interatomic interaction functions. However, a strictly pair-wise potential is not valid for metals since the total energy of these crystals contains other interaction terms due to electron-electron interactions etc.. The aim of this research is therefore to gain some insight into the influence of more-body interactions. The pair-wise interactions are taken into account to predict the order - disorder and order - order transformations in the f.c.c. part of the ternary system CuNiZn using CVM. This diagram is particularly interesting since in the f.c.c. part two ordered structures occur.

    Detailed information about the coherent phase diagram of the CuNiZn system, calculated using the CVM, will be published elsewhere. In the following, we will restrict ourselves mainly to the pseudopotential model for the stoichiometric Cu_2NiZn composition and the final result obtained using the cluster variation method.

2. THEORETICAL FRAMEWORK

    The basic idea of a pseudopotential is to replace the strong potential inside the core of an ion by a weaker potential which eliminates the bound core states but has the same energy eigenvalues for the valence state. The pseudopotential method has been extensively discussed by several authors [2] and for a comprehensive survey reference should be made to the review of Heine and Weaire [3]. We have applied the method of Heine and Abarenkov [4] for obtaining a model pseudopotential for metals which is an extension of the quantum defect method [5]. The model

potential has been used by Animalu [6] to calculate the phonon
dispersion curves for a number of noble and transition metals.

The transition-metal model potential of the Heine-
Abarenkov type reads :

$$w_0 = - \frac{Z}{R_m} - \Sigma^\ell_{\ell_0} \; \theta(R_m-r) \, [A_\ell(E) - \frac{Z}{R_m}] \, P_\ell \tag{1}$$

where P_ℓ is the projection operator that picks out the ℓ-th
angular momentum component of an incident one - electron wave
function. $A_\ell(E)$ is the energy-dependent depth of the core potential
well for a given model radius R_m. Z represents the chemical valence
of the isolated ion of the metal. $\theta(R_m-r)$ is the step function :

$$\theta(R_m-r) = \begin{cases} 1 & \text{if} \quad r \leq R_m \; , \\[2mm] 0 & \text{if} \quad r > R_m \; . \end{cases} \tag{2}$$

The model potential is, in fact, a non local operator
which depends on the energy. This dependence is determined first
by considering the free ion, described by the Schrödinger equation
containing the eigen-function $\psi(r)$, eigenvalue E and potential
$w_0(E,r)$. Since we are dealing with metals, the one-electron model
Hamiltonian has to be modified including the potential seen by an
electron due to the other conduction electrons : $V_{SC}(r)$.
The modified Hamiltonian is :

$$[\frac{1}{2} p^2 + \Sigma_i w_0(E'_k,\vec{r}-\vec{R}_0) + V_{SC}(r)] \, \psi_k(\vec{r}) = E_k \psi_k(\vec{r}), \tag{3}$$

where $\vec{R}_i$ is the position vector of the i-th ion.
In order to find the relation between E_k and E'_k, consider the
Schrödinger equation in a cell around one ion at $\vec{R}_0$:

$$[\frac{1}{2} p^2 + w_0(E'_k,\vec{r}-\vec{R}_0)] \, \psi_k(\vec{r}) =$$

$$[E_k - \underset{i\neq 0}{\Sigma} \; w_0(E'_k,\vec{r}-\vec{R}_i) - V_{SC}(r)] \, \psi_k(\vec{r}) \tag{4}$$

Eq. (4) is identical to the Schrödinger equation in the case of a
free ion for r in the cell around $\vec{R}_0$ to the extent that the core
shift Δ_k is independent of r :

$$\Delta_k = \underset{i\neq 0}{\Sigma} \; w_0(E'_k,\vec{r}-\vec{R}_i) + V_{SC}(r) \tag{5}$$

As a result $E'_k = E_k - \Delta_k$ and the value for $A_\ell(E)$ in equation (1)
should be taken at $E = E'_k$. Because Δ_k (eq.5) depends on $(\vec{r}-\vec{R}_i)$ we
have approximated Δ_k by taking the mean value over the unit cell.

This allows one to consider the change Δ_k on alloying for dilute and also for concentrated alloys without the necessity of knowing the structure factors in the alloy. This assumption has been made also for liquid alloys [7] and leads to the following coupled equations in the case of pure metals :

$$\Delta_k = \frac{3Z}{2R_A} - \frac{1}{3} k_F^2 - \phi(k_F),$$ (6)

$$E_k = \frac{1}{2} k - \frac{1}{3} k_F^2 + \phi(k) - \phi(k_F),$$ (7)

where R_A is the radius of the Wigner-Seitz cell, k_F the Fermi wave vector and $\phi(k)$ is given by :

$$\phi(k) = \sum_{\ell=0}^{\ell} <\vec{k}|w_0(E',r) + \frac{Z}{R_m}|\vec{k}> .$$ (8)

For binary and ternary alloy systems similar equations can be written where Δ_k depends on k.

One of the parameters required for the calculation of the total interionic interaction function is the effective valence Z. In order to obtain a concentration dependent potential for an alloy we take $\overline{Z}$ equal to $\Sigma_i c_i Z_i$ where c_i represents the concentration of the i-th kind of atom. In addition, k_F depends on the mean atomic volume $\overline{\Omega}$. To a first approximation, we take $\overline{\Omega}$ equal to $\Sigma_i c_i \Omega_i$. In that way and using the equivalence of equation (6) and equation (7) the interactions are not the same in the alloy as in the pure metals.

The values of $A_\ell(E)$ at several values of energy E are obtained from spectroscopic data for free atoms and then extrapolated to the energy of the solid (see ref.[6]). This does not represent an experimental pseudopotential, since experimental information is introduced at the most fundamental level. It is more akin to the a priori calculation but has the advantage of including experimental corrections for correlation and exchange between conduction and core electrons.

3. RESULTS AND DISCUSSION

In figure 1 the ordering energies W_{CuNi}, W_{NiZn} and W_{CuZn} as a function of r_j are depicted. r_j is the distance to the j-th site and the ordering energy $W_{ij}(k)$ is defined by :

$$W_{ij}(k) = V_{ij}(k) - (V_{ii}(k) + V_{jj}(k))/2$$ (9)

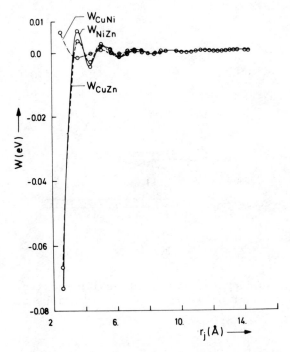

<u>Fig.1</u> : The ordering energies W_{CuNi}, W_{NiZn} and W_{CuZn} as a function of r_j.

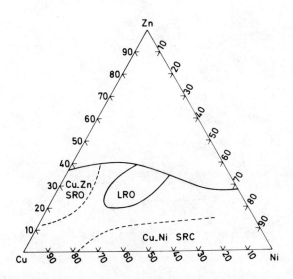

<u>Fig.2</u> : Experimental phase diagram of the CuNiZn system.

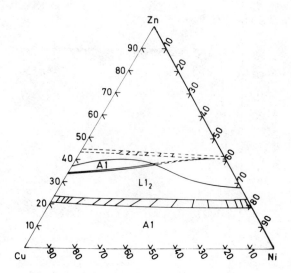

Fig.3 : Calculated phase diagram of CuNiZn ordered region taking
        into account only pair-wise interactions at 685 K.

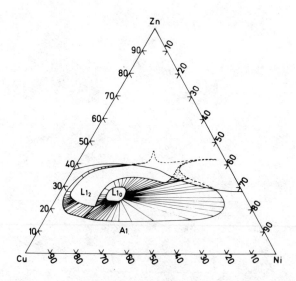

Fig.4 : Coherent phase diagram of CuNiZn (A_1 : disordered f.c.c.,
        Ll_2 : ordered A_3B, Ll_0 : ordered A_2BC).

where $V_{ij}(k)$ is the pairwise interaction energy between the k-th neighbouring atom i and j. The experimental phase diagram [8] [9] [10] has been depicted in figure 2.

The phase diagram calculated using CVM and the ordering energies of the first shell (figure 1) is shown in figure 3. The calculated phase diagram is not in agreement with the experimental phase diagram. Around the composition $Cu_{50}Ni_{25}Zn_{25}$ two different types of ordering are found experimentally: one from disorder to A_3B ($L1_2$) below 770 K and the other from A_3B to A_2BC ($L1_0$) in CuNiZn below 700 K. More-body forces α_{ijkl} have to be introduced because the asymmetry in the binary systems. Indicating Cu by 1, Ni by 2 and Zn by 3, the following more-body forces were taken in the calculation of the phase diagram as shown in figure 4 :
α_{1113} = -0.075, α_{2223} = -0.15, α_{1123} = 0.3, α_{1223} = -0.2, α_{1112} = -1. However, the chosen set of parameters is certainly not unique. In general, all pair- and more-body interactions are varied in the CVM to obtain a best fit with the experimental phase diagram. Here, the pair-wise interactions result from basic principles calculations and are kept fixed. This leaves the many-body interactions for fitting. It is rather our aim to vary only some of these more-body parameters within a limited range in order to investigate how the coherent phase diagram approaches the experimentally known diagram. Mainly due to the high value of α_{1123} the critical temperature is about 180 K too high as compared with the experimental observation. At any rate, around the composition 50/25/25 the A_2BC disorder is more stable, caused by α_{1123}, than the A_3B-disorder which is in occurance with experiments.

In conclusion we may state that using pair-wise interactions from transition metal model potentials and more-body forces as independent variables in the CVM tetrahedron approximation we are able to predict a phase diagram of the ternary system which is in reasonable agreement with experiments. This is in sharp contrast to the Bragg-Williams-Gorski approximation which predict phase diagrams which are not even topologically correct. Moreover, the applicability of the transition-metal model potential supports our guiding consideration [11] of the existence of a unifying interaction function which does not only predict phase diagrams but also other structural aspects such as the separation between the Shockley partial of $1/2<110>$ superlatice dislocations in ternary systems like CuNiZn [12] [13] .

Acknowledgement - This work is part of the research program of the Foundation for Fundamental Research on Matter (F.O.M. - Mt VI-2, Utrecht).

REFERENCES

[1] R. Kikuchi, Physical Review, 81, 988 (1951).

[2] W.A. Harrison, Pseudopotentials in the Theory of Metals,
     (Benjamin, Reading) 1966, p.5.

[3] V. Heine, D. Weaire, Solid State Physics (Academic Press,
     London) 1970, p.249.

[4] V. Heine, I. Abarenkov, Philosophical Magasine, 9, 451 (1964).

[5] F.S. Ham, Solid State Physics, Academic Press, New York, 1,
     127 (1955).

[6] A.O.E. Animalu, Physical Review, B8, 3542 (1973).

[7] M.J. Huyben, W. van der Lugt, W.A.M. Reimert, J.Th.M. de
     Hosson, C. van Dijk, Physics 97 B+C, 338 (1978) and references.

[8] H. Thomas, Zeitschrift für Metallkunde, 63, 106 (1963).

[9] W. Köster, R. Störing, Zeitschrift für Metallkunde, 51, 182
     (1972).

[10] A. Kuzsmann, H. Wollenberger, Zeitschrift für Metallkunde, 50,
     94 (1959).

[11] J.Th.M. de Hosson, American Institute of Physics, 53, 146
     (1979).

[12] G.J.L. van der Wegen, P.M. Bronsveld, J.Th.M. de Hosson,
     Scripta Metallurgica, 13, 303 (1979).

[13] J. de Groot, P.M. Bronsveld, J.Th.M. de Hosson, Physica
     Status Solidi (a), 52, 635 (1979).

ELECTRONIC TRANSPORT IN INTERMEDIATE VALENCE COMPOUNDS

H.J. Leder
Institut für Theoretische Physik
Universität Köln
D-5000 Köln 41, West-Germany

G. Czycholl
Institut für Physik
Universität Dortmund
D-4600 Dortmund 50, West-Germany

To describe transport properties of intermediate valence
compounds, we study the periodic Anderson model with off-site
hybridization. A particle density operator which is consistent
with the model assumptions is derived, and the current operator
is determined via the continuity equation. The model is treated
within the alloy analog approximation, using the CPA to solve
the alloy problem. The static conductivity and other transport
quantities are then calculated within the CPA theory of electronic
transport.

1. INTRODUCTION

Intermediate valence compounds are of immediate
experimental and theoretical interest because of their quite
unusual physical properties[1,2]. Besides the single particle
quantities like magnetic susceptibility and specific heat, also
transport quantities of mixed valence systems have extensively
been investigated experimentally[3-6]. The temperature dependences
of the resistivity, the thermopower, the thermal conductivity,
etc. all show certain interesting anomalies, which are regarded
as characteristic features for the mixed valent state, and which
are not yet understood theoretically. The resistivity of $CePd_3$[4,6],

for instance, shows an increase proportional to T^2 for low temperatures, a broad maximum between 100 K and 200 K and a decrease for higher temperatures.

It is widely accepted[2,7] that the periodic Anderson model (PAM) may serve as a simple starting point for the theoretical description of intermediate valence systems. So far, there are only a few very recent attempts to study the transport properties of the PAM[8,9]. These works investigate the importance of phonon scattering[9] and of alloying[8] on the resistivity which is calculated within the memory function approach. However, the strong Coulomb correlation of the localized f-electrons is treated only in Hartree-Fock approximation. On the other hand, the alloy analog approximation (AAA), which is certainly superior to a Hartree-Fock treatment, has recently been applied to the PAM[10-12]. In particular, it has been shown by the authors[11] that this approach gives good agreement with experiment for single-particle quantities. Therefore, one should try to describe also the transport properties of the PAM within an approach which is consistent with the AAA.

In this paper, we present preliminary results of an application of the alloy analogy to the calculation of transport quantities for the PAM. Here we consider off-site hybridization, which is only allowed from inversion symmetry, and determine the current operator, consistent with the model Hamiltonian, via the continuity equation. The current operator necessarily contains mixing terms between band (d) and localized (f) states because hopping from a Wannier d state into a nearest neighbour f state and vice versa obviously contributes to the current. By applying the AAA to this model, the many particle Hamiltonian is replaced by the sum of two single particle "alloy" Hamiltonians the "concentrations" of which are just the f-electron occupation numbers and must, therefore, be determined self-consistently[11]. This "alloy" problem is solved within the CPA[13] as a second approximation. Then the CPA transport theory[14] is used for the calculation of the transport quantities of the model. This treatment ensures that the approximation for two-particle Green functions is consistent with the approximation for one-particle Green functions in the sense that Ward identities remain valid. As in the original CPA[14] the vertex corrections for the current operator vanish even in this two-band problem.

There already exist some applications of the CPA transport theory to two band models in connection with real alloys (with fixed concentrations)[15,16]. It has also been proposed[10,17,18] to study the influence of local Coulomb correlations on transport properties within the AAA for hybridized two band models. However, in none of the cited works an off-site hybridization and a current containing explicitely hybridization terms has been considered, and no practical calculation of the temperature dependence of actual transport quantities has been performed.

2. THE MODEL, HAMILTONIAN AND CURRENT

We study the model Hamiltonian

$$H = H^d + H^f + V^{fd} \tag{1}$$

with

$$H^d = \sum_{k\sigma} \varepsilon_k \, d^+_{k\sigma} d_{k\sigma} = \sum_{i \neq j, \sigma} t_{ij} \, d^+_{i\sigma} d_{j\sigma} \tag{2}$$

$$H^f = \sum_{i\sigma} (E_o f^+_{i\sigma} f_{i\sigma} + \frac{U}{2} f^+_{i\sigma} f_{i\sigma} f^+_{i-\sigma} f_{i-\sigma}) \tag{3}$$

$$V^{fd} = \sum_{i \neq j\sigma} V_{ij} (f^+_{i\sigma} d_{j\sigma} + d^+_{i\sigma} f_{j\sigma}) = \sum_{k\sigma} V(k) (f^+_{k\sigma} d_{k\sigma} + hc) \tag{4}$$

The three terms describe the conduction (d-band), a lattice of localized f-levels including the Coulomb repulsion U between electrons of different spin-direction at the same lattice site, and a hybridization between localized f-levels of a site i and d-levels of an alternative site j. Thus the hybridization becomes explicitly k-dependent. Making the tight binding assumptions

$$t_{ij} = \{ \begin{array}{l} t \text{ for } i, j \text{ nearest neighbours} \\ 0 \text{ otherwise} \end{array} \tag{5}$$

and

$$V_{ij} = \{ \begin{array}{l} V \text{ for } i, j \text{ nearest neighbours} \\ 0 \text{ otherwise} \end{array} \tag{6}$$

thereby choosing the d-band center to zero, we get obviously

$$V(k) = \frac{V}{t} \cdot \varepsilon(k) = \tilde{V} \, \varepsilon(k) \tag{7}$$

With the same assumptions, which necessarily underly the model (1 - 4), one gets for the Fourier transformed density operator in the limit of small q[19]

$$\rho(q) = \sum_{k\sigma} [d^+_{k+q,\sigma} d_{k\sigma} + f^+_{k+q,\sigma} f_{k\sigma} + 0(q^2)] \tag{8}$$

By means of the continuity equation $i[H,\rho(q)] = i \, q \, j \, (q)$ the longitudinal part of the electrical current in the limit $q \to 0$[19] is obtained as

$$j = e \lim_{q \to 0} \frac{qj(q)}{|q|} = e \sum_{k\sigma} \frac{\partial \varepsilon}{\partial k} [d^+_{k\sigma} d_{k\sigma} + \tilde{V}(d^+_{k\sigma} f_{k\sigma} + hc)] \qquad (9)$$

The static conductivity according to Kubo's Formula is then given by

$$\sigma = \lim_{\omega \to 0} \frac{Im \ \chi_{jj}(\omega + i0)}{\omega + i0} \qquad (10)$$

with

$$\chi_{jj}(z) = -\frac{1}{\Omega} i \int_0^\infty <[j(t),j]>_0 \ e^{izt} dt \qquad (11)$$

being the current-current response function.

3. APPROXIMATIONS, AAA AND CPA

Within the alloy analog approximation (AAA), the two-particle part H^f of our model Hamiltonian (1) is replaced by the one-particle "alloy"-Hamiltonian[11]

$$\tilde{H}^f = \sum_{i\sigma} E_i f^+_{i\sigma} f_{i\sigma} = \sum_\sigma \tilde{H}^f_\sigma \qquad (12)$$

with

$$E_i = \begin{cases} E_o & \text{with probability } 1 - \bar{n}^f_{-\sigma} \\ E_o + U & \text{with probability } \bar{n}^f_{-\sigma} \end{cases} \qquad (13)$$

Here the "alloy concentration" $\bar{n}^f_\sigma$ is the averaged f-electron occupation number of spin-σ-electrons. The averaged a-electron occupation number (a $\in \{f,d\}$) is given by

$$\bar{n}^a_\sigma = \int dE \ \rho^a_\sigma(E) \ f(E) = -\frac{1}{\pi} \int dE \ Im \ \bar{G}^{aa}_\sigma(E+i0) f(E) \qquad (14)$$

$\rho^a_\sigma(E)$ and $\bar{G}^{aa}_\sigma(E)$ denote the averaged density of states and single particle Green function of a_σ-electrons, $f(E) = (\exp(E-\mu)/kT+1)^{-1}$ is the Fermi-Function and μ is the chemical potential fixed by the total number of electrons per lattice site

$$n_o = \sum_\sigma (\bar{n}^f_\sigma + \bar{n}^d_\sigma) \qquad (15)$$

To treat the "alloy"-problem (12,13), i.e. to determine the Green functions

$$\bar{G}_\sigma^{aa}(z) = <a_\sigma|\bar{G}(z)|a_\sigma> \text{ for given } \bar{n}^f_{-\sigma}$$

the coherent potential approximation (CPA)[13] is used. The self-energy operator M(z) is defined by

$$\bar{G}_\sigma(z) = (z-H_\sigma^d - V_\sigma^{fd}-M_\sigma(z))^{-1} \tag{16}$$

where $G_\sigma(z) = (z-H_\sigma)^{-1}$ is the resol-ent operator and the bar denotes the configurational average. Within the CPA we have simply

$$M_\sigma(z) = \sum_i \Sigma_\sigma(z)\ f_{i\sigma}^+ f_{i\sigma} \tag{17}$$

and the f-electron self-energy $\Sigma_\sigma(z)$ is determined by the solution of the CPA-self-consistency equation

$$\overline{(E_{i\sigma}-\Sigma_\sigma(z))\ (1-\bar{G}_\sigma^{ff}(E_{i\sigma}-\Sigma_\sigma(z)))^{-1}} = 0 \tag{18}$$

The static conductivity (10) for the effective single-particle model (12,13) can be written[13] as

$$\sigma = \frac{2}{\pi\Omega} \int dE(-\frac{df}{dE})\ \overline{Tr(j\ ImG(E+i0)\ j\ ImG(E+i0))} \tag{19}$$

where again a configurational average has been performed. Introducing the current vertex operator Γ by

$$\overline{G(z)\ j\ G(z')} = \bar{G}(z)(j+\Gamma(z,z'))\bar{G}(z') \tag{20}$$

the CPA transport theory[14] can be applied to get an approximation for Γ which is consistent with the approximation for the self-energy in the sense that Ward identities hold. Within the CPA Γ is site-diagonal like the self energy, i.e.

$$\Gamma(z,z') = \sum_i \Gamma_i(z,z') \tag{21}$$

As in the usual CPA the current vertex operator vanishes identically[19]

$$\Gamma(z,z') = 0 \tag{22}$$

because of the symmetry properties of $\varepsilon(\underline{k})$.

Thus we get the CPA-conductivity

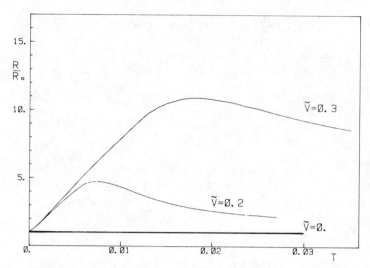

<u>Fig.1</u> : Temperature dependence of the resistivity for $E_o = 0$, U = 3, and $n_o = 2$.

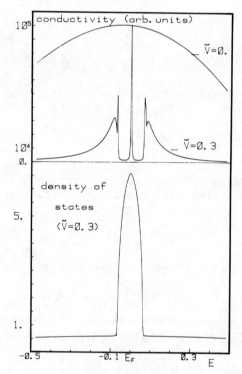

<u>Fig.2</u> : Total density of states and static conductivity as a function of the band energy E for $E_o = 0$, U = 3, $\tilde{V} = 0.3$, $n_o = 2$.

$$\sigma = \frac{2e^2 \cdot N}{\pi\Omega} \int dE (-\frac{df}{dE}) \frac{1}{N} \sum_k (\frac{\partial\varepsilon}{\partial k})^2 \cdot$$

$$\cdot \{ (Im\bar{G}_{k\sigma}^{dd}(E))^2 + 4\tilde{V} \, Im \, \bar{G}_{k\sigma}^{dd}(E) \, Im \, \bar{G}_{k\sigma}^{df}(E) + 2\tilde{V}^2[(Im\bar{G}_{k\sigma}^{df}(E))^2 + Im \, \bar{G}_{k\sigma}^{dd}(E) \, Im \, \bar{G}_{k\sigma}^{ff}(E)] \}$$

$$(23)$$

4. RESULTS

We use a semi-elliptical model density of states for the unperturbed d band of width 2, i.e. we assume

$$F_o(z) = \frac{1}{N} \sum_k \frac{1}{z - \varepsilon_k} = \frac{2}{\pi} (z - \sqrt{z^2 - 1}) \qquad (24)$$

Thus, energies are measured in units of half the band width. For the evaluation of the k-summations in (23) one needs a further model assumption consistent with (24). Following Ref.14, we assume

$$R(z) = \frac{1}{N} \sum_k \frac{1}{z - \varepsilon_k} (\frac{\partial\varepsilon}{\partial k})^2 = 2(z^2 - 1)^{3/2} - z(2z^2 - 3) \qquad (25)$$

For comparison with Ce Pd$_3$ measurements we choose the following model parameters : A localized level near the band center of the conduction band, i.e. $E_0 = 0$, a Coulomb repulsion U larger than the d-band width, e.g.~U = 3, a hybridization small compared to the d-band width i.e. V << U, and a total number of $n_o = 2$ electrons per lattice site so that the occupation number for the f states is less than one and the conduction band is partly filled. This corresponds roughly to a Ce compound with a valence between 3+ and 4+.

In figure 1 we show the resistivity as a function of the temperature. As our effective Hamiltonian describes a disordered system there is a small finite residual resistance at zero temperature which may be regarded as an artifact of the AAA treatment. What is more important, we observe a strong increase with temperature, a broad maximum in the intermediate temperature range, and a decrease of the resistivity for higher temperatures. This behaviour is qualitatively in good agreement with experimental results obtained for Ce Pd$_3$[6]. Figure 2 shows the effective density of states and conductivity as a function of energy. Near the local level E_o both quantities have pronounced structures which are strongly reflected in the temperature dependence of the conductivity, as the Fermi energy lies close to E_o.

A more detailed version of this work as well as further results will be published elsewhere[19].

REFERENCES

1. For a review on the experimental situation see : D. Wohlleben, Journ. de Phys. Coll. 37, C4-231 (1976).

2. For a review on the theoretical situation see : N. Grewe, H.J. Leder, P. Entel, to be published in Festkörperprobleme XX, ed. J. Treusch (Vieweg, Braunschweig 1980).

3. T. Penney, F. Holtzberg, Phys. Rev. Letters 34, 322 (1975).

4. P. Scoboria, J.E. Crow, T. Mihalisin, J.Appl.Phys. 50, 1895 (1979).

5. W. Franz, A. Grießel, F. Steglich, D. Wohlleben, Z. Physik B31 7 (1978).

6. H. Schneider, Diplomarbeit Köln (1980).

7. J.H. Jefferson, K.W. Stevens, J.Phys. C11, 3919 (1978).

8. Y. Ono, S.G. Mishra, Solid State Commun. (1980), to be published.

9. P. Entel, B. Mühlschlegel, Y. Ono, Z. Physik B (1980), to be published.

10. O. Sakai, S. Seki, M. Tachiki, Journ. Phys. Soc. Japan 45, 1465 (1978).

11. H.J. Leder, G. Czycholl, Z. Physik B35, 7 (1979).

12. M.R. Martin, J.W. Allen, J. Appl. Phys. 50, 7561 (1979).

13. B. Velicky, S. Kirkpatrick, H. Ehrenreich, Phys. Rev. 175, 747 (1968).

14. B. Velicky, Phys. Rev. 184, 614 (1969).

15. F. Brouers, A.V. Vedyayev, Phys. Rev. B5, 348 (1972); F. Brouers, M. Brauwers, Journ. de Physique Lett. 36, L-17 (1975).

16. P.N.Sen, Phys. Rev. B8, 5613 (1973).

17. K. Elk, J. Richter, V. Christoph, J. Phys. F9, 307 (1979).

18. S.K. Ghatak, M. Avignon, K.H. Bennemann, Journ. de Phys. Coll. 37, C4-289 (1976); J. Phys. F6, 1441 (1976).

19. G. Czycholl, H.J. Leder, to be published.

DEVIATIONS FROM MATTHIESSEN'S RULE DUE TO

SURFACE SCATTERING: ALUMINIUM

J. van der Maas and R. Huguenin
Institut de Physique Expérimentale de l'Université
de Lausanne, CH-1015 Lausanne, Switzerland

C. Rizzuto
Istituto di Scienze Fisiche and Gruppo Nazionale di
Struttura della Materia del C.N.R., Genova, Italy

An analysis of experimental data from various authors on the electrical resistivity of aluminium foils and wires shows that the DMR due to surface scattering varies linearly with the residual surface resistivity $\rho_o^s = \rho_o - \rho_o$ (bulk) and is the same for residual bulk resistivities ranging from 0.1-1 nΩcm. The temperature dependence of this DMR is consistent with a T^2-dependence below 20 K, and we estimate the coefficient of T^2 to be of the order of $10^{-3}\rho_o^s\Omega$cmK^{-2}. Sample dependent anomalies in this temperature dependence considerably complicate an estimate of the size effect below 4 K.

INTRODUCTION

The low temperature electrical resistivity of bulk aluminium has been studied extensively and it has been found that distinct scattering mechanisms such as impurities[1,2] and dislocations[3] can change the temperature dependent resistivity $\rho - \rho_o$ by a substantial amount. Surface scattering also, is known to produce deviations from Matthiessen's rule (DMR) but the available theories cover only approximately the experimental results[2,4].

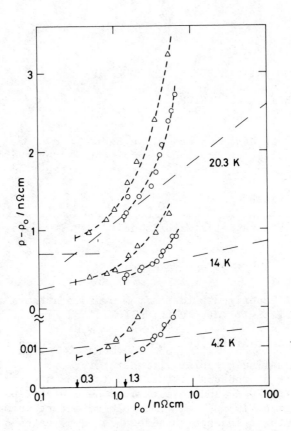

Fig.1 : Temperature dependent part of the resistivity, $\rho - \rho_o$, as a function of the residual resistivity ρ_o on a logarithmic scale, of aluminium foils and wires for three representative temperatures; data from ref. 5.

Dashed curves correspond to $\rho - \rho_o \propto \rho_o$.

Straight dashed lines represent the estimated impurity induced DMR $|1|$.

($\triangle$) foils[1], thickness 3,6-204 µm, ρ_o^{∞} = 0.3 nΩcm

($\bigcirc$) wires[1], diameter 13-820 µm, ρ_o^{∞} = 1.3 nΩ cm.

It is therefore interesting to see how deviations produced by surface scattering compare with those produced by other defects. This can best be done for aluminium because of the availability in the literature of a large number of studies of size dependent DMR in this metal.

PRESENTATION OF THE DATA ANALYSIS

In figure 1 we have plotted the temperature dependent part of the resistivity, $\rho - \rho_o$, of aluminium foils and wires from ref.5 as a function of residual resistivity, ρ_o, on a logarithmic scale. This is done for three representative temperatures. The residual resistivity of a specimen, ρ_o, can be expressed as the sum of the residual resistivity of the bulk, ρ_o^∞, and of a residual surface resistivity, ρ_o^s; $\rho_o = \rho_o^\infty + \rho_o^s$. The triangles and circles in figure 1 correspond to foils with $\rho_o^\infty = 0.3$ nΩcm and to wires with $\rho_o^\infty = 1.3$ nΩcm respectively. The straight dashed lines represent the estimate of Caplin and Rizzuto[1] for impurity induced DMR. Indeed, the thickest most bulk-like samples fall so close to these lines that no distinction can be made between a DMR due to impurity or to surface scattering. However, when the specimens with a higher value of ρ^s are taken into account the size induced DMR clearly appears, varying linearly and not logarithmically with ρ_o^s. This linearity was suggested earlier by Bass[2].

Figure 2 presents most published size effect data on aluminium and shows at 14 K and 20 K how the temperature dependent resistivity depends on ρ_o^s. To facilitate intercomparison of the data, we take as a reference the wire data of Aleksandrov[6] which are represented by circles in figure 2a. The solid lines represent a low estimate of the slope $\Delta(\rho-\rho_o)/\Delta\rho_o^s$ at 14 K and 20 K, and are given in Figs.2b-f by the dashed lines.

The scatter in the data is sometimes quite large, as in Reich's data[7] (Fig.2a, squares), and for higher surface resistivity saturation seems to take place (Fig.2f). However, at least for $\rho_o^s \lesssim 1$nΩcm, the various series of data are consistent with $\rho - \rho_o$ varying linearly with ρ_o^s, with a temperature dependent coefficient.

The fact that the data points are systematically grouped along the dashed lines (from ref.6) suggests that the surface resistivity of the different sets of samples would have

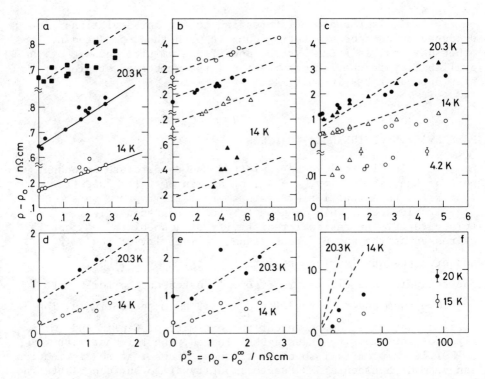

Fig.2 : Temperature dependent part of the resistivity as a function
of the residual surface resistivity, $\rho_o^s = \rho_o - \rho_o^\infty$, on a
linear scale at temperatures 14 K and 20.3 K; data from
ref.5-11.
The solid lines through the data of ref. 6 in Fig.2a
correspond to a low estimate of the slope $\Delta(\rho-\rho_o)/\Delta\rho_o^s$
at the temperature 14 K and 20.3 K, and are represented
in Figs.2b-f by dashed lines.

Fig.2a : $(0,\bullet)$ wires[6], diameter d = 0.2-3.6 mm, ρ_o^∞ =
0.09 nΩcm; ($\blacksquare$) wires[4], d = 0.3-4 mm, ρ_o^∞ = 0.1 nΩcm;

Fig.2b : oriented single-crystal films[8] :
(0) series A, ($\bullet$) series B, (Δ) series C, thickness
t = 30-140 μm; ($\blacktriangle$) series D, t = 63 $\pm$ 4 μm, for different
orientations, ρ_o^∞ = 0.11 nΩcm;

Fig.2c : same data as in figure 1.

Fig.2d : films[9], t = 17-950 μm, ρ_o^∞ = 0.12 nΩcm;

Fig.2e : single-crystal plates[10], t = 15-80 μm, ρ_o^∞ = 0.15
nΩcm;

Fig.2f : single-crystal films[11], t = 0.16-1.7 μm, ρ_o^∞ =
1.35 nΩcm.

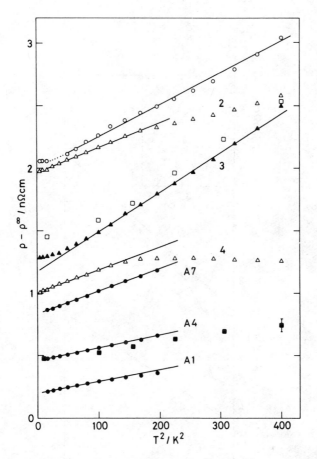

Fig.3 : The surface resistivity, $\rho - \rho^{\infty}$, as a quadratic function
of temperature for a selection of film samples; data
from ref.5-11.
Straight lines have been drawn through the data points
to serve as a guide for the eye.

($\bullet$) foils[8]; sample A_1 : t = 124 μ; sample A_4 : t = 62 μ;
sample A_7 : t = 32 μ, ρ_o^{∞} = 0.11 nΩcm.

($\triangle$,$\blacktriangle$) foils[10]; sample 4 : t = 35 μ; sample 3 : t = 24 μ;
sample 2 : t = 18 μ; ρ_o^{∞} = 0.15 nΩcm.

($\blacksquare$,$\square$) foils[9]; ($\blacksquare$) sample D, t = 117 μ; ($\square$) sample A,
t = 17 μ; ρ_o^{∞} = 0.12 nΩcm.

(O) foil[5], t = 15 μ, (shifted up by 0.4 nΩcm); ρ_o^{∞} = 0.3
nΩcm.

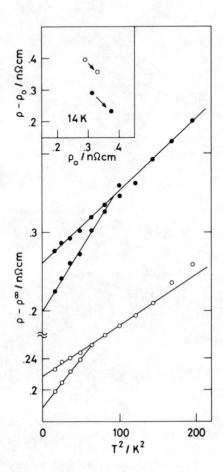

Fig.4 : As in figure 3.
Risnes and Sollien[12] measured their specimens twice and
observed a splitting of the curves, which they attribute
to a variation in the dislocation configuration.
The insert shows the result of the splitting on a DMR-
plot at 14 K.
(O) sample 10A1, t = 101 μm;
(O) sample 10B2, t = 98 μm.

the same temperature dependence. To investigate this, we determined the slope $\Delta(\rho-\rho_o)/\Delta\rho_o^s$ and found it to vary linearly with T^2 for temperatures between 4 K and 15 K. The detailed behaviour of the surface resistivity as a function of temperature is shown in figure 3 where $\rho - \rho^\infty$ is plotted against T^2 (ρ^∞ is the total bulk resistivity). Apart from the generally observed linearity with T^2, the particular samples exhibit a variety of behaviours : a) the T^2-variation may extend to higher (up to 20 K) or to lower (down to 2 K) temperatures, b) $\rho - \rho^\infty$ may level off at high and/or at low temperatures.

A striking anomaly is illustrated in figure 4. Risnes and Sollien[12] measured their samples twice and observed a splitting of the resistivity vs. temperature plots. When we plot the surface resistivity as a function of T^2 a kink is appearing at some temperature. They found[8] that during the annealing process, some strain could accidentally be introduced. This shows that kinks in the function $(\rho-\rho^\infty)$ (T^2) may be associated with the configuration of dislocations.

In the insert in figure 4, it is shown that at fixed temperature an increase in the residual resistivity is coupled to a decrease in $\rho - \rho_o$. This is the same behaviour as is observed in studies of the DMR due to dislocations[9].

Forgetting about the observed anomalies we estimate the coefficient of T^2 to be of the order of $10^{-3}\rho_o^s \Omega\text{cmK}^{-2}$. With $\rho \times \ell$ being $6 \times 10^{-12}\Omega\text{cmK}^{-2}$ in aluminium a wire sample of 1 mm diameter would have a coefficient of the order of $10^{-3}\rho\ell/d = 10^{-1}$ $\text{p}\Omega\text{cmK}^{-2}$ for residual bulk resistivities of 0.1-1 nΩcm.

DISCUSSION

There are several reasons why surface scattering is expected to contribute to the temperature dependence of the resistivity[2,4]. The most obvious is the fact that the electron distribution function depends on the position in the sample. Another mechanism suggested by Olsen[13] is small-angle electron phonon scattering, which is not an effective mechanism to produce resistivity in the bulk, but which in thin samples can scatter electrons to the surface where they suffer diffuse reflection. An approximate theory[14] for wires, based on Olsen's idea, predicts a DMR which varies as $(\rho\ell/d)^{2/3}T^{7/3}$. In our empirical approach we find that the data are well described by a similar expression $(\rho\ell/d)T^2$.

Diffuse scattering at surfaces can in principle enhance the effectiveness of all scattering processes which otherwise would generate little resistivity in the bulk, as is the case for electron-electron scattering[17] and small angle electron-dislocation scattering[12]. Therefore investigations of dislocation resistivity in wires[3] will not be representative for the bulk. Finally other authors conclude that the observed DMR are only related to a varying anisotropy of the scattering over the Fermi surface[8,12,15].

Nevertheless we have extended our analysis of size effect data to other metals, and observe that the DMR from surface scattering is qualitatively the same, which may point to a more universal mechanism.

In conclusion we have shown that surfaces can produce a T^2-resistivity which does not depend on the bulk mean free path but is only determined by the surface resistivity. The magnitude of this contribution is comparable to the observed T^2-resistivities of wire samples below 2 K, and which is presently attributed to electron-electron scattering[16]. In view of the observed anomalies, we suggest that a contribution from the size effect can not be ruled out completely.

Acknowledgement - The authors would like to thank the Fonds National Suisse pour la Recherche Scientifique for financial support and one of us (JvdM) is most grateful for a Council of Europe Fellowship of the Italian Government in 1977, during which part of this work was conceived.

REFERENCES

1. M.R. Cimberle, G. Bobel ånd C. Rizzuto, Advances in Physics 23, 639-71 (1974); A.D. Caplin and C. Rizzuto, Journal of Physics C : Solid State Physics 3, L117-20 (1970).

2. J. Bass (review article on DMR), Advances in Physics 21, 431-604 (1972).

3. J.A. Rowlands and S.B. Woods, Journal of Physics F : Metal Physics 8, 1929-39 (1978); T. Fujita and T. Ohtsuka, Journal of Low Temperature Physics 29, 333-44 (1977).

4. See the following review articles of size effects :
   G. Brändli and J.L. Olsen, Materials Science and Engineering 4,
   61-83 (1969); D.C. Larson, Physics of Thin Films, Vol.6
   (New York : Academic Press) 81-149 (1971).

5. J.B. van Zytveld and J. Bass, Physical Review 177, 1072-82
   (1969).

6. B.N. Aleksandrov, Zh.Eksp. i Teor.Fiz. 43, 399-410 (1962);
   Soviet Physics JETP 16, 286-94 (1963).

7. R. Reich, Thesis (1965) Paris.

8. R. Risnes, Philosophical Magazine 21, 591-97 (1970).

9. I. Holwech and J. Jeppesen, Philosophical Magazine 15, 217-28
   (1967).

10. Yu N. Chiang, V.V. Eremenko and O.G. Shevchenko, Zh.Eksp. i
    Teor.Fiz. 54, 1321-32 (1968); Soviet Physics JETP 27, 706-12
    (1968).

11. A. von Bassewitz and E.N. Mitchell, Physical Review 182,
    712-16 (1969).

12. R. Risnes and V. Sollien, Philosophical Magazine 20, 895-905
    (1969).

13. J.L. Olsen, Helvetica Physica Acta 31, 713-26 (1958).

14. F.J. Blatt and H.G. Satz, Helvetica Physica Acta 33, 1007-20
    (1960).

15. P. Cotti, E.M. Fryer and J.L. Olsen, Helvetica Physica Acta 37,
    585-88 (1964).

16. J.C. Garland and D.J. van Harlingen, Journal of Physics F :
    Metal Physics 8, 117-124 (1978); J.M.J.M. Ribot, J. Bass,
    H. van Kempen and P. Wyder, Journal of Physics F : Metal
    Physics 9, L117-122 (1979).

17. F.J. Blatt, Physics of Condensed Matter 9, 137 (1969).

LONG WAVELENGTH OPTICAL LATTICE VIBRATIONS AND STRUCTURAL PHASE TRANSITION IN MIXED a-$In_2S_{3-x}Se_x$ CRYSTALS

K. Kambas and J. Spyridelis

First Laboratory of Physics

University of Thessaloniki, Greece

The FIR and Raman measurements of lattice vibrations in a-$In_2S_{3-x}Se_x$ mixed crystals are reported. For $0 < x < 2.5$ the crystals have a cubic defect spinel structure and one mode behavior is evident. For $x > 2.5$ the a-In_2Se_3 hexagonal phase appears and the crystals exhibit a mixed one- and two mode behavior. This abrupt change in the optical spectrum evidence a structural phase transition occurring for a concentration near $x \sim 2.5$.

INTRODUCTION

In_2S_3 exists in three crystallographic modifications. The room temperature modification, β-In_2S_3 crystallizes in a defect spinel lattice with a high degree of ordering of tetrahedral and octahedral vacancies[1]. This results in a lattice with a tetragonal supercell consisting of three spinel blocks.

Above 420°C a disordering of the tetrahedral vacancies takes place and a new cubic modification appears, the α-In_2S_3 with a defect spinel lattice in which one third of tetrahedral metal positions remain empty in random arrangement. Under certain conditions[2] the transition into the β-phase can become entirely blocked and α-In_2S_3 is considered stable at room temperature.

In_2Se_3 also exists in many modifications. The room temperature phase crystallizes in a hexagonal wurtzite-like layer structure[3] which is clearly two-dimensional and exhibits highly anisotropic mechanical properties. The exact structure of this compound is not yet completely determined and in the literature

73

one can find quite a lot of controversial results.

In this work an attempt was made of study the long-wavelength optical phonons in the $In_2S_{3-x}Se_x$ system which is presented for the first time. The fact that the end members crystalize in different structures implies a structural phase transition in the mixed system in the intermediate range of composition. Our experimental results confirm that this composition is about $x \simeq 2.5$. Thus this system crystalizes in the structure of $\alpha-In_2S_3$ for $x < 2.5$ whereas for $x > 2.5$ another single phase region exists which is isotypic with the non completely known structure of $\alpha-In_2Se_3$.

FAR-INFRARED SPECTRA

The Infrared reflectivity measurements for the range of composition $0 < x < 2,5$ are shown in figure 1. The spectrum of $\alpha-In_2S_3$ ($x = 0$) has five distinct bands, two of which are broad dominant peaks at ~ 227 and 328 cm^{-1}.

The reflection spectra of the solid solutions maintain the general character of the host crystal but we must note the following features. For all solid solutions two maxima have been observed analogous to those of pure crystal. It is clear that as the composition varies, these two maxima are continously shifted toward low frequencies with almost the same mode strength. This is to be expected since the atomic mass of Se is larger than that of S. The two modes exist throughout the whole range of composition and are monotonically shifted. So we can say that the system for $x < 2.5$ exhibits an "one-mode" behavior.

An interesting feature of these restrahlen bands appears from the comparison with the spectra given in figure 2. This figure contains the IR spectrum of the $\alpha-In_2S_3$ quenched sample (b) together with the spectrum with the same sample after two months annealing (a). We observe that the reflectivity near the low frequency range becomes lower for the annealed sample and the spectrum shows a fine structure compared with the disordered structure of the quenched sample. The lines become sharper and new lines appear. This can be attributed to the cation ordering which produces a change inside the unit cell of the spinel lattice and also noticeable changes in the infrared spectrum[4,5]. Generally, ordering of cations results in a lowering of the overall symmetry which means an increasing number of IR active vibrations, which are indeed observed. $\alpha-In_2S_3$ has the space group O_h^7 and it was

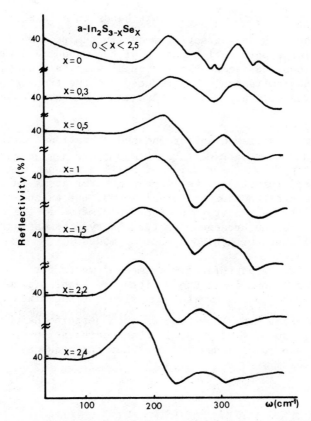

Fig.1 : Infrared reflectivity spectra for $In_2S_{3-x}Se_x$ with concentrations $0 < x < 2.4$.

found that the lattice remains cubic after annealing and after the substitution of S-Se in the mixed crystals. According to group theoretical considerations[6] two possible symmetry changes can occur, $0_h^7 \leftrightarrow 0_h^7$, $0_h^7 \leftrightarrow 0^7$ (Spinel $\leftrightarrow$ 1:3 order on B sites). Because of the increasing of the number of bands[7] after annealing, the symmetry is lowered by the cation ordering and we conclude that the second change occurs. The annealed $In_2S_{3-x}Se_x$ specimens did not show any variation in the IR spectrum. The absence of ordering can be explained by the presence of Se atoms by which the 1:3 order is prevented by normal-inverse disorder. So the structures remain disordered spinel (normal, inverse or statistic) indicating an exchange of cations and vacancies between the octahedral and tetrahedral sublattices. Whereas it is known that the alloys have mainly the same crystal structure as the parent crystal of quenched α-In_2S_3, the distribution of the constituent ions over the sublattices is not known. The fact that in the spectra of the alloys there exist only two very broad bands centered almost on

the same position as the two dominant bands in $\alpha-In_2S_3$ can be
related to the high degeneracy of modes in the reduced Brillouin
zone of the large unit cell. Because of the random distribution
of S and Se ions and vacant sites a quasi-unit cell containing
many $In_2S_{3-x}Se_x$ units can be introduced. So the translational
symmetry of the crystal is broken and the $\bar{q} = 0$ selection rules
are no longer conserved.

A way to view disorder induced vibrational activity is
to realize that there are more q = 0 modes for semi-crystalline
solids resulting from the folding of Brillouin zone. Because of
this situation, inspite of the large number of atoms in the unit
cell giving a great number of normal modes, the extend of optical
spectrum is extremely narrow and what we observe is the envelope
of all bands and the broadness of the observed bands corresponds
to frequency ranges with high number of modes.

The infrared reflectivity measurements for compositions
2.5 < x < 3 are shown in figure 3. Three restrahlen bands were
observed for pure In_2Se_3 crystal and five for the mixed compounds
with x = 2.9, 2.8, 2.6. The three of them pertain the characteristic
features of those of host crystal. The incorporation of S atoms in
In_2Se_3 gives two local modes lying higher in frequency than the

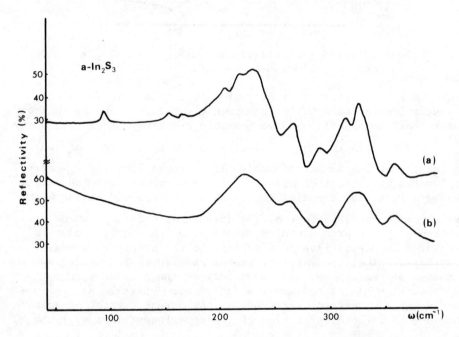

Fig.2 : Infrared reflectivity spectra for different degree of
annealing in $\alpha-In_2S_3$.

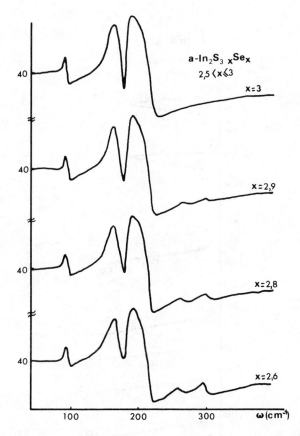

Fig.3 : Infrared reflectivity spectra for $In_2S_{3-x}Se_x$ with concentrations 2.5 < x < 3.

LO-mode frequencies of pure In_2Se_3 whereas when Se replaces S in In_2S_3 neither a resonance nor a gap mode is yielded. So two-mode behavior is expected to generate from the local modes in In_2Se_3 alloys and one-mode behavior in In_2S_3 (x < 2.5). If we use the mass criterion of Chang and Mitra[8] we find also that two-mode behavior is expected. Actually as can be observed from figure 4 in which we plot the frequencies variations for x > 2.5, both one and two-mode behavior exists. This behavior cannot be explained by the present criteria concerning the behavior of mixed crystals. We can expect that a layer-mixed crystal is not a conventional case for any of the criteria because of its extreme anisotropy. For such an example a relation of masses alone cannot be sufficient. So we have here, say, an unusual one- and two-mode behavior at optical modes in the center of Brillouin zone. This duality in behavior has been observed in other crystals containing more than one optical bands[9-14].

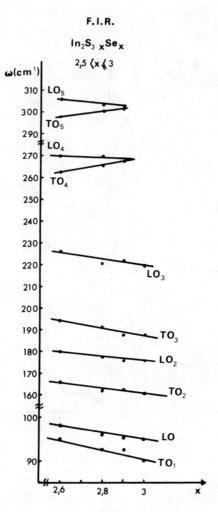

Fig.4 : Frequencies of TO and LO modes for $In_2S_{3-x}Se_x$ as a function of composition 2.5 < x < 3.

An interesting characteristic feature concerning $\alpha-In_2Se_3$ and its alloys is the difference between the reflection spectra of quenched and annealed samples. Figure 5 compares the spectra of the same sample taken by quenching (a), and after two months (b) annealing. We observe that all bands were modified after annealing but they remained in the same position with regard to the frequency and new bands did not appear. We also observe a very high reflectivity at low frequencies for the quenched phase.

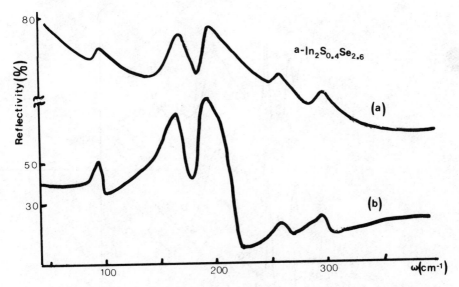

Fig.5 : Infrared reflectivity spectra for different degree of annealing in $\alpha\text{-}In_2S_{0.4}Se_{2.6}$.

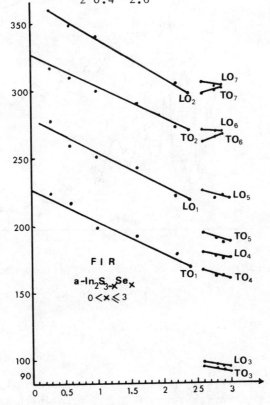

Fig.6 : Frequency dependence of $In_2S_{3-x}Se_x$ for the whole concentration range $0 < x \leqslant 3$.

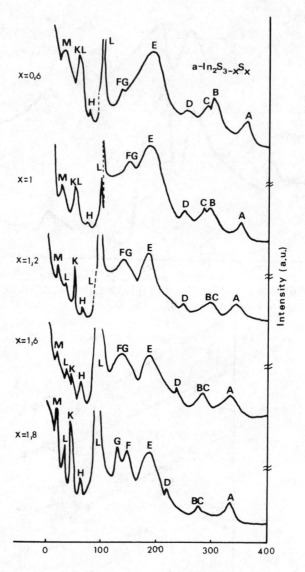

<u>Fig.7</u> : Raman spectra of $In_2S_{3-x}Se_x$ alloys for different
concentration 0 < x < 2.5.

        Finally figure 6 shows the frequency distribution for
the whole concentration range 0 < x < 3 in which one can see the
discontinuity at x ≃ 2.5 which is due to a structural phase
transition.

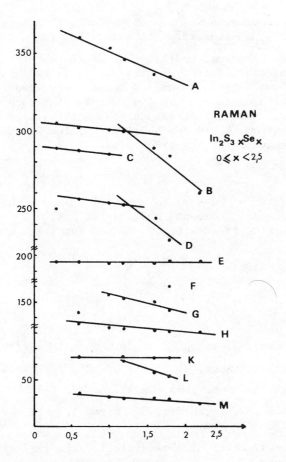

Fig.8 : Frequency dependence of Raman active modes of the alloys
$In_2S_{3-x}Se_x$ with $0 < x < 2.5$.

RAMAN SCATTERING

        The results of our Raman-Scattering experiments, taken
at room temperature, are shown in figure 7, for $0 < x < 1.8$.
Unpolarized excitation and detection was performed for measuring
these spectra, by using a Yag Laser ($\lambda = 1.06 \mu$) and the anti-
Stokes components of the scattered radiation. A large number of
peaks are observed corresponding to a set of normal mode
frequencies. Some lines shift continously with composition, in
some others the shift is discontinous. Also there are lines which
appear or disappear as a function of the concentration. The
variation of the frequencies of the Raman active modes vs the
composition is shown in figure 8. The great number of peaks
observed, do not occur at the same frequency as in the IR spectra
because having inversion symmetry the exclusion principle is
obeyed in In_2S_3 and modes that are Raman active will be IR inactive
and vice versa.

In conclusion we have observed in this type of alloys a very unusual and diversified behavior. For In_2S_3 we observe a great number of normal modes. As we increase the Se concentration, the infrared spectra show only two restrahlen bands giving two sets of TO and LO modes whose frequencies shift continously with concentration and exhibit "one mode" behavior. For $x \simeq 2.5$ we observe a structural phase transition and above this concentration "one mode" behavior for low frequencies and "two mode" behavior for higher ones.

Acknowledgements - We want to thank Professor M. Balkanski for his very helpful comments and suggestions. Also thanks are due to Dr. G. Kanellis and Dr. M. Massot for their help in this work.

REFERENCES

1. C.J. Rooymans, J. Inorg. Nuclear Chem. 11, 78 (1959).

2. R. Diehl, R. Nitsche, J. Cryst. Gr. 28, 306 (1975).

3. S.A. Semiletov, Sov. Phys. Cryst. 6, 158 (1960).

4. B. Deangelis, V. Keramidas, N. White, S. Sol. St. Chem. 3, 358 (1971).

5. V.A. Brabers, Phys. St. Sol. (a), 12, 629 (1972).

6. C. Haas, J. Phys. Chem. Solids, 26, 1225 (1965).

7. W.I. White, B.A. Deangelis, Spectrochim. Acta, 23A, 985 (1967).

8. I.F. Chang, S.S. Mitra, Phys. Rev. 172, 924 (1968).

9. A.S. Barker, J.A. Ditzenbarger, H.J. Guggenheim, Phys. Rev. 175, 1180 (1968).

10. C.H. Perry, N.E. Torenberg, Phys. Rev. 183, 95 (1969).

11. M. Hayek, O. Brafman, R. Lieth, Phys. Rev., B8, 2772 (1973).

12. N.A. Bakhysov, N.M. Gasanly, B. Yavadof, V. Tagiror, S. Efendiev, Phys. St. Sol. (b) 91, K1 (1979).

13. M.K. Teng, M. Massot and M. Balkanski, Phys. Rev. 17, 3695 (1978).

14. K. Wakamura, T. Arai, S. Onari, K. Kudo and T. Takahashi, J. Phys. Soc. Japan 35, 1430 (1973).

ELECTRONIC STRUCTURE AND MAGNETIC PROPERTIES

OF THE TiFe$_x$Co$_{1-x}$ SYSTEM

J. Giner
Institut de Physique
Université de Liège, 4000 Sart Tilman/Liège 1, Belgium

F. Gautier
Laboratoire de Structure Electronique des Solides
Université Louis Pasteur, 67000 Strasbourg, France

We investigate the density of states and the spin susceptibility
of TiFe$_x$Co$_{1-x}$ in the paramagnetic phase. The partial densities
of states are determined using an extension of the coherent
potential approximation. The spin susceptibility of each component
is calculated taking into account the electron-electron interaction via the molecular field. The results show that the antistructure atoms for nearly ordered states can carry a local moment
and could be the elements which determine ferromagnetism in the
system.

1. INTRODUCTION

        The electronic properties of the pseudobinary alloys
TiFe$_x$Co$_{1-x}$ are known to be composition dependent. In particular,
the boundary binary alloys TiFe and TiCo are paramagnetic whereas
TiFe$_x$Co$_{1-x}$ presents a ferromagnetic phase in the range
$0.4 \lesssim x \lesssim 0.8$ with a maximum Curie temperature around 63 K[1,2,3].
In addition, experimental data show that the magnetism is
homogeneous for the Co-rich alloys and more localized for the
Fe-rich alloys[4,5]. The magnetic properties of the Co-rich alloys,
excepting the disappearence of ferromagnetism for $x \lesssim 0.4$, are
well interpreted in terms of a weak itinerant ferromagnet model[2,3,6].
As far as the evolution from the local to the homogeneous behaviour
of the magnetism is concerned, it has been qualitatively interpreted

83

in a model of coupling between magnetic moments localized on the Co atoms[4]. However, no complete theoretical explanation of the origin of the magnetism in this system has been proposed up to now.

In this paper, we report the results of a calculation which show that the antistructure Fe and Co atoms could play an important part in the appearence of ferromagnetism at least in the Fe-rich alloys.

2. ATOMIC ARRANGEMENT AND ELECTRONIC STRUCTURE

The $TiFe_xCo_{1-x}$ alloys have the CsCl structure over the whole range of compositions, the Ti atoms sitting on one simple cubic sublattice (hereafter referred as n°2) and the Fe and Co atoms sitting at random on the other one (here after referred as n°1). In a previous paper[8] we proposed a simple band model for the description of ordering effects in relation with the electronic structure of the components of pseudo-binary alloys like $TiFe_xCo_{1-x}$, VMn_xFe_{1-x}, etc. In this model, the partial and average densities of states (DOS) are determined using the coherent potential approximation (CPA) generalized to account for the long range order. The main conclusions concerning the $TiFe_xCo_{1-x}$ alloys was :

i) the average DOS at the Fermi level increases from the Fe-rich alloys to the Co-rich alloys in qualitative agreement with the experimental data;
ii) the general shape of the partial DOS corresponding to structure (s) atoms (i.e. atoms in their own sublattice) is not very sensitive to the composition x; these partial DOS are characterized by a typical valley in the middle region due to the CsCl symmetry[9];
iii) the partial DOS corresponding to antistructure (A) atoms (i.e. atoms not in their own sublattice) have a well-marked peak in the middle region. The Fermi level lies in the high density region of the Fe and Co A-atoms suggesting that these atoms could play an important part in the occurence of ferro-magnetism.

In order to clarify the role of the A-atoms we have studied the composition dependence of the static paramagnetic susceptibility of $TiFe_xCo_{1-x}$ assuming a small number of A-atoms (and no vacancies) in the system. To be consistent with the experimental data, and in particular with the neutron-diffraction ones, we have done the calculations for the equilibrium temperature

at which the total concentration of A-atoms (Fe+Co+Ti) in our model[8]
is about 1.5 % for TiFe$_{0.5}$Co$_{0.5}$. We show in figure 1 the calculated
equilibrium concentration of antistructure Fe and Co atoms versus
the composition. As one can see the concentration of antistructure
Fe atoms is larger than the concentration of antistructure Co
atoms except for the phases near to the TiCo. It is also
interesting to note that the total concentration of A-atoms is
more or less constant for the whole range of compositions. This
is not surprising insofar as Fe and Co are very similar in nature
(they have similar electronegativities and sizes).

3. COMPOSITION DEPENDENCE OF THE STATIC PARAMAGNETIC SPIN SUSCEPTIBILITY

Neglecting the dependence of the susceptibility of an
atom upon its environment (which is consistent with the CPA
assumptions), one can write the susceptibility of an atom i
sitting on the site λ as follows :

$$\chi_\lambda^i = \chi_\lambda^{oi} + \Sigma_\mu \Sigma_j \chi_{\lambda\mu}^{oij} I_\mu^{eff} \chi_\mu^j \qquad (1)$$

$$(i,j = Ti,Fe,Co)$$

where : i) the first summation is extended over the whole lattice;
ii) χ_λ^{oi} is the susceptibility calculated without taking into
account the electron-electron interaction (i.e. the Hartree-Fock
susceptibility) and is given by the relation

$$\chi_\lambda^{oi} = 2 \mu_B^2 g_\lambda^i (E_F) \qquad (2)$$

where $g_\lambda^i(E_F)$ is the corresponding local density of states at the
Fermi level; iii) $\chi_{\lambda\mu}^{oij}$ is a coupling term between sites λ and μ
and results from a local contribution describing the
magnetization of the atom at   from a magnetic field localized
on the atom at   and a non local term describing the change of
the effective medium with the magnetic field (see Brouers et al[10]
for the details of formalism in the case of a completely disordered
alloy; iv) I_μ^{eff} is an effective intra-atomic Coulomb interaction
between electrons at site μ. In the present case, the system of
equations (1) reduces to a system of 6 linear coupled equations.

The main results can be summarized as follows : i) the
susceptibility χ_λ^i diverge for critical values of $I_\lambda^{eff} \equiv I^{eff}$

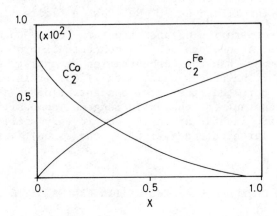

<u>Fig.1</u> : Equilibrium concentration of antistructure Fe atoms (C_2^{Fe})
and Co atoms (C_2^{Co}) in $TiFe_xCo_{1-x}$. The corresponding
concentration for the antistructure Ti atoms is given by
$C_1^{Ti} = C_2^{Fe} + C_2^{Co}$.

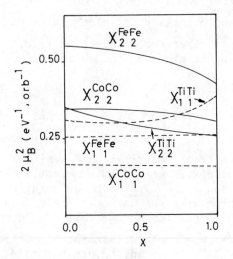

<u>Fig.2</u> : Composition dependence of the various Hartree-Fock
susceptibilities $\chi_{\lambda \lambda}^{oi\ i}$.

around 2 eV (for simplicity, we assumed I_λ^{eff} independent of the
whole range of compositions the susceptibility χ_2^{Fe} corresponding
to Fe atoms sitting on the Ti sublattice increases much more
sharply with I^{eff} than the other susceptibilities; ii) the same
results are obtained in a simpler approach, i.e. in a local
approximation which consists in neglecting all the $\chi_\lambda^{oi}\,{}_\mu^j$ if
$(i,\lambda) \neq (j,\mu)$. In that case, the condition for the appearence
of a localized moment is given by the relations

$$1 - I^{eff}\,\chi_\lambda^{oi}\,{}_\lambda^i = 0 \tag{3}$$

we show in figure 2 the composition dependence of the 6 terms
$\chi_\lambda^{oi}\,{}_\lambda^i$. As one can see $\chi_2^{oFe}\,{}_2^{Fe}$ is the dominant term and increases
from the Fe-rich alloys to the Co-rich alloys like the average
DOS at the Fermi level (see § 2).

4. DISCUSSION

Our results indicate that the antistructure Fe atoms
have a net tendency to carry a local moment for a minimum value
of X. We believe that these atoms could in turn determine local
moments on the neighbouring Fe and Co atoms giving rise to
magnetic clusters coupled via the conduction electrons. In that
case the exact range of appearence of ferromagnetism as well as
the value of the magnetization would depend on 2 factors : i) the
concentration of local moments, ii) the coupling between them.

As far as the first point is concerned, our model
predicts correctly the disappearence of ferromagnetism in TiCo
because the concentration of antistructure Fe atoms goes to 0
when X goes to 0 (see Fig.1). As for the second point, let us
say that the magnetic couplings depends sensitively on the
electronic structure of the alloy and that our band model is
too simple to give a correct description.

To end, let us point out one experimental fact which
seems to confirm the importance of the A-atoms : the Curie
temperature T_{Curie} of the off-stoechiometric CsCl phases $Ti_y(FeCo)_{1-y}$
$(0.4\ 97 \leqslant y \leqslant 0.52)$ varies sharply with the composition y. In fact,
T_{Curie} increases when the number of antistructure Fe and Co atoms
is increased, i.e. for $y < 0.5$, and decreases when the number of
antistructure Ti atoms is increased, i.e. for $y > 0.5$.

Let us recall also that the electronic structure of the VMn_xFe_{1-x} system is very similar to that of the $TiFe_xCo_{1-x}$ system so that the former could present such magnetic properties.

Acknowledgement - We thank Dr. J. van der Rest for many interesting discussions and for the critical reading of the manuscript.

REFERENCES

1. Desavage B.F. & Goff J.F., J. Appl. Phys. 38, 1337 (1967).

2. Asada Y. & Nose H., J. Phys. Soc. Japan 35, 409 (1973).

3. Hilscher G., Buis N. & Franse J.J.M., Physica 91 B, 170 (1977).

4. Beille J., Bloch D., Towfiq F. & Voiron J., J. Magn. and Magn. Mat. 10, 265 (1979).

5. Buis N., Ph. D. Thesis Amsterdam Univ. (1979).

6. Hilscher G. & Gratz E., Phys. Stat. Solidi (a) 48, 473 (1978).

7. Pickart S.J., Nathans R. & Menzinger F., J. Appl. Phys. 39, 2221 (1968).

8. Giner J. & Gautier F., J. de Physique 38, C7-301 (1977).

9. Yamashita J. & Asano S., Prog. Theor. Phys. 47, 2119 (1972).

10. Brouers F., Gautier F. & van der Rest J., J. Phys. F 5, 975 (1975).

THE DETERMINATION OF Mg SURFACE ENRICHMENT IN HEAT TREATED

AlMgSi ALLOYS USING THE SXES METHOD

L. Kertész, J. Kojnok, and A. Szász

Institute for Solid State Physics Eötvös University

Budapest, Muzeum krt.6-8, Hungary

The heat treatments result Mg loss by diffusion from AlMgSi alloys. This was identified by many authors using different bulk measurements. Our results show, that the Mg concentration essentially increases during the heat treatment near to the surface.

It has been known for quite a time that longer heat treatment of AlMgSi alloys is producing Mg enrichment on the surface[1-4].

Most of the investigations referring to this did not contain concrete data concerning the amount of Mg in the surface layer. Bulk type measurements were performed, with the exception of the[4] measurement, extrapolating on the surface the Mg concentration.

In our present paper we attempted to determine the Mg concentration enriched in the 200 nm thick surface layer.

For the investigation of the surface concentration of Mg we used the soft X-ray emission spectroscopy method /SXES/. We performed the measurements using the transitions of the Al $L_{2,3}$ levels.

On the basis of the calculation of Segall[5] and Rooke[6] was obtained Al $L_{2,3}$ SXES curve, using the[7] band structure calculation.
The symmetry points of the Brillouin zone are : L_2'/68,6 eV/,

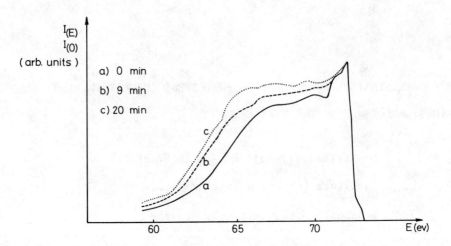

Fig.1 : SXES curves of AlMgSi alloys after annealing in vacuum
   a. 0 min
   b. 9 min
   c. 20 min

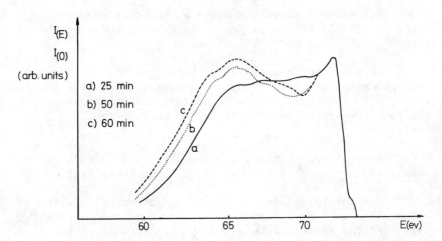

Fig.2 : SXES curves of AlMgSi alloys after annealing in vacuum
   a. 25 min
   b. 50 min
   c. 60 min

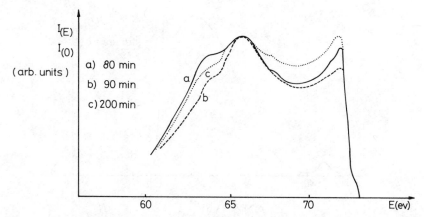

Fig.3 : SXES curves of AlMgSi alloys after annealing in vacuum
     a. 80 min
     b. 90 min
     c. 200 min

L_1/69,0 eV/, X_4/70,1 eV/, X_1/71,3 eV/, K_1/71,4 eV/, W_3/72 eV/,

K_1/72,7 eV/. The emission edge characteristic of the Al Fermi

surface appears. at 72,8 eV. The type of equipment was RSM 500/
Burewestnik, Leningrad/[8]. The resolving power in the used range
was 0,07 nm /0,3 eV/. The composition of the specimens was the
following : Al matrix, Mg 0,58 wt % /0,65 at %/, Si 0,35 Wt %/,
0,34 at %/,Fe 0,14 wt %/, 0,068 at %/, Cu 0,01 wt %/, Ti 0,03 wt %/.

     The specimen was annealed for 45 min at 800 K in air and
was quenched in water at room temperature.

     The final treatment of the specimens was done "in situ"
with an exciting electron beam /4 kV, 0,3 mA/ in a vacuum not

worse than 10^{-4} Pa. The temperature of the specimen during this
heat treatment was 530 K $\pm$ 30 K. It was measured by the new
method[9].

     For the determination of the Mg concentration we used
H. Neddermayer's measurements[10].

     As a starting point and for comparison we used the plot
of AlMgSi taken after the annealing and quenching of the specimen
/figure 1, curve a/. The curve in its main lines is in agreement
with the pure Al $L_{2,3}$ SXES curve, with the exception that the fine

structure is less pronounced.

     The plot shown in figure 1 curve b was taken after
9 min in situ heat treatment, essential variation has not been

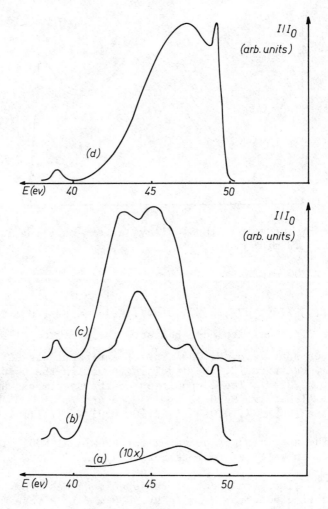

<u>Fig.4</u> : SXES curves, Mg L$_{2,3}$ peak of AlMg alloy before and after
annealing in vacuum

a. before annealing
b. after annealing

found yet. The situation after 20 min is shown in figure 1 curve c
on which the appearence of a new local maximum can be observed at
66 eV.

   With the further increase of the time of heat treatment,
this peak emerges more vigonrously and at the same time the
maximum of the large peak below the emission edge gradually
decreases. /Fig.2 curves : a,b,c/. After 80 and 90 min a new

local peak appears at 63 eV in figure 3 curves a,b, curve c
represents the emission spectrum after 200 min heat treatment,
it does not change with further heating. Similar experiments were
carried out using Al-5 wt % Mg alloy. The SXES received from the
sample without any preliminary heat treatments is in the figure 4
curve a. The next curve /Fig.4, curve b/ was made after 530 K
heat treatment for 100 minutes, and the surface was cleaned by
electron beam.

On the basis of[10] Neddermayer's work on the various
composition of AlMg alloys, the spectra presented on Figure 1
curves a,b,c correspond to 0,6 at %, 10 at % + 5 %, 20 at % + 10 %
concentration. Our experimental results on figure 2 curves a,b,c
correspond to 30 at % + 10 %, 40 at % + 10 %, 50 at % + 10 %
concentrations. The emission spectra on figure 3 curves a,b,c
correspond to 60 at % + 10 %, 70 at % + 10 %, 40 at % + 10 % of
Mg concentrations.

Thus the maximum Mg concentration in the surface layer
under given conditions is about 70 %. It is most likely that the
Mg atoms are outdiffusing from the matrix, and they evaporate
from the surface during the same period.

In accordance with the described results it can be
ascertained that due to the effect of heat treatment for longer
periods of time /90 min/ at 500 K and higher temperatures, the
Mg on the surface layer vigonrously enriches and instead of the
24 % obtained through extrapolation with the[4] measurement,
according to our measurements the concentration increases to 70 %.

The behaviour concerning the SXES of the Mg $L_{2,3}$ line
has been studied on AlMg alloy so far. According to the
SXES measurements low concentration alloy gives a very small
effect. The same alloy shows a significance increase of Mg
amount about 13 times greater in the surface layer after 100 min.
heat treatment at 530 K.

REFERENCES

1. Chatterjee D.K. and Entwistle K.M. : J. Inst. Met. 101, 53
   (1973).

2. Kovács I., Lendvai J., Ungáar T. : Mat. Sci. Eng. 21, 169 (1975).

3. Hidvégi E. and Kovács-Csetényi E. : Mat. Sci. Eng. 27, 39 (1977).

4. Csanády A., Stefániay V., Beke D. : Mat. Sci. Eng. 38, 55 (1979).

5. Segall B. : Phys. Rev. 124, 1797 (1961).

6. Rooke G.A. : J. Phys. C. $\underline{1}$, 767 (1968); J. Phys. C. $\underline{1}$, 776 (1968).

7. Singhal S.P. and J. Callaway : Phys. Rev. $\underline{16}$, 1744 (1977).

8. Lukirskii A.P., Brytov I.A. and Komyak N.I. : in book : Methods and apparats in X-ray spectroscopy 2. Ed. Leningrad p.2, 1967.

9. Kertész L., Szász A., Kacnelson A.A. : to be published.

10. H. Neddermayer : in book : Band Structure Spectroscopy of Metals and Alloys. Eds.:D.J.Fabian and L.M.Watson, Academic Press, p.153, 1973.

THE EFFECT OF HEAT TREATMENT ON THE ELECTRON

STRUCTURE OF DILUTE AlMgSi TYPE ALLOYS

L. Kertész, J. Kojnok, and A. Szász
Institute for Solid State Physics
Eötvös University, Hungary

A.S. Sulakov
Institute for Solid State Physics
A.A. Zdanov State University, Leningrad, USSR

The changes of the electron density of states of dilute AlMgSi alloys is investigated by soft X-ray Emission Spectroscopy /SXES/ method. The states of the alloy was also measured by differential thermoanalysis /DTA/.

      The AlMgSi alloys have been investigated by a number of people and by different methods[1]. It is known that directly after quenching[2] and as the temperature increases, a transformation process takes place, after which a medium temperature G.P. zone and an Si separation can be observed, and later on the solution of this Si takes place at medium temperatures from 480 K to 680 K. At higher temperatures a separation, after which the repeated solution of all the Mg and Si can be observed, from temperatures between 680 K and 800 K in solid phase.

      In our present work our purpose was to perform measurements which give direct information concerning the effects influencing the electron structure of the individual processes which take place in the alloys.

      We performed the tests on the alloys with the help of soft X-ray emission spectroscopy /SXES/ and the differential thermoanalysis /DTA/.

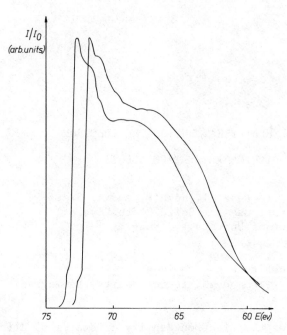

<u>Fig.1</u> : The SXES plots of the pure Al /99,99 %/ /curve a/ and
of AlMgSi alloy /curve b/ measured immediately after the
quenching.

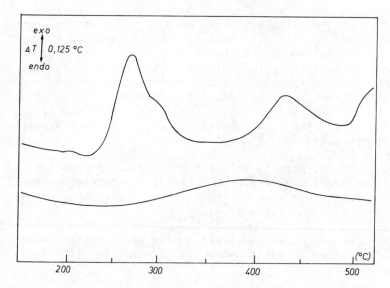

<u>Fig.2</u> : The DTA curve of AlMgSi alloy measured immediately after
the quenching.

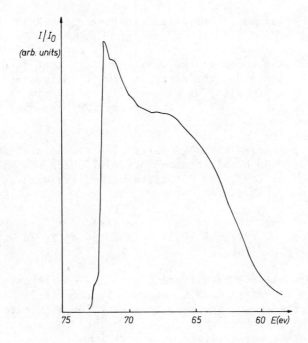

Fig.3 : The SXES curve of AlMgSi annealed for 6 hours at 443 K.

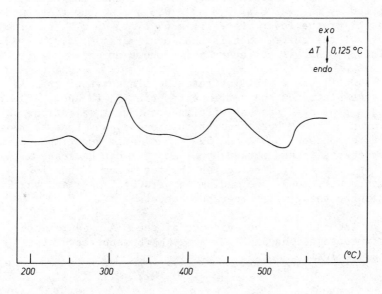

Fig.4 : The DTA curve of the alloy annealed for 6 hours at 443 K.

We obtained the basis information with SXES from the surface layer of a thickness 200 nm.

The main point of the measurement is that it gives direct information about the electron density of states[3]. The concentration of the alloying elements is relatively low, so we paid attention first of all to the energy range characteristic of the Al matrix, to the Al $L_{2,3}$ level.

We performed the SXES tests with the help of the type RSM 500 device, the detailed description of which was given by Lukirszkij and Britov[4]. The DTA tests were given with the help of the DTA equipment[5].

The composition of the tested specimens was the following: Mg 0,58 wt % /0,65 at %/, Si 0,35 wt % /0,34 at %/ Fe 0,14 wt %/ 0,068 at %/, Cu 0,01 wt %, Ti 0,03 wt %.

We achieved the different physical conditions in the alloys with a two-stage heat treatment. In the first stage each specimen received resolving heat treatment for 45 minutes at 800 K temperature, after this it was quenched in water at room temperature,which was followed by either immediate testing, or the specimen was subjected to another heat treatment.

On the first figure we show the SXES plot of pure Al /99,99 %/ and of AlMgSi /curve / alloy taken of the specimen measured immediately after the resolving heat treatment and quenching. The most striking observation is that the Fermi level has shifted by the value of 0,8 eV $\pm$ 0,3 eV, at the time increases the density of the lower states from 67 eV to 70 eV. The fine structure of the curve, that is the local maximum points are more blurred, that is not every point shows up.

On figure 2 we illustrate the DTA curve of the alloy of the same state. On the curve up to point M, to about 480 K, there is a prolonged process with inherent heat rejection, which is followed by the M heat absorption process, and then at medium temperatures between 520 K and 590 K two large superimposed exotherm processes /peaks marked by N_1 and N_2/ can be seen, among others. We can easily distinguish the formation of the low temperature G.P. zone, the medium temperature zones and the Si separation on the given temperature points.

According to the evidence of the DTA measurements, the specimen tested by SXES was an optimally supersaturated metastable solution.

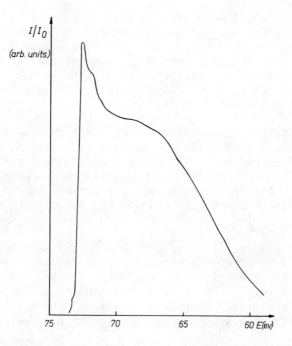

Fig.5 : The SXES curve of AlMgSi annealed for 40 min at 573 K.

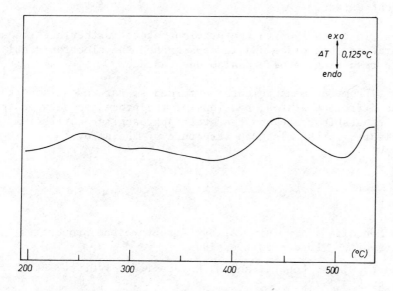

Fig.6 : The DTA curve of AlMgSi annealed for 40 min at 573 K.

On figure 3 we plotted the SXES curve of the alloy annealed for 6 hours at 443 K temperature. The Fermi level corresponds to that of the pure Al. The fine structure characteristic of the Al curve can be approximately found in this measurement too and this indicates that the band structure and density of state of the Al matrix are not essentially affected by the alloying elements put into Al.

On figure 4, on the DTA curve of this same state, the low temperature exotherm and the changes indicated by M and N_1 practically disappeared, a large scale medium temperature zone formation took place /around 540 K/. This process partially decreased the alloying content of the solution and this adequately coincides with the results of the emission measurements also.

The SXES curve taken of the sample annealed for 40 minutes at 573 K, after the resolving heat treatment and quenching is shown on figure 5. The Fermi level /72,8 eV/ corresponds to that of the pure Al, within the measuring accuracy. An essential change can be seen in the slight decrease of the relative peak height connected to the K symmetry point.

On figure 6 we plotted the DTA curve of this same state of alloy. Comparing this with the curve plotted of the alloy having the previous state, it can be ascertained that up to 590 K practically no effects show up. These conditions indicate that the Mg and Si content of the Al matrix is starved to a great extent.

In comparison to the previous state further significant Si separation took place and in consequence the electron density of state resulting in peak K also decreased.

From the measurements performed so far we can ascertain that the different states, sections of the processes between the solute metastable state and the state approaching equilibrium, are reflected in the electron structure of the alloy.

REFERENCES

1. L.F. Mondolfo : Aluminium Alloys structure and properties, Butterworths, London-Boston, 1976.

2. L. Kertész, Cs. Lénart, and M. Kovacs-Treer : Cryst. Lattice Def. 8, 99, 1979.

3. G.A. Rooke : J. Phys. C. 1, 767, 1968.

4. A.P. Lukirskii, I.A. Brytov, N.I. Komiak : in book :
   Apparatura i metody rentgenovskovo analiza, vyp : 2,
   Leningrad, 1967, p.4.

5. Hungarian Patent, Budapest MA-2720/254, 172.496/619.

ELECTRONIC DENSITIES OF STATES AND SURFACE

SEGREGATION IN ALLOYS

P. Lambin and J.P. Gaspard

Université de Liège, B5, Institut de Physique

4000 Sart Tilman/Liège 1, Belgium

A theory of surface segregation in transition metal alloys is investigated. In this theory, the segregation energy is deduced from the electronic d densities of states. The alloy is described in a tight binding approximation using the moments method. A particular application to CuNi alloys is presented. A systematic prediction of surface segregation is made on the basis of the d electrons per atom ratio of the components of the alloy. A good agreement with experimental results is found except for the 3-d series. The addition of the correlation energy removes the discrepancy.

The theoretical study of surface segregation in alloys has received a new interest since experimental data on surface composition are available. Experimental techniques like UPS, AES, LEIS, have shown that the surface composition of alloys can be very different from that of the bulk[1]. Recent theories which assume pairwise interaction between atoms[2] on which are based on semi empirical calculations of the surface tensions[3,4] give a certain number of correct explanations of the experimental results. In the case of transition metal alloys, quantum mechanical calculations have shown that the d electrons play an important role in the segregation mechanism[5,6]. The aim of this paper is to give a simple formulation to the electronic forces responsible for segregation at the surface of transition (and noble) metal alloys. We shall see that the band fillings of the components of the alloy are relevant parameters for the prediction of the surface segregation.

In this paper, a monolayer approximation is used, i.e., the compositions of the first, second, ..., layers parallel to the surface are equal to the bulk composition. The driving force for the segregation of A at the surface is the change ΔE in the internal energy when an atom A in the bulk exchanges positions with an atom B at the surface. ΔE can be written as the difference between the surface energies of the components A and B

$$\Delta E = \gamma_A - \gamma_B \ . \tag{1}$$

We assume that the predominant contribution to the surface energies are due to d electrons[7]. The surface energies are evaluated at OK, in the Hartree approximation, as

$$\gamma_A = \int^{E_{F,A}^s} n_A^s(E)E\,dE - \int^{E_{F,A}^b} n_A^b(E)E\,dE \ , \tag{2}$$

where $n_A^b(E)$ and $n_A^s(E)$ are the partial d densities of states on an atom A, respectively in the bulk and at the surface. The charge transfer and the charge oscillations near the surface has been neglected. The surface charge oscillations near the surface produce a more or less rigid transition of surface band in order to equalize the Fermi level with that of the bulk[8]. In this paper, the Fermi levels $E_{F,A}$ are calculated from the d band filling assuming a rigid shift of the surface band.

It is shown in the appendix that the change ΔF in the free energy when a bulk atom A exchanges positions with a surface atom B is equal to zero when the alloy is in equilibrium. This means that

$$\Delta E = T\Delta S \tag{3}$$

where T is the absolute temperature and ΔS, the change in entropy. It is assumed that the temperature dependance of the internal energy ΔE is negligible, and we only account for the configuration entropy as the relevant contribution to ΔS. Thus, Eq.(3) yields

$$\Delta E = k_B T \ln \frac{c_A^b \, c_B^s}{c_A^s \, c_B^b} \tag{4}$$

where k_B is the Boltzmann constant, c_A^b and c_A^s being respectively the bulk and surface atomic concentration of the component A. Eq.(4) allows to calculate the surface composition of the alloy for different temperatures. The segregation energy ΔE is computed

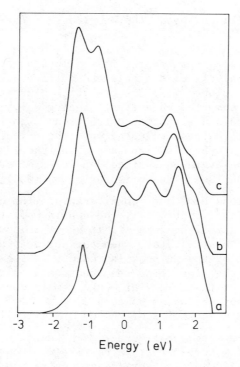

Fig.1 : (001) surface densities of states of the $Cu_{0.1}Ni_{0.9}$ alloy for surface concentrations of Cu equal to 10 % (a), 40 % (b) and 70 % (c). The curves are normalized to ten electrons per atom.

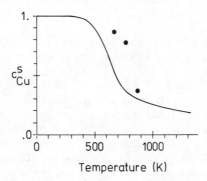

Fig.2 : Surface concentration of $Cu_{0.1}Ni_{0.9}$ alloy in equilibrium as a function of the temperature. The black dots represent experimental results.

from Eq.(1) and (2) and needs the knowledge of the densities of states.

Preliminary results for CuNi alloys have been presented[6] and a detailed analysis of the bulk and surface densities of states will be published[9,10]. The densities of states of CuNi alloys have been computed in a tight binding Hamiltonian using the moments method and a continued fraction expansion for the Green function[11]. In these calculations, the off-diagonal disorder as well as the s-d hybridisation have been ignored. As an example, figure 1 shows the averaged surface densities of states $Cu_{0.1}Ni_{0.9}$. The surface plane orientation is (001) and various surface compositions have been considered. In the case of the non-enriched surface (figure 1a), the density of states bears resemblance to that of the pure Ni; it shows in addition a narrow minority peak of Cu at -1.2 eV. When the surface is enriched with Cu, the magnitude of the Cu peak increases. At the same time, the structure of the Ni side of the spectrum is altered, but the energy positions of the Cu and Ni peaks remain constant, in agreement with photo-emission spectra[12].
The (001) surface composition of the $Cu_{0.1}Ni_{0.9}$ alloy in equilibrium is presented on figure 2 as a function of the temperature. At low temperature, a nearly 100 % enrichment of Cu at the surface is found. The surface concentration of Cu decreases abruptly between 500 K and 800 K. The black dots on figure 2 represent experimental results[13,14]. A complete agreement between the theory and the experimental results is not possible at the present time. However, this type of calculations show that the Cu segregation at the surface of CuNi alloys can be understood within the quantum mechanical description of the metallic cohesion. An enrichment of Cu at the surface is found because Cu has the lower surface energy, in agreement with pair bonding calculations[15].

In order to bring out the relevant parameters in the electronic contribution to the surface segregation, a systematic prediction can be made on the basis of a two moment theory. Within this model, the d partial densities of states are assumed symmetrical with respect to the atomic energies and rectangular. The width of the rectangular bands are proportional to the square root of the second centered moment of the partial densities of states. In a diagonal approximation, the second centered moment is proportional to the coordination number. Thus, the bands are narrower at the surface than in the bulk. From Eq.(1) and (2), one obtains

$$\Delta E = \frac{W}{20} \left(1 - \sqrt{\frac{z^s}{z^b}}\right)(N_A - N_B)(10 - N_A - N_B) , \qquad (5)$$

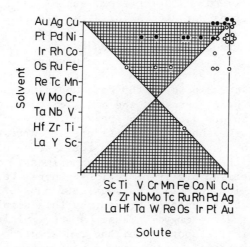

Fig.3 : Solvent solute diagram. The abscissas and ordinates give
the number of d electrons per atom ranging from 0 to 10.
See text.

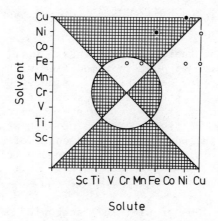

Fig.4 : Same as Fig.3, for 3-d transition metal alloys when the
d-d electron correlations are included in the theory.
The parameter U/W has been taken equal to 0.5.

where W is the bulk bandwidth, Z^b and Z^s are the coordination
number in the bulk and at the surface, and N_A (or N_B) is the
number of d electrons per atom A (or B). A negative value of ΔE
corresponds to the segregation of A at the surface. In this
simplified model, the sign of ΔE depends only on the number
of electrons N_A and N_B. The results are presented on figure 3
and are compared with experimental results on dilute alloy B(A) :
the open circles correspond to the case where the segregation of
the solute A has been found and the black circles represent the
cases where the solute does not segregate[4]. The diagram is divided
in four regions, the shaded areas (figure 3) correspond the
positive values of ΔE (Eq.5), i.e. to the absence of segregation
of A. Of course, there are not enough experimental data to see a
net separation between different regions. There are two observed
cases of segregation, namely, the segregation of Cr in Fe(Cr) and
Mn in Fe(Mn) which are neither predict by this simple theory nor
by the elastic strain energy[16]. These can be explained, however,
if one includes in the above model the d-d electron correlations,
which give an important contribution to the surface energy in the
3-d series. Following Friedel and Sayers[17], a new expression of
the surface energy is obtained in which the Coulomb interaction
is treated at the second order in U/W, where U is an average
Coulomb interaction. The resulting expression of the segregation
energy is

$$\Delta E = \frac{W}{20} \left(1 - \sqrt{\frac{Z^s}{Z^b}}\right) (N_A - N_B)(10 - N_A - N_B)$$

(6)

$$\cdot \left\{1 + 0.09 \left(\frac{U}{W}\right)^2 [(N_A - 5)^2 + (N_B - 5)^2 - 50]\right\} \; ;$$

when $U/W < \sqrt{2}/3$, the factor in parenthesis $\{...\}$ is always
positive and the sign of ΔE is not modified with respect to Eq.5.
But, in the case of the 3-d series, a value of U/W equal to 0.5
is possible[17,18] and its effects are shown in figure 4. The
segregation can now occur in two quaters of circle in the otherwise shaded regions of figure 3. A value of U/W equal to 0.5 is
just suffisant to include the observed cases of segregation in
Fe(Cr) and Fe(Mn). This value seems therefore fully justified
for the 3-d series.

Acknowledgement - One of us (P.L.) acknowledges the Belgian
FNRS for financial support.

APPENDIX

When the alloy is in equilibrium, the variation of the free energy $F(C_A^s, C_A^b)$ for an arbitrary small variation of the concentrations C_A^s and C_A^b, must be equal to zero, with the constraint that the total number of atoms A

$$n_A = C_A^s n^s + C_A^b n^b \tag{7}$$

is constant; n^b and n^s are the numbers of atomic sites in the bulk and at the surface respectively. This gives the condition

$$\frac{1}{n^s} \frac{\partial F}{\partial C_A^s} - \frac{1}{n^b} \frac{\partial F}{\partial C_A^b} = 0 \ . \tag{8}$$

On the other hand, the change ΔF in the free energy when an atom A in the bulk exchanges positions with an atom B at the surface is

$$\Delta F = (\frac{n_A^s + 1}{n^s}, \frac{n_A^b - 1}{n^b}) - F(\frac{n_A^s}{n^s}, \frac{n_A^b}{n^b})$$

$$= \frac{\partial F}{\partial C_A^s} \frac{1}{n^s} - \frac{\partial F}{\partial C_A^b} \frac{1}{n^b} \ , \tag{9}$$

where n_A^b and n_A^s are respectively the number of bulk and surface sites which are occupied by the atoms A. Eq. 8 and 9 show that ΔF is equal to zero when the alloy is in equilibrium.

REFERENCES

1. W.M.H. Sachtler and R.A. Van Santen, Appl. Surf. Sci. 3, 121 (1979).

2. F.F. Abraham, N.H. Tsai and G.M. Pound, Surf. Sci. 83, 406 (1979).

3. A.R. Miedema, Z. Metallkde 69, 455 (1978).

4. J.C. Hamilton, Phys. Rev. Lett. 42, 989 (1979).

5. G. Kerker, J.L. Moran-Lopez and K.H. Bennemann, Phys. Rev. B15, 638 (1977).

6. Ph. Lambin and J.P. Gaspard, Solid State Commun. 28, 123 (1978).

7. F. Cyrot-Lackmann, J. Phys. Chem. Solids 29, 1235 (1968).

8. M.C. Desjonqueres and F. Cyrot-Lackmann, J. Phys. F : Metal Phys. 5, 1368 (1975).

9. Ph. Lambin and J.P. Gaspard, J. Phys. F : Metal Phys., under press.

10. Ph. Lambin and J.P. Gaspard, to be published.

11. J.P. Gaspard and F. Cyrot-Lackmann, J. Phys. C : Solid St. Phys. 6, 3077 (1973).

12. K.Y. Yu, C.R. Helms, W.E. Spicer and P.W. Chye, Phys. Rev. B15, 1629 (1977).

13. K. Watanabe, M. Hashida and T. Yamashina, Surf. Sci. 69, 721 (1977).

14. H.H. Brongersma, M.J. Sparnaay and T.M. Buck, Surf. Sci. 71, 657 (1978).

15. F.L. Williams and D. Nason, Surf. Sci. 45, 377 (1974).

16. J.J. Burton and E.S. Machlin, Phys. Rev. Lett. 37, 1433 (1976).

17. J. Friedel and C.M. Sayers, J. Physique 38, 697 (1977).

18. F. Ducastelle, to be published.

METALLURGICAL EFFECTS ON MAGNETIC PROPERTIES

OF THE ALLOY 15.7 at.% Fe-Au

J. Lauer, W. Keune, and T. Shigematsu

Laboratorium für Angewandte Physik, Universität Duisburg

D-4100 Duisburg 1, Germany

The Mössbauer effect was used to determine the influence of different metallurgical treatments on the magnetic properties of 15.7 at.% Fe-Au. The spectra of an as-quenched sample at 4.2 K can be explained as consisting of two magnetic hyperfine patterns with different average hyperfine fields due to the presence of mictomagnetic and ferromagnetic clusters. The mictomagnetic freezing temperature for this specimen was determined by the appearance of a central doublet spectrum for temperatures above ∿40 K. Aging of the sample at 22°C or plastic deformation result in changes of the transition temperatures, and plastic deformation drastically reduces the relative amount of the mictomagnetic phase in the sample.

INTRODUCTION

Gold-iron alloys are particular interesting systems since long-range ferromagnetic order evolves from spin glass with increasing iron concentration. The first infinite iron clusters with long-range ferromagnetic order (at T = 0) form at 15.5 at.% Fe by percolation along a network of ferromagnetically coupled nearest-neighbor Fe links /1,2/. The magnetic properties near the percolation concentration are extremely sensitive to the metallurgical state of the sample /3-6/.

EXPERIMENTAL

        The alloy was prepared by Argon arc melting of high
purity Au (5N) and Fe (5N, 50 % enriched in 57-Fe). The alloy
was homogenized by annealing in an evacuated quartz capsule
at 950°C for 24h and water-quenched. We prepared a quenched
sample by re-annealing of a 2 μm thick foil at 750°C in an
evacuated quartz capsule for 48h followed by a water-quench.
Mössbauer transmission measurements (57-Co - Rh source) were
performed with this quenched foil immediately after preparation
as well as after storing the same foil at 22°C for six months
(aged sample). Another specimen was obtained by cold-rolling a
100 μm thick foil down to 2 μm thickness (cold-rolled sample).
Mössbauer spectra were taken immediately after its preparation.

RESULTS AND DISCUSSION

        The Mössbauer spectrum taken with the as-quenched
specimen at 4.2 K without application of an external magnetic
field is shown in figure 1 (top). Magnetic ordering occurs at
this temperature as revealed by magnetic hyperfine-splitting,
but the spectrum is clearly not simple. The line width at 4.2 K
is relatively large indicating a distribution of hyperfine fields.
Clearly a shoulder can be seen at about -4 mm/s (marked with an
arrow) due to broad satellite line in addition to the main left
line at about -3.5 mm/s. A similar satellite line with an intensity
depending on the Fe concentration has been observed as well in
Mössbauer spectra of Au-Fe alloys in the mictomagnetic concentration
region and above /7-9/. The complicated Mössbauer spectra at 4.2 K
is the result of the large number of possible (geometrical) local
atomic configurations at higher Fe concentrations. A shifting to
lower velocities of the six-line hyperfine pattern associated
with a particular local configuration with increasing numbers of
Fe neighbors has been inferred (-0.04 mm/s isomer shift (I.S.)
per Fe neighbor), combined with an increase of the magnetic
hyperfine field with increasing Fe neighbors /9/. A detailed
computer analysis of our spectrum (figure 1 top) is very difficult.
However, the overall shape of this spectrum can be qualitatively
explained by the superposition of two different magnetic hyperfine
spectra. Both indicated subspectra I and II in figure 1 (top) should
be considered as rough averages, giving an approximate limit to
the peak positions in the hyperfine field, isomer shift and
quadrupole splitting distribution curves.

        Qualitatively we can interpret the spectrum of our
quenched sample at 4.2 K (figure 1) in the following way. Spectrum
I with the larger average hyperfine field of ∿313 kOe and the more
negative I.S. is associated with infinite Fe clusters where the

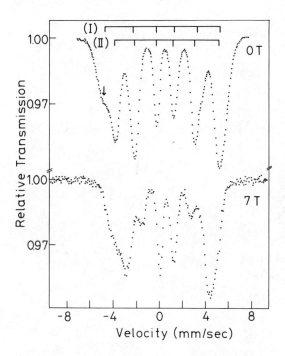

Fig.1 : top : Mössbauer spectrum of 15.7 at.% Fe-Au (as-quenched specimen) at 4.2 K (external field H_{ext} = 0)

bottom : as above, but H_{ext} = 7 Tesla.

number of Fe neighbors surrounding a central Fe atom is relatively high and where nearest-neighbor coupling is ferromagnetic. Consequently, spectrum II (with the smaller average hyperfine field of ∿280 kOe and the more positive isomer shift) corresponds to finite Fe clusters ("cluster glass"), i.e. to regions where the number of Fe neighbors is relatively small. The intensity ratio of the left line of spectrum II (at -3.5 mm/s) and the satellite line (at -4.5 mm/s) indicates that a substantial number of finite Fe clusters exist in our as-quenched specimen.

The influence of an external field of 7 Tesla (oriented parallel to the γ-direction) with the as-quenched sample at 4.2 K is shown in figure 1 (bottom). Such a strong field certainly aligns the spins of infinite ferromagnetic Fe clusters along the field direction resulting in total polarization, i.e. disappearance of the ($\Delta m = 0$) - lines near -1.5 and 3.0 mm/s in the corresponding

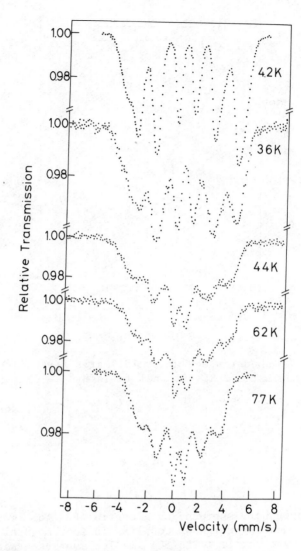

<u>Fig.2</u> : Temperature dependence of Mössbauer spectra for 15.7 at.%
        Fe-Au (as-quenched specimen).

spectrum combined with a reduction of the total splitting. These
effects are observed in figure 1 (bottom). However the
polarization effect is not complete in the sense that some
intensity remains at the ($\Delta m = 0$) line positions. This latter
fact is due to the presence of finite Fe clusters (cluster glass)
which are frozen in a spin-glass like manner at 4.2 K, and which
do not align completely even in the strong applied field /10/.

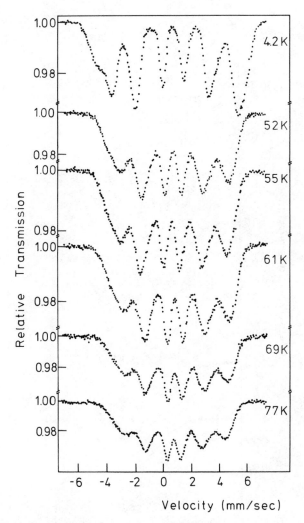

Fig.3 : Temperature dependence of Mössbauer spectra for 15.7 at.%
Fe-Au aged for 6 months.

        The temperature dependence of the Mössbauer spectra for
15.7 at. % Fe-Au (as-quenched) is shown in figure 2. The most
important feature of these spectra is the clear appearance of a
doublet between 36.2 K and 44.0 K which is superimposed on a broad
magnetic hyperfine spectrum. The appearance of this probably
superparamagnetic doublet and the relative sharpness of the
transition (at 40 + 4 K) is clear Mössbauer evidence for the
"melting" of the finite Fe-cluster glass, since the magnetic phase
diagram for Au-Fe alloys predicts a cluster freezing temperature of

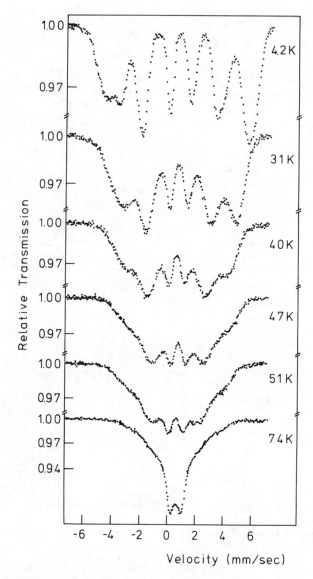

Fig.4 : Temperature dependence of Mössbauer spectra for 15.7 at.%
         Fe-Au (cold-rolled specimen).

∿40 K̊ for 15.7 at.% Fe /1,2/. The superimposed broad magnetic
hyperfine spectrum in figure 2 is caused by the infinite Fe
clusters which are ferromagnetically ordered with an apparent
ordering temperature above 77 K. Preliminary Mössbauer spectra
taken above this ordering temperature in the presence of applied
fields indicate superparamagnetic behavior of Fe-clusters /11/.

Mössbauer spectra of our aged specimen at various temperatures are shown in figure 3. No measurable difference in the 4.2 K - spectra and 300 K - spectra (not shown) of the aged and as-quenched sample could be found. The doublet spectrum, however, which we have associated with finite clusters above their freezing temperature appears at 60 K, superimposed on the magnetically - split spectrum of infinite clusters. Thus the cluster glass freezing temperature has increased as compared to the as-quenched specimen. On the other hand a comparison of the average hyperfine fields at a similar temperature for the infinite-cluster spectra in figure 2 and 3. (e.g. at 62 K and 69 K, respectively) shows that the hyperfine field is larger for the aged sample ($\sim$227 kOe at 69 K) than for the as-quenched sample ($\sim$192 kOe at 62 K). We conclude that the ordering temperature of the infinite clusters also has increased by the aging process. Both, an increase of the cluster glass freezing temperature and of the ferromagnetic transition temperature can be understood on the basis of the magnetic phase diagram for Au-Fe alloys near our composition if atomic clustering is assumed as a result of the aging process, probably via quenched-in vacancy diffusion. Qualitatively atomic clustering has the effect of changing the local composition, i.e. spatial regions of lower local Fe concentrations (mictomagnetic regions) in the sample will be more diluted from Fe and other regions of higher local Fe concentration (infinite cluster regions) will be more concentrated in Fe. A slightly lower Fe content for the mictomagnetic phase accompanied by a slightly higher Fe concentration for the ferromagnetic phase shift the respective transition temperature upwards.

A drastic change in the Mössbauer spectra as compared to the as-quenched alloy was observed for the as-rolled specimen (figure 4). At 4.2 K (figure 4 top) the satellite line to the left (at about -4 mm/s) now has almost equal intensity with the former main line at about -3.5 mm/s. According to our interpretation this means that the fraction of infinite (or large) ferromagnetic clusters in the sample has increased as compared to the fraction of finite (mictomagnetic) clusters as result of plastic deformation. It seems as if movement of dislocations not only divides larger ferromagnetic clusters into smaller fractions (still large enough to be "infinite"), but also may cause linking of existing mictomagnetic finite Fe clusters /12/, which as a consequence become ferromagnetic. The almost complete collapse of the hyperfine field at 74 K indicates a drastic lowering of the apparent ferromagnetic transition temperature, and the unresolved, broad structure of the spectra below that temperature may be a result of a distribution of transition temperatures due to inhomogeneities in the sample. The mictomagnetic freezing temperature can be seen to be near 47 K and thus is slightly higher than for the as-quenched case, since the typical doublet spectrum appears for the first time at 47 K (cf. figure 4).

Acknowledgements - We are very grateful for many helpful
discussions with Drs. R.A. Brand and H.-D. Pfannes. This work
was supported by the Ministerium für Wissenschaft und Forschung
des Landes Nordrhein - Westfalen.

REFERENCES

1. B.R. Coles, B.V.B. Sarkissian and R.H. Taylor, Phil.Mag.B37,
   489 (1978).

2. B.H. Verbeek and J.A. Mydosh, J. Phys. F : Metal Phys. 8,
   L 109 (1978).

3. R.J. Borg, D.Y.F. Lai and C.E. Violet, Phys. Rev. B 5, 1035
   (1972).

4. I. Maartense and G. Williams, Phys. Rev. B 17, 377 (1978).

5. G. Zibold, J. Phys. F : Metal Phys. 9, 917 (1979).

6. S. Crane and H. Claus, Solid State Commun, will be published.

7. R.J. Borg, Phys. Rev. B 1, 349 (1979).

8. M.S. Ridout, J. Phys. C (Solid St.Phys.) 2, 1258 (1969).

9. B. Window, Phys. Rev. B 6, 2013 (1972).

10. G. Chandra and J. Ray, J. Physique 39, C 6-914 (1978).

11. J. Lauer, T. Shigematsu and W. Keune (in preparation).

12. U. Gonser, R.W. Grant, C.J. Meechan, A.H. Muir, Jr. and
    H. Wiedersich, J. Appl. Phys. 36, 2124 (1965).

ROTOPLASMONIC EXCITATIONS IN THE ELECTRON GAS

A.K. Theophilou and C. Papamichalis

Nuclear Research Center "Demokritos"

Athens, Greece

By means of group theory we show that collective excitations of the electron gas which carry certain angular momentum in addition to the linear momentum exist. The energy E_q^m of such excitations has been calculated. For q tending to zero E_q^m goes to E_q^o, i.e. it coincides with the usual plasmon energy.

Hitherto many types of electronic excitations have been observed in solids. Such are e.g. particle-hole pairs, plasmons, excitons, electronic polarons[1,2].

In this paper we shall show theoretically that other types of excitations exist and calculate the excitation spectrum of a type of such excitations, which will be named "rotoplasmonic excitations" as their excitation spectrum is similar to that of plasmons for small q and carry certain angular momentum in addition to the linear momentum $\vec{q}$.

The many electron Hamiltonian in the jellium model approximation (in the second quantization formalism) is

$$\hat{H} = \sum_{\vec{k}} \frac{k^2}{2} a_{\vec{k}}^{+} a_{\vec{k}} + \sum_{q} {}' \frac{2\pi}{q^2} \rho_{\vec{q}}^{+} \rho_{\vec{q}} - \sum_{q} {}' \frac{2\pi}{q^2} \hat{\rho}_{o} \tag{1}$$

where the prime at the summation means that the term in q = 0 is not included; this is because the positive background cancels this term. By $a_{\vec{k}}^{+}$, $a_{\vec{k}}$ we denote the well-known creation and annihilation operators, respectively, in the $\vec{k}$ representation.

The operator $\rho_{\vec{q}}^{+}$ is

$$\rho_{\vec{q}}^{+} = \sum_{\vec{k}} a_{\vec{k}+\vec{q}}^{+} \, a_{\vec{k}} \, .$$

This Hamiltonian is invariant under the full space group I_3^3 in the three dimensional space which included all translations and rotations[3]. Up to now only the translation group has been taken into account. This gave rise to excitations characterized by the irreducible representations of this group. However, the irreducible representations of the symmetry group of the jellium Hamiltonian, I_3^3, are labelled by two quantum numbers m and $|\vec{q}|$, where m is the eigenvalue of the component of the angular momentum about the $\vec{q}$ direction, and $\vec{q}$ is the linear momentum[3]. Thus the wavefunction which transforms according to the irreducible representations of I_3^3 are labelled as $|\psi_{\vec{q}}^{m}\rangle$. The eigenstates of the Hamiltonian belonging to the irreducible subspace $S_{|\vec{q}|}^{m}$ have the same energy irrespective of the direction of $\vec{q}$[3,4]. The study of these representations is simplified when one uses the algebra of the generators of the group, where one can use the Casimir operators for characterizing the irreducible representations[3].

In order to derive the rotoplasmonic states and the corresponding excitation energies, we follow a procedure similar to that of Pines in his derivation of the plasmon excitations[5]. It must be noted that this is different from the one that Bohm and Pines developed initially for the R.P.A.[6,7,8].

In this derivation one makes use of the approximate relation,

$$[H, \, a_{\vec{k}+\vec{q}}^{+} \, a_{\vec{k}}] \simeq (\frac{q^2}{2} + \vec{q} \cdot \vec{k}) a_{\vec{k}+\vec{q}}^{+} a_{\vec{k}} + \frac{4\pi}{q^2} (n_{\vec{k}} - n_{\vec{k}+\vec{q}}) \rho_{\vec{q}}^{+} \qquad (2)$$

where $n_{\vec{k}}$ is the Fermi distribution function at T = 0 and constructs an operator $b_{\vec{q}}^{+}$ of the form

$$b_{\vec{q}}^{+} = \sum_{\vec{k}} f(\vec{k}) a_{\vec{k}+\vec{q}}^{+} a_{\vec{k}} \qquad (3)$$

by the requirement

$$|\hat{H}, b_{\vec{q}}^{+}| = \omega \, b_{\vec{q}}^{+} \qquad (4)$$

where ω is to be calculated. Because of the relations (2), (3), (4) it is possible to expand f(k) in a series of the form

$$f(k) = C' \sum_{n=o}^{\infty} \left(\frac{\frac{q^2}{2} + \vec{q}.\vec{k}}{\omega} \right)^n . \tag{5}$$

By making now an approximation for ω and b_q^+, say a first order approximation, one obtains $\omega = \omega_p = aq^2$

$$b_q^+ = C[(\omega_p - aq^2)\rho_q^+ + \frac{q^2}{2} \rho_q^+ + \sum_k \vec{q}.\vec{k} \; a_{k+q}^+ \vec{a_k}] \tag{6b}$$

where terms of order $\sum_k (\vec{q}.\vec{k}) a_{k+a}^+ \vec{a_k}$ have been omitted.

By examining the operator b_q^+, from the group theoretical point of view, one finds that this transforms according to the q-irreducible representation of the translation group, which is an invariant subgroup of I_3^3. As with respect to I_3^3, b_q^+ transforms according to the $\Gamma_{|q|}^{(o)}$ irreducible representation. In order to develop excited states which transform according to $\Gamma_{|q|}^{(1)}$, i.e. states which carry an angular momentum, we construct the operator

$$B_q^{(1)} = m_{q-q'}^+ b_{q'}^+, \tag{7}$$

where $\vec{q}' = \lambda\vec{q}$, λ is a real number, and $m_{q-q'}^+$, is defined by the relation :

$$m_q^+ \equiv \sum_k e^{i\phi_k} \vec{q}.\vec{k} \; a_{k+q}^+ \vec{a_k} \tag{8}$$

The angle $\phi_{\vec{k}}$ is the azimuthal angle of $\vec{k}$ with respect to the $\vec{q}$ axis.

In taking the commutation relation of the operator $m_{\vec{q}}^+$ with H, terms different from $m_{\vec{q}}$ appear on the right hand side of the equation. These are neglected by the same arguments used in the approximations made before, i.e. the definition (8) is in accordance to the approximations we have carried out up to now. Finally, the following relation holds

$$[H, B_q^{(1)}] = \varepsilon_{|q|}^{(1)} B_q^{(1)} \tag{9}$$

where the excitation energy $\varepsilon_{|q|}^{(1)}$ is given by the formula

$$\varepsilon_{|q|}^{(1)} = \omega_p + \frac{a}{1 + 2a} q^2 \tag{10a}$$

where

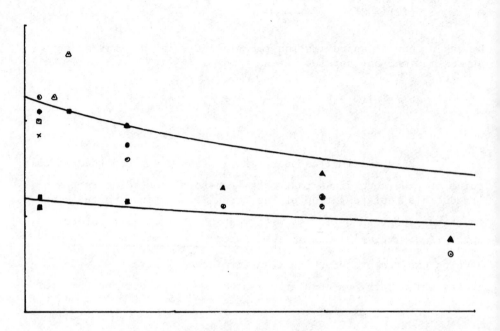

Fig.1 : Figure 1 shows the coefficient, $a^{(m)}$, of q^2 in the
dispersion relation $E_{|\vec{q}|}^{(m)} = \omega_p + a^{(m)} q^2$ vs r_s. The
scaling of axes is as follows :

Horizontal axis 2-5 : atomic units
Vertical axis 0-0.6 : atomic units

Solid lines : present paper theoretical results
Upper line corresponds to m = 0
Lower line corresponds to m = 1.

⊙ Ref. 9
△ Ref. 10
⊡ Ref. 11
× Ref. 12
▲ Ref. 13
⊠ Ref. 14
● Ref. 15
■ Ref. 16

$$\omega_p = (4\pi\rho_o)^{1/2} \tag{10b}$$

ρ_o is the ground state density and

$$a = \frac{3K_F}{10\omega_p} \tag{10c}$$

In figure 1 we plot the experimental results of various authors[9-16] for the coefficient of q^2 in the dispersion relation $\omega = \omega_p + aq^2$ and our own theoretical results for m = 0, m = 1. The m = 0 is the usual plasmon coefficient for q^2 as given by the R.P.A. As it is seen the experimental results show a dispersion from the usual plasmon spectrum. These deviations can be interpreted as due to the rotoplasmonic excitations. Most of these lie between the m = 0 and m = 1 curves.

From the above we conclude that it is worthwhile for experimentalists to carry out experiments which take into account the polarisation of the x-ray beam.

We also remark that the excitations presented here are not the only type of rotoplasmonic excitations.

REFERENCES

1. Kittel, Quantum theory of solids, John Wiley and Sons Inc.

2. Ziman, Principles of the theory of solids, Cambridge University Press.

3. Hamermesh, Group theory, Addison-Wesley Publishing Company.

4. Tinkham, Group theory and quantum mechanics, McGraw-Hill Book Company.

5. Lectures on the many body problem, vol.2 1964, Academic Press.

6. D. Bohm and D. Pines, Phys. Rev. 82, 625 (1951).

7. D. Bohm and D. Pines, Phys. Rev. 85, 338 (1952).

8. D. Bohm and D. Pines, Phys. Rev. 92, 609 (1953).

9. Kloos T., Z. Phys. 265, 225 (1973).

10. Zeppenfeld K., Z. Phys. 223, 32 (1969).

11. Petri E. - Otto A., Phys. Rev. Lett. 34, 1283 (1975).

12. Geiger J., Z. Naturf. 17a, 696 (1962).

13. Kunz C., Z. Phys. 196, 311 (1966).

14. Gibbons P.C. et al., Phys. Rev. B13, 2451 (1976).

15. Chen C.H., J. Phys. C9, L321 (1976).

16. K. Sturm, Solid State Comm. 27, 645 (1978).

THEORY OF EXCITED STATES FOR MANY-ELECTRON SYSTEMS BY MEANS

OF VARIATIONAL PRINCIPLES FOR SUBSPACES

Andreas K. Theophilou

Nuclear Research Center "Demokritos"

Athens, Greece

In this paper the excitation spectrum of many-electron systems is discussed in terms of variational principles for subspaces. It is shown how these principles can be applied to derive Hartree-Fock type equations for excited states. In this scheme fractional occupation numbers appear. The theory of variational principles for subspaces is used to derive Slater's transition state theory. This is achieved by showing that there is a one to one correspondence between the subspace spanned by the M lowest energy eigenstates and the sum of the densities of these eigenstates. Slater's transition state theory comes out as a generalisation of the H-K-S theory for excited states. In this theory too fractional occupation numbers appear.

In a paper to be referred to as 1 the author has introduced the variational principles for subspaces[1,2]. In terms of these principles the H-K-S theory[3,4] was generalized for excited states. For special cases, like e.g. systems with certain symmetry, simplifications result. In this paper we shall deal with special cases of the general theory[1,2].

The essence of paper 1 is the definition of functionals with variable a subspace. Thus if A is an operator and S an M-dimensional subspace S of the Hilbert space the following functional is rigorously defined[1]

$$G_A(S) = \frac{1}{M} \sum_{i=1}^{M} \langle \Psi_i | A | \Psi_i \rangle \qquad \langle \Psi_i | \Psi_i \rangle = \delta_i , \qquad (1)$$

Let us consider now the many-electron Hamiltonian,

$$H = H_o + \int \hat{\rho}(\vec{r}) V(\vec{r}) d^3 r \tag{2}$$

where H_o is the sum of the kinetic energy and electron-electron interaction operator, $\hat{\rho}(\vec{r})$ is the electron density operator in the second quantisation representation and $V(\vec{r})$ the external potential. As it was shown in Ref.1 the functional $G_H(S)$ assumes its minimum value when S is the subspace spanned by the M-lowest energy eigenstates of H. Further it was shown that there exists a one to one correspondence between the subspace density $\rho(r)$

$$\rho(\vec{r}) = \frac{1}{M} \sum_{i=1}^{M} \langle \Psi_i | \hat{\rho}(\vec{r}) | \Psi_i \rangle \tag{3}$$

and the subspace S.

From the above it follows that when the ground state is M-fold degenerate the development of the H-K-S theory for excited states[1,2] is directly applicable. This case may be considered academic as most physical many electron systems are considered to have their ground state nondegenerate.

Degeneracy however is not accidental but derives from the fact that the Hamiltonian of the physical system is invariant under a group of transformations. This group may be of geometric or dynamical origin or both. If such a group exists then by using group theoretic techniques one can simplify the problem. As it will be shown in the following such simplifications result in the scheme developed in Ref.1. In order to realise the necessity of such simplification consider the electron-hole excitation in a solid. These excitation energies are a function of the "continuous" variable $\vec{k}$ which characterises the irreducible representation of the translation group. Then the direct application of the scheme developed in Ref.1 is practically impossible. However, by taking into account group theory the problem simplifies a lot.

Let us consider the set of Hamiltonians which are invariant under a group of transformations G. Then, as is well known from group theory[5,6] one can classify the eigenstates of H according to the irreducible representation of the group G. The energy eigenstates $|\Psi_\gamma^{(\Gamma)}\rangle$ belonging to certain energy level, $E^{(\Gamma)}$ form an irreducible subspace $S_M^{(\Gamma)}$ of dimension M. The direct sum of such subspaces forms an infinite dimensional space H^Γ which is a subspace of the Hilbert space. This subspace, being an invariant subspace of G, is also an invariant subspace of the

Hamiltonian H because if $|\phi^{\Gamma}_{\gamma}>$ belongs to H so does the state $H|\phi^{\Gamma}_{\gamma}>$. Then the variational principles and theorems proved in Ref.7 hold equally well when restricted to the subspace S^{Γ}. Thus instead of the definition of $G_H(S)$ in S one uses the definition

$$G_H(S^{\Gamma}_M) = \frac{1}{M} \sum_{i=1}^{M_{\Gamma}} \sum_{\gamma=1} <\Psi^{\Gamma}_{\gamma,i}|H|\Psi^{\Gamma}_{\gamma,i}> . \tag{4}$$

Obviously the dimension of S_M must be an integral multiple of the dimension of the irreducible representation considered, i.e. $M = \ell M_{\gamma}$. For the special case that $M = M_{\gamma}$ one finds the lowest energy and subspace density of the irreducible representation Γ.

The procedure for the development of the present scheme is to correspond to the subspace $S^{(\Gamma)}$ of the interacting system a subspace $S'^{(\Gamma)}$ of the "noninteracting system" in such a way that the two subspace densities coincide[1,2]. The simplest case of the subspace $S'^{(\Gamma)}$ is when this is spanned out of states of the form

$$|\phi^{\Gamma}_{\gamma,i}> = a^+_{\gamma,i}|\phi_o> \tag{5a}$$

where $|\phi_o>$ is a state of N-1 particles which transforms according to the identity representation of the group G,

$$|\phi_o> = a^+_1 \ldots a^t_{N-1}|0> \tag{5b}$$

The $a^+_{\gamma,i}$ are creation operators for the one particle states $\phi^{\Gamma}_{\gamma,i}(\vec{r})$. The subspace density

$$\rho(r) = |\sum_{j=1}^{N-1} |\phi_j(r)|^2_+ \sum_{i,\gamma} \frac{1}{M} |\phi^{\Gamma}_{\gamma,i}(\vec{r})|^2 \tag{6}$$

The difficult problem, as in the HKS theory is to device a good approximation for $E_{xc}(\rho)$, the exchange and correlation energy. The lowest order approximation is of the form[1,7]

$$E_{xc}(\rho) = c_{xc} \int \rho^{4/3}(\vec{r})d^3r \tag{7}$$

and the exchange and correlation potential is

$$V_{xc}(\vec{r}) = \frac{4}{3} C_{xc}\rho^{1/3}(\vec{r}) \tag{8}$$

The kinetic energy is that of S' plus a correction term[1].
After variation of the energy expression $E(S')$ the resulting
equations for $\phi_i(r)$, $\phi^{\Gamma}_{\gamma,i}$, are like those of the H-K-S theory,
i.e.

$$-\frac{1}{2} \nabla \phi_i(\vec{r}) + (V(\vec{r}) + V_{el}(\vec{r}) + V_{xc}(\vec{r}))\phi_i(r) = \varepsilon_i \phi_i(r) \quad (9)$$

The same equations hold for $\phi^{\Gamma}_{\gamma,i}$. These equations must be solved
self-consistently, i.e. the density resulting from the solution
must satisfy Eq.8 and Poisson's equation for the electrostatic
potential $V_{el}(\vec{r})$. Such equations have been solved self-
consistently by Slater and his co-workers for certain atoms[7,8].
In fact the theory developed here is Slater's transition state
theory derived from variational principles[7].

For the lowest energy state of a crystal metal belonging
to the k-representation of the translation group one has to
consider the state

$$|\Phi_k\rangle = a^+_{\vec{k},1} a^+_{\vec{k}_1} a^+_{-\vec{k}_1} \ldots a^+_{\vec{k}_{n-1}} |0\rangle \quad (10)$$

Thus all the $a_{\vec{k}}$ with $\phi_{\vec{k}}$ below a certain energy level enter and in
addition $\phi_{\vec{k},1}$ whose energy is above that level. Thus, one has to
do a H-K-S calculation in which the density expression is of
different form. Note that functional occupation numbers appear
in the density expression, Eq.6. The spin variable is supressed
throughout in this paper.

In many cases it is not possible to form $|\Phi^{\Gamma}_\gamma\rangle$ states of
the form given by Eq.5 but one has to use linear combinations of
Slater determinants. This is a topic for further development of
the present theory. A theory that deals with such cases was
developed by von Barth[9]. This theory uses states instead of
variational principles.

If in the present theory one considers the subspace of
$|\Phi^{\Gamma}_\gamma\rangle$ states, $^{\Gamma}_\gamma$, then the correspondence theorems between S^{Γ}_γ
and S'^{Γ}_γ also hold. However, this theory gives nonsymmetric
potentials and this makes the numerical calculations difficult.
In the case of 1-dimensional space one derives Gunnarson's e.a.
theory[10].

For developing the Hartree-Fock approximation for excited
states one has to consider subspaces of the form

$$\phi = a_N^+ a_{N-1}^+ \cdots a_1^+ |0>$$

$$\phi_1 = a_{N+1}^+ a_{N-1}^+ \cdots a_1^+ |0>$$

and minimise the functional

$$E(S') = <\Phi_0|H|\Phi_0> + <\Phi_1|H|\Phi_1>$$

in this set of subspace. The energy resulting from the minimisation of $E(S')$ is an approximation for the sum of the ground state and first excited state energies. The equations for the one particle states $\phi_i(\vec{r})$ are like those of the H-F approximation with $1/2$ occupation numbers for $\phi_N(\vec{r})$ and $\phi_{N+1}(\vec{r})$. Thus, the theory of Slater about fractional occupation numbers[7] can be derived from variational principles.

REFERENCES

1. A.K. Theophilou, J. Phys. C : Solid State Phys. _12_, 5419 (1979).

2. A.K. Theophilou, submitted to J. Phys. C : Solid State Phys.

3. P. Hohenberg and W. Kohn, Phys. Rev. _B136_, 864 (1964).

4. W. Kohn and L.J. Sham, Phys. Rev. _A140_, 1133 (1965).

5. M. Tinkham, Group Theory and Quantum Mechanics. McGraw-Hill, New York (1964).

6. M. Hamermesh, Group Theory. Addison-Wesley, London (1964).

7. J.C. Slater, Quantum Theory of Molecules and Solids, vol.4, McGraw-Hill, New York (1974).

8. J.C. Slater and J.H. Wood, Int. J. Quantum Chem., _45_:3 (1971).

9. U. von Barth, Phys. Rev. _A20_, 1693 (1979).

10. O. Gunnarsson and B.I. Lundqvist, Phys. Rev. _B13_, 4274 (1976).

STATIC ELECTRIC FIELD PROFILE IN A METAL AND

ELECTROREFLECTANCE

R. Garrigos, R. Kofman, and P. Cheyssac

Laboratoire d'Electrooptique - L.A. 190
Associé au CNRS - Parc Valrose

06034 Nice Cédex, France

Numerical resolution of Poisson-Boltzmann's equation has been carried out to determine the charge distribution induced by a strong static electric field just within the surface of a metal. This distribution has been fitted by an analytical model and then introduced in the wave equation of the light-electric field. The resolution of this last equation with the help of the Fourier transform gives the optical surface impedance of the metal as a function of the excess surface charge σ. The logarithmic derivative of the reflectance with respect to σ is then calculated and successfully compared to the experimental electroreflectance spectrum for the (111) face of a gold single crystal at the potential of zero charge.

1. INTRODUCTION

A strong static electric field ($\sim 10^{10}$ V/m) applied at the surface of a metal induces a change in its optical properties. Such a field is obtained by applying a small d.c. bias (~ 1 V) between the metallic sample and an appropriate electrolytic solution[1].

This electroreflectance effect has been investigated by many workers especially from an experimental point of view[1-9].

The aim of this paper is to give a microscopic interpretation of metallic electroreflectance by considering that a static electric field induces a noticeable change in the electron density just within the surface of the metal.

The determination of this charge distribution is done in Section 2. The calculation of the optical surface impedance in terms of the excess surface charge σ is given in Section 3. In Section 4 the logarithmic derivative of the reflectance with respect to σ is established; the result of the theoretical calculation is then compared with experiment and discussed.

2. STATIC ELECTRIC FIELD AND POTENTIAL IN A METAL

Let $E_{stat}(x,\sigma)$, $\psi(x,\sigma)$ and $N(x,\sigma)$ be the electric field, the potential and the electron density in the metal respectively; σ is the excess surface charge given by

$$\sigma = - e \int_{0}^{\infty} \{N(x,\sigma) - N_{o}\}dx \qquad (1)$$

e is the absolute value of the electron charge, N_{o} is the electron density of the undisturbed metal, and the positive x direction, normal to the surface, is into the metal.

The electron density $N(x,\sigma)$ fulfils the Boltzmann's distribution[10] :

$$N(x,\sigma) = N_{o} \exp \{e\psi(x,\sigma)/k_{B}T\} \qquad (2)$$

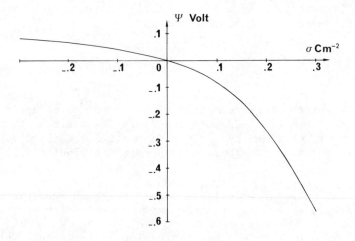

Fig.1 : Electric potential at the surface of the metal (ψ = 0 in the bulk). Note that ψ is asymmetrical with respect to the sign of σ.

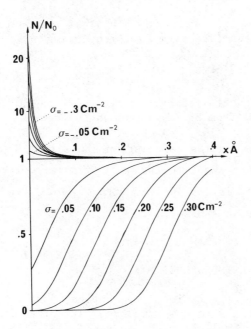

Fig.2 : Electron density in the metal when a static electric field
is applied. N is asymmetrical with respect to the sign of
σ and is different from N_0 only within a depth of a
fraction of an Angström (typically the Thomas–Fermi
screening length).

The electric potential $\psi(x,\sigma)$ satisfies Poisson's
equation :

$$\frac{e}{\varepsilon_0} \{N(x,\sigma) - N_0\} = \frac{d^2\psi}{dx^2} (x,\sigma) \tag{3}$$

written in the international unit system; ε_0 is the dielectric
permittivity of the vacuum.

A first integration of Eq.(3) yields :

$$\frac{2eN_0}{\varepsilon_0} \{\frac{k_B T}{e} [\exp \frac{e\psi(x,\sigma)}{k_B T} - 1] - \psi(x,\sigma)\} = [\frac{d\psi}{dx} (x,\sigma)]^2 \tag{4}$$

The electric potential $\psi(o,\sigma)$ at the surface of the metal (fig.1) is derived from Eq.(4) by using :

$$E_{stat}(o,\sigma) = - \frac{d\psi}{dx}(o,\sigma) = - \frac{\sigma}{\varepsilon_o} \qquad (5)$$

Now $\psi(x,\sigma)$ can be obtained by a numerical integration of Eq.(4). $N(x,\sigma)$ calculated from Eq.(2) is shown in figure 2.

As a remark the calculation shows that $E_{stat}(x,\sigma)$ is asymmetrical with respect to the sign of σ (e.g. for x = 0.1 Å the value of E_{stat} is 1.5 10^9 V/m with σ = -0.3 C/m^2 but is -2.4 10^{10} V/m with σ = 0.3 C/m^2.

3. OPTICAL SURFACE IMPEDANCE

Let Z be the ratio of the optical surface impedance on the metal side to the vacuum impedance. At normal incidence the wave equation of the light electric field is :

$$\frac{d^2E}{dx^2}(x,\sigma) + \frac{\omega^2}{c^2}\varepsilon_B E(x,\sigma) = - \frac{i\omega}{\varepsilon_o c^2} J(x,\sigma) \qquad (6)$$

ω represents the light wave angular frequency and c its celerity; $\varepsilon_B = 1 + \chi_B$ is the interband transitions contribution to the dielectric susceptibility of the metal. The current density of conduction electrons is J = ΓE where the electrical conductivity Γ is deduced from Drude–Lorentz model :

$$\Gamma(x,\sigma) = \{N(x,\sigma)e^2 \tau/m\} (1 - i\omega\tau)^{-1} \qquad (7)$$

m and τ are the mass and the relaxation time of the electron. Eq.(6) can be solved with the help of the Fourier transform leading to the following expression of $Z(\sigma)$:

$$Z(\sigma) = \varepsilon_o c E(o,\sigma)/H(o,\sigma) = Z_o \{1 - i \chi_f Z_o F(\sigma)\}^{-1} \qquad (8)$$

$$Z_o = \frac{1}{\tilde{n}} \; ; \; F(\sigma) = \frac{\omega}{c} \int_0^\infty [\frac{N(x,\sigma)}{N_o} - 1] \exp \frac{i \omega \tilde{n} u}{c} du \qquad (9)$$

H is the light magnetic field; $\tilde{n}$ is the complex optical index of the metal; $\chi_f = -\omega_p^2 \{\omega(\omega+i/\tau)\}^{-1}$ the free electron contribution to the susceptibility of the metal; ω_p the free electron plasma angular frequency.

        The determination of $Z(\sigma)$ requires a tedious
mathematical development which, for this paper, seems irrelevant.
The principles of such a calculation have already been used[11,1,12].

        Instead of performing a numerical calculation of $F(\sigma)$
it seems more interesting to express it analytically. For this,
$N(x,\sigma)$ has been fitted by an analytical model, using both the
calculated value of $N(o,\sigma)$ and Eq. (1) (fig.3). Within a good

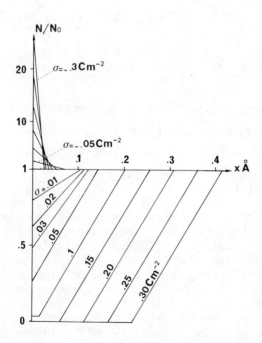

<u>Fig.3</u> : Analytical model for the electron density in the metal
        when a static electric field is applied.

approximation it can be shown that[12] :

$$Z(\sigma) = (\widetilde{n} + i\alpha\sigma)^{-1}; \qquad \alpha = \omega \chi_f / N_o \, e \, c \qquad (10)$$

4. ELECTROREFLECTANCE SPECTRUM

At normal incidence the complex reflectivity of the metal is :

$$r = \sqrt{R} \exp i\phi = \frac{Z(\sigma) - 1/n_E}{Z(\sigma) + 1/n_E} \qquad (11)$$

The (real) refractive index of the electrolytic solution n_E has been assumed independant of σ.

Derivation of Eqs. (11) and (10) with respect to σ leads to :

$$\frac{1}{R} \frac{dR}{d\sigma} \simeq - \, \mathcal{R}e \, \frac{4 \, i \, n_E \, \alpha}{n_E^2 - (\widetilde{n} + i\alpha\sigma)^2} \qquad (12)$$

In the vicinity of the "point of zero charge" (p.z.c., i.e. $\sigma = 0$)

$$\frac{1}{R} \frac{dR}{d\sigma} \simeq \frac{8 \, \omega \, \chi_f'}{N_o \, e \, c} \frac{n_E \, n \, k}{(n_E^2 - n^2 + k^2)^2 + 4n^2 \, k^2} \qquad (13)$$

n and k are the optical constants of the metal; χ_f', the real part of χ_f, is always negative and its absolute value much greater than the imaginary part. These conditions are always satisfied for metals and confirm both the negative sign of R^{-1} (dR/dσ) and its expression at the p.z.c.

Note that Eq. (10) represents exactly the expression of $Z(\sigma)$ obtained by taking for the charge distribution[12]; $N(x,\sigma) = N_o - 2 \sigma \, \delta(x)/e$ where $\delta(x)$ is the "Dirac function".

The electroreflectance spectrum of a gold single crystal calculated from Eq. (12) is found in good agreement with the experimental one for the (111) face at the p.z.c. (fig.4). It is worth pointing out that for the first time such a quantitative agreement has been obtained without requiring any scaling factor.

Similar agreement has been obtained for Silver. An extension of this work to other metals can be considered if the two following conditions are simultaneously satisfied :

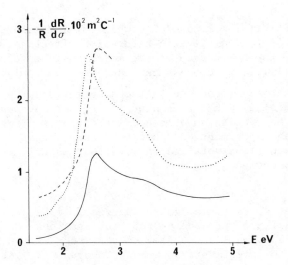

Fig.4 : Comparison between theoretical and experimental spectra :
- solid line : theoretical calculation using the optical
  constants of gold found in the literature[13]
- dashed line : theoretical calculation using the optical
  constants of the sample under study
- dotted line : experimental spectrum for the (111) face,
  of a gold single crystal, at $\sigma = 0$.

   i/ existence of a noticeable variation of the reflectance
of the metal in the spectral range under study (visible range here)
   ii/ existence of a double layer region for the metal
electrolyte interface.

   These two conditions do not seem to stand for all metals;
for instance platinum which nevertheless exhibits a double layer
region has a reflectance so smooth that the electroreflectance
effect is very weak and, by the way, difficult to bring in evidence.

   Outside the p.z.c. it is not possible using Eq.(12) to
find the changes observed in the experimental spectrum in terms
of σ. These observed changes are much more important than those
predicted by the theory.

   As a consequence it seems that the present theory is
valid as far as the static electric field does not disturb the
dielectric properties of the electrolytic solution, i.e. as far

as $|\sigma|$ is small ($|\sigma| < 0.05$ C/m^2). When $|\sigma|$ increases, the electrostrictive and dielectric saturation effects of the water dipoles close to the surface of the metal must be considered. Then, in Eq.(10), $1/n_E$ depends on σ and must be replaced by Z_E the optical surface impedance on the solution side. A first tentative calculation has been done taking into account for the terms of σ. The results so obtained are found in a rather good agreement with experiment and show that such effects cannot be ignored. This will be detailed in a further paper.

REFERENCES

1. R. Kofman, R. Garrigos and P. Cheyssac, to be published in Surface Science. Proceedings of "International Conference on non traditional approaches to the study of the Solid - Electrolyte interface". Snowmass, Colorado (U.S.A.) September 24-27 (1979).

2. J. Feinleib, Phys. Rev. Lett. 16, 1200 (1966).

3. W.N. Hansen, A. Prostak, Phys. Rev. 174, 500 (1968).

4. J.D.E. Mac-Intyre, Surf. Sci. 37, 658 (1973).

5. R. Kofman, R. Garrigos and P. Cheyssac, Surf. Sci. 44, 170 (1974).

6. R.M. Lazorenko-Manevitch, E.B. Brick and Y.M. Kolotyrkin, Electrochim. Acta 22, 151 (1977).

7. T.E. Furtak, D.W. Lynch, Nuovo Cimento, 39 B, 346 (1977).

8. C.N'Guyen Van Huong, C. Hinnen, J. Lecoeur and R. Parsons, J. Electroanal. Chem. 92, 239 (1978).

9. H. Gerischer, D.M. Kolb and J.K. Sass, Advances in Physics, 27, 437 (1978).

10. R. Enderlein, A.J. Hunt and B.O. Seraphin, Phys. Rev.B18, 2472 (1978).

11. K.L. Kliewer, R. Fuchs, Phys. Rev. 172, 607 (1968).

12. R. Kofman, R. Garrigos and P. Cheyssac, to be published.

13. G.P. Pells, M. Shiga, Private Communication.

THE FICTITIOUS EQUILIBRIUM ENSEMBLE IN THE THEORY OF QUENCHED DISORDERED SYSTEMS

Albrecht Huber

Institut für Theoretische Physik und Sternwarte
Christian-Albrechts-Universität

D-2300 Kiel, Germany

A method proposed by Morita for calculating the thermodynamic properties of a quenched disordered system by equating it to a suitably chosen fictitious equilibrium ensemble in an enlarged phase space, is re-examined. It is shown that Morita's prescription, in its original form, is incomplete in that it leaves the free energy of the fictitious ensemble entirely undefined. A missing postulate, named the "zero mean potential condition", is found which must supplement the Morita condition for a full specification of the fictitious ensemble to be achieved. The formulation thus obtained is shown to be thermodynamically self-consistent up to first order in the derivatives of the free energy, and to be equivalent to the maximum entropy formalism proposed by Mazo. The resulting free energy is, except for an irrelevant entropic contribution, equal to the Brout-Vuillermot "quenched" free energy.
In the calculation of fluctuations, fictitious ensemble formulation and maximum entropy procedure are both shown to be inconsistent : the formulae based on the second derivatives of the free energy give the correct result, while the direct calculation of fluctuations yields an incorrect extra contribution. The origin of this behaviour is discussed.

1. INTRODUCTION

In the theory of disordered systems – or, more generally, of incompletely specified systems – two kinds of variables are needed to describe microscopic behaviour. First, there are the "configurational" variables, κ, knowledge of which would completely

specify the constitution of the system in question. As an example,
one may think of κ as giving the lattice locations of all impurity
atoms in a randomly dilute alloy. Secondly, there are the dynamic
variables proper, i.e. the observables which characterize the
microstates of the system once the impurity configuration is given;
they will be denoted by σ. The dynamical variables may be either
classical or quantum mechanical in character. However, for the
sake of simplicity I shall always write Σ_{σ} for a sum over states;
this is to be read as the appropriate trace in the quantum case.

The disorder is called quenched if the configurational
variables do not take part in the thermal motion of the system.
In this case, the configuration κ is a fixed random variable subject
only to a temperature- and field-independent a-priori probability
distribution $p(\kappa)$ with

$$p(\kappa) \geqslant 0 \quad \text{for all } \kappa, \ \Sigma_{\kappa} \ p(\kappa) = 1 \ . \tag{1}$$

According to a by now famous argument due to Brout[1], the quantity
of interest in such situations is the "quenched" free energy
density f_q, which is obtained from the Hamiltonian $H(\kappa,\sigma)$ by
first computing the partition function

$$Z(\kappa) = \Sigma_{\sigma} \ e^{-\beta H(\kappa,\sigma)} \ , \quad \beta = \frac{1}{k_B T} \ , \tag{2}$$

and then calculating the average

$$f_q = - \frac{1}{\beta N} <\ln Z> \ . \tag{3}$$

Here, N is the number of lattice sites; the angular brackets
denote the configurational average formed with the probability
distribution (1).

Brout's argument indicates that the true free energy
density of a quenched sample, as a random variable, converges
in probability to f_q when the sample size goes to infinity. For
a certain class of bond-diluted random spin systems, the argument
has been turned into a rigorous proof by Vuillermot[2]. Occasionally,
one encounters the opinion that Brout's argument fails near a
critical point where the coherence length becomes infinite[3,4,5].
However, as can be seen from Brout's original discussion and is
borne out by Vuillermot's analysis, the range of the dynamical
correlations plays no role in establishing the argument : it is
only for the dynamical interactions, and for the correlations
between impurity locations, that finite ranges need to be assumed.
If these two are of finite range, then the Brout argument is fully
valid right up to the center of criticality.

The question with which the present contribution is concerned goes back to a reformulation of the quenched free energy problem due to Morita[6], who proposed using a "fictitious" equilibrium ensemble to simulate the effect of configuration averaging. The idea – which has since been rediscussed by Falk[7] and by Sobotta and Wagner[8] – is to introduce a kind of "grand system" whose dynamical variables are now the pairs (κ, σ) instead of the σ alone, and to endow it with a "fictitious potential" $\Psi(\kappa)$ chosen such that the new system with the total Hamiltonian

$$\mathcal{H}(\kappa, \sigma) = H(\kappa, \sigma) + \Psi(\kappa) \tag{4}$$

will have thermodynamic equilibrium properties identical to the "quenched" thermodynamic properties of the disordered system. If this can be achieved, then – so at least one hopes – the full machinery of equilibrium statistical mechanics can be applied to the calculation of the thermodynamic properties of quenched (i.e. non-equilibrium) impurity systems. In what follows, an appraisal of this idea is attempted.

2. FICTITIOUS POTENTIAL, THERMODYNAMIC SELF-CONSISTENCY, AND THE ZERO MEAN POTENTIAL CONDITION

By evaluating the equilibrium expectation value of an arbitrary observable $A(\kappa, \sigma)$ in the fictitious ensemble and equating it to the configuration average

$$<A> = \Sigma_\kappa \ \{p(\kappa) \ \frac{1}{Z(\kappa)} \ \Sigma_\sigma \ A(\kappa, \sigma) \ e^{-\beta H(\kappa, \sigma)}\} \ , \tag{5}$$

Morita established the following requirement for the fictitious potential : The quantity $\Psi(\kappa)$ must be chosen such that

$$\frac{p(\kappa)}{Z(\kappa)} = \frac{1}{Z} \ e^{-\beta\Psi(\kappa)} \tag{6a}$$

where

$$Z = \Sigma_\kappa \ Z(\kappa) \ e^{-\beta\Psi(\kappa)} \ . \tag{6b}$$

Morita did not advance an explicit expression for $\Psi(\kappa)$, but contented himself with the remark that (6) "determines $\Psi(\kappa)$ uniquely except for an arbitrary additive constant". Indeed, it follows from (6) that $\Psi(\kappa)$ must be of the form

$$\Psi(\kappa) = -\frac{1}{\beta} \ln \frac{p(\kappa)}{Z(\kappa)} + \Phi \ , \tag{7}$$

where Φ is arbitrary, and is a constant in the sense that it does not depend on κ.

In order to check whether Morita's condition (6) - which is clearly a necessary condition on the fictitious potential - is actually sufficient to establish equivalence with the properties of the quenched system, one has to calculate the fictitious ensemble free energy

$$F = - \frac{1}{\beta} \ln \sum_{\kappa,\sigma} e^{-\beta\{H(\kappa,\sigma) + \Psi(\kappa)\}} \tag{8}$$

with $\Psi(\kappa)$ given by (7). The result is simply

$$F = \Phi \tag{9}$$

which shows that Φ - far from being an arbitrary and irrelevant constant - is in fact solely responsible for the way in which the free energy of the fictitious ensemble is going to depend on temperature, external fields, interaction parameters and impurity concentration.

Thus, choosing $\Psi(\kappa)$ so as to make the fictitious ensemble equilibrium probabilities

$$p(\kappa,\sigma) = \frac{1}{Z} e^{-\beta H(\kappa,\sigma)} \tag{10}$$

equal the "quenched" joint probabilities

$$P_q(\kappa,\sigma) = p(\kappa) \frac{1}{Z(\kappa)} e^{-\beta H(\kappa,\sigma)} \tag{11}$$

- which is exactly what Morita's condition accomplishes - is not sufficient for establishing an equivalent fictitious ensemble, because this leaves the free energy of the latter totally un-specified. An additional requirement must be imposed upon the fictitious potential for the resulting free energy actually to describe the thermodynamics of the disordered system.

One rather direct way of doing this would be to demand that F equal the Brout free energy, i.e. that

$$\Phi = - \beta^{-1} <\ln Z(\kappa)> \quad . \tag{12}$$

Unfortunately, this would cause an inconsistency in that the two alternative entropy expressions

$$k_B^{-1} S_{probabil.} = - \sum_{\kappa,\sigma} p(\kappa,\sigma) \ln p(\kappa,\sigma) \tag{13}$$

and

$$k_B^{-1} S_{thermodyn.} = \ln Z - \beta \frac{\partial(\ln Z)}{\partial\beta} \tag{14}$$

- which are of course identical for any physical equilibrium ensemble with a temperature-independent Hamiltonian - would no

longer agree : The "probabilistic" entropy (13) would be larger
than the "thermodynamic" entropy (14) by the amount

$$k_B^{-1} S_o = - \sum_{\kappa} p(\kappa) \ln p(\kappa) . \tag{15}$$

The difference is just the "entropy of mixing" obtained by Mazo[9]
which describes the uncertainty about the impurity configuration
itself. Although (15) is only a constant entropy contribution, the
discrepancy would have the unpleasant consequence of leading, via
the relation

$$E = F + TS , \tag{16}$$

to two conflicting answers for the internal energy E of the
fictitious ensemble which would differ by a term linear in T.

Since such a behaviour does not appear easily acceptable,
it seems a sounder procedure to secure at the outset a certain
measure of thermodynamic self-consistency for the fictitious
ensemble, and to try then to adapt the ensemble thus obtained,
as best possible, to the properties of the quenched system. With
the help of (6) and (10), it is easily shown that the two expressions
(13) and (14) agree if an only if

$$\frac{\partial <\Psi>}{\partial \beta} = 0 . \tag{17}$$

Furthermore, for each extensive macroscopic observable $M(\kappa,\sigma)$
governed by a conjugate intensive "field" h (in the sense that
the original Hamiltonian contains a term $-hM$), there exists a
self-consistency condition requiring that the two alternative
expressions for the mean value of $M(\kappa,\sigma)$, viz

$$<M>_{\text{probabil.}} = \sum_{\kappa,\sigma} M(\kappa,\sigma) p(\kappa,\sigma) \tag{18}$$

and

$$<M>_{\text{thermodyn.}} = \beta^{-1} \frac{\partial (\ln Z)}{\partial h} , \tag{19}$$

give the same result. Again, for physical equilibrium ensembles
(18) and (19) would automatically be identical, while for the
fictitious ensemble it follows from (6) and (10) that (18) and
(19) agree if and only if

$$\frac{\partial <\Psi>}{\partial h} = 0 . \tag{20}$$

Thus, the fictitious ensemble is thermodynamically self-consistent
in first order - i.e. all standard thermodynamic relations
containing no higher than first-order derivatives of the thermodynamic potential are satisfied - provided that the configuration

average of the fictitious potential is independent of temperature and independent of all fields, chemical potentials, etc. which occur in the original Hamiltonian.

When the conditions (17) and (20) are satisfied, the internal energy of the fictitious ensemble is unambiguously given by

$$E = <\frac{1}{Z(\kappa)} \Sigma_\sigma H(\kappa,\sigma) \ e^{-\beta H(\kappa,\sigma)}> + <\Psi> \ . \tag{21}$$

Since the first term by itself already equals the "quenched" internal energy, the second term must be made equal to zero, i.e. the fictitious potential must be made to satisfy the condition

$$<\Psi> \equiv 0 \ . \tag{22}$$

This is the "zero mean potential condition" by which Morita's condition (6) must be supplemented. Together, the two conditions completely specify the fictitious potential, which now reads

$$\Psi(\kappa) = - \beta^{-1} \{\ln \frac{p(\kappa)}{Z(\kappa)} - <\ln \frac{p(\kappa)}{Z(\kappa)}>\} \ , \tag{23}$$

The free energy and entropy of the fictitious ensemble are thus given by

$$F = - \beta^{-1} <\ln Z(\kappa)> - TS_o \tag{24}$$

and

$$S = <S(\kappa)> + S_o \ , \tag{25}$$

respectively. These expressions agree with the formulae obtained by Mazo[9] and by Sobotta and Wagner[8], who treat the problem of quenched disorder by maximizing the entropy functional

$$S = - k_B \Sigma_{\kappa,\sigma} \ p(\kappa,\sigma) \ \ln p(\kappa,\sigma) \tag{26}$$

under the subsidiary conditions

$$\Sigma_\sigma \ p(\kappa,\sigma) = p(\kappa) \tag{27}$$

and

$$\Sigma_{\kappa,\sigma} \ H(\kappa,\sigma) \ p(\kappa,\sigma) = E \ . \tag{28}$$

The results (24) and (25) deviate from the correct Brout–Vuillermot expressions by a constant term in the entropy and by a corresponding term linear in T in the free energy. These discrepancies are the unavoidable consequence of the use of an equivalent equilibrium ensemble. Since they do not affect measurable quantities, they need not be considered as major

defects in the equivalent ensemble formalism. However, they are a
first indication of the fact that the fictitious equilibrium
ensemble cannot be perfectly adapted to the quenched impurity
problem.

Furthermore, it should be noted from (23) that a complete
determination of the fictitious potential requires the calculation
of the Brout free energy, i.e. the very calculation that the
fictitious ensemble was meant to facilitate. This somewhat circular
feature creates obvious problems for the practical application of
the fictitious ensemble formalism, which can, however, at least in
principle be overcome within a suitable scheme of successive
approximations for the fictitious potential as used e.g. by
Morita[6].

3. FLUCTUATIONS AND RESPONSE

Additional difficulties are encountered when one
considers fluctuations in a quenched system. As has been pointed
out by Leff[10], there exist several candidates for the right
expression to describe fluctuations in a system with quenched
disorder. To discuss them, the thermal average, as calculated
for fixed κ, shall be denoted by a bar and the configuration
average, as before, by angular brackets. Then the mean fluctuation
of the observable $M(\kappa,\sigma)$ may be considered as given by the
expression

$$\delta M^2 = <\overline{M^2} - \overline{M}^2> , \tag{29}$$

which is what one is led to by attaching importance to the thermodynamics of the individual disordered system. Alternatively, one
could form the expression

$$\Delta M^2 = <\overline{M^2}> - <\overline{M}>^2 ; \tag{30}$$

this would appear logical in approaches like those of Mazo[9] and of
Sobotta and Wagner[8], where the variables κ and σ are treated on
the same footing. One might call (29) the "small" and (30) the
"large" fluctuation because, evidently, the latter is larger than
the former by the amount

$$\Delta M^2 - \delta M^2 = <\{\overline{M} - <\overline{M}>\}^2> , \tag{31}$$

i.e. by the configurational fluctuation of the thermal mean
value of M.

The fictitious ensemble formalism is incapable of
resolving this ambiguity since it, too, produces both alternatives.

If one calculates the "generalized susceptibility" by twice differentiating F with respect to the conjugate field h, one obtains

$$- \beta^{-1} \frac{\partial^2 F}{\partial h^2} = \delta M^2 \quad , \tag{32}$$

i.e. the small fluctuation. On the other hand, calculating the fluctuation directly yields

$$\sum_{\kappa,\sigma} M^2(\kappa,\sigma)p(\kappa,\sigma) - \left\{ \sum_{\kappa,\sigma} M(\kappa,\sigma)p(\kappa,\sigma) \right\}^2 = \Delta M^2 \quad , \tag{33}$$

which is the large fluctuation. (The same ambiguity would be found in the maximum entropy formulations of Mazo and of Sobotta and Wagner). For any physical equilibrium ensemble, of course, the left-hand sides of (32) and (33) would be equal by virtue of the fluctuation-response-theorem. However, the fictitious Hamiltonian contains h not only in the term -hM, but also implicitly through Ψ, and this non-standard h-dependence causes the fluctuation-response-theorem to be violated. (Analogous remarks apply to energy fluctuations and specific heat). Thus, although care has been taken to secure thermodynamic self-consistency for the fictitious ensemble in first order, self-consistency is definitively lost in second order, and there is no way of restoring it.

Which of the two answers, (29) or (30), is, then, the correct expression for the fluctuation? A Brout-Vuillermot type argument shows that the "quenched" fluctuation (29) is correct in the sense that it represents (if suitably normalized with regard to sample size) the limit in probability of the true configuration-dependent fluctuation, as the sample size goes to infinity. Thus, as already emphasized by Leff, in systems with quenched disorder the small fluctuation is the quantity of interest.

However, the large fluctuation, too, has a meaning. It estimates the width of the range within which an onlooker interested in predicting the result of the observation of M can reasonably expect the measured value to lie. When the experiment is repeated many times, the distribution of the values obtained will show a spread which will not agree with that estimate, but which, in general, will be smaller (its average being the small fluctuation). This is so because the impurity configuration is kept fixed when the experiment is repeated.

The uncertainty about the actual impurity configuration gives rise to the extra terms in entropy and fluctuations documented in (15) and (31). These extra terms are relevant for prediction, but not for repetition and thus not for thermodynamics. In a correct thermodynamic description of systems with quenched disorder they must be absent. If each repetition of the measurement of M were to be performed with a new sample from the sample reservoir, then the extra entropy term (15) would indeed apply, and the large fluctuation (30) would indeed be observed in the distribution of the values of M. However, this is not how experiments with randomly dilute alloys are normally done.

For the reasons just outlined, the concept of a "mixture" of random experiments, which at first sight seems so appropriate for the thermodynamic description of incompletely specified systems, is in fact inadequate.

Acknowledgement — The author wishes to thank D. Wagner and G. Sobotta for having introduced him to the subject, and for several stimulating discussions.

REFERENCES

1. R. Brout, Phys. Rev. 115, 824 (1959).

2. P.A. Vuillermot, J. Phys. A 10, 1319 (1977).

3. G. Sobotta and D. Wagner, J. Phys. C 11, 1467 (1978).

4. C. De Dominicis, in : Dynamical Critical Phenomena and Related Topics (Lecture Notes in Physics vol.104), ed. by C.P. Enz (Springer, Berlin, 1979), pp.251-279, v. p. 254.

5. T.C. Lubensky, in : Ill-Condensed Matter (Les Houches Session XXXI, 1978), ed. by R. Balian, R. Maynard and G. Toulouse (North-Holland, Amsterdam, 1979), pp.405-475, v. p. 411.

6. T. Morita, J. Math. Phys. 5, 1401 (1964).

7. H. Falk, J. Phys. C 9, L 213 (1976).

8. G. Sobotta and D. Wagner, Z. Physik B 33, 271 (1979).

9. R.M. Mazo, J. Chem. Phys. 39, 1224 (1963).

10. H.S. Leff, J. Chem. Phys. 41, 596 (1964).

LOCAL ORDER IN MOLTEN $Se_{1-x}Te_x$

R. Bellissent and G. Tourand

Service de Physique du Solide et de
Résonance Magnétique

CEN-Saclay, BP n° 2, 91190 Gif-sur-Yvette, France

In this paper we present a study of the short range order and of the coordination number in liquid Selenium-Tellurium systems.

The first part deals with neutron diffraction measurements of the structure factors of liquid $Se_{1-x}Te_x$, in the whole concentration range, at 475 C, performed at El_3 reactor in Saclay using a 640 cell multidetector. From these data we have calculated the radial distribution functions from which were deduced the corresponding coordination numbers.

In a second part we present a structural model based on random chains for Selenium and on a quasicrystalline behaviour of Tellurium. For Se-rich melts we assume that Tellurium enters the Selenium chains by substitution. In the Te-rich range we assume that the local order if represented by substituted SeTe chains in a Tellurium matrix. This model provides us with a good representation of the various structure factors. Moreover we have calculated the coordination number for each concentration in the model and the results are consistent with our experimental data.

The 2 fold coordination of Se and the 3 valency of Te in the liquid state are emphasized and they can be associated with the metallisation of liquid Tellurium whereas Selenium remains a semiconductor.

1. INTRODUCTION

Although Selenium and Tellurium belong to the same class of lattice in the crystalline state they do exhibit very different local orders above the melting point. Liquid Selenium[1] remains divalent and its resistivity[2] decreases only very slowly with increasing temperature. On the contrary liquid Tellurium becomes trivalent just above the melting point [3] and its resistivity[4] decreases very quickly down to a metallic state. It therefore seemed to us interesting to under take a determination of local order in the whole concentration range of the liquid SeTe mixtures in order to study the variation of the coordination number.

2. EXPERIMENTAL RESULTS

Structure factors determination at 475 C from neutron diffraction measurements have been carried out at the El_3 reactor in Saclay for various $Se_{1-x}Te_x$ melts with 10 values of x the whole concentration range. The use of a 640-cell multidetector provides us with very high counting rates for short times of measurement. The liquid samples were put in cylindrical containers of vitreous silicium of 10 mm diameter and 0.5 mm thickness. The furnace consisted of a Vanadium cylinder 30 mm in diameter and 0.1 mm thick. The mean scattered intensities for the furnace and for the container were respectively about 5 % and 12 % of the intensity scattered by the SeTe samples. After geometrical correction for absorption[5] and Placzek type correction for inelasticity[6] the scattered intensity has been normalized in order to obtain the structure factors. The mean relative uncertainty in these structure factors is estimated to be less than 2 %.

We give in figure 1 and figure 2 a plot of the radial distribution function g(r) for the various studied concentrations.

The most important feature of these curves in the Se-rich range consists of sharp peaks for first and second neighbour distances and very diffuse order elsewhere. For Te concentrations higher than 60% the second peak becomes very large and finally parts into two distances. More over an additional peak arises between 6 and 6.25 Å.

The distances of first and second neighbours for all concentrations increase very slowly, and there is a very weak smoothing of the corresponding peaks for increasing Te concentration. Such a progressive change in the radial distribution function suggests that the same kind of structural model, based on

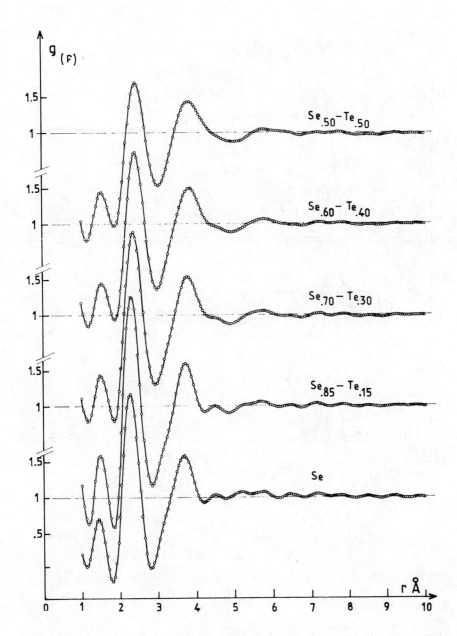

Fig.1 : Pair correlation function, g(r), of liquid $Se_{1-x}Te_x$ mixtures ($0 \leqslant x \leqslant 0.5$).

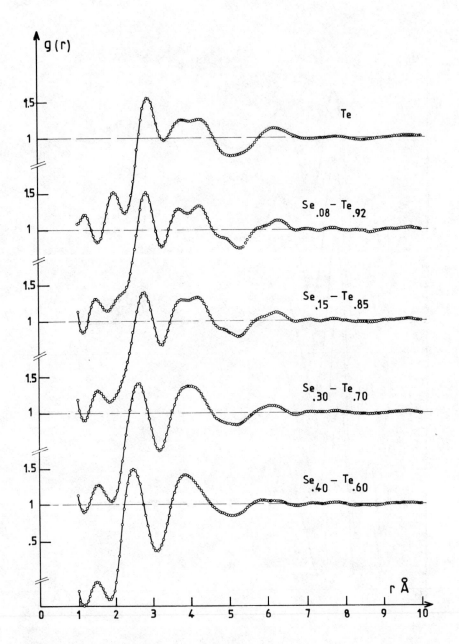

<u>Fig.2</u> : Pair correlation function, g(r), of liquid $Se_{1-x}Te_x$
         mixtures $(0.6 \leqslant x \leqslant 1)$.

the first and second neighbours distances can be used for both
liquid Selenium and Se-rich mixtures. On the contrary, in the
Te-rich range more characteristic distances must be involved in
a representation of local order.

3. STRUCTURAL MODEL

The structural model chosen to give representation of
the various structure factors in the Selenium rich range is based
on tye Nearly Free Rotating Chains (NFRC) model that we have already used in the case of pure Selenium[7].

It consists of chains of atoms in which the distances d_1
and the angles φ of the covalent bonds are conserved but in which
the dihedral angle has a nearly random distribution, as shown in
figure 3.

By a nearly random distribution we mean that we have
forbidden in our model that any distance be smaller than the nearest
neighbour distance the fact that the g(r) is very close to zero
after the first peak is an experimental evidence that the distribution
functions for higher order have no overlap with the first neighbours
distribution. In order to evaluate the chain parameters d_1 and φ
for SeTe melts we assume that Tellenium enters by substituion in
Selenium chains.

The mean distance d_1 of nearest neighbours has been
evaluated respectively by using, for a couple of atoms i and j
with concentrations and scattering length $x_i x_j$ and $b_i b_j$; a
statistical weight ω_{ij} such as : $\omega_{ij} = x_i x_j b_i b_j$.

To compute the mean value of the covalent bond φ we
consider second neighbour atoms i and j covalently bonded to the
same third atom k. The mean value for the second neighbour
distance d_2 is calculated with a statistical weight : $\omega_{ijk} = x_i x_j x_k b_i b_j$ and we deduce the mean value of φ by the relation :

$$\sin \frac{\varphi}{2} = \frac{d_2}{2d_1}$$

The structural model used to fit the structure factor in the
Tellurium rich range is based on a quasicrystalline behaviour of
pure liquid Tellurium. In a previous paper[8] we had made an attempt
to represent the local order by using as a starting point the
hexagonal structure of Tellurium. Here following Friedel[9] we have
used the solid state structure of Arsenic as a structural basis in
order to obtain a better representation of trivalent liquid Te.

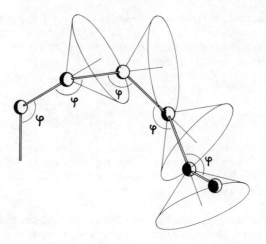

Fig.3 : Structural basis of the NFRC model.

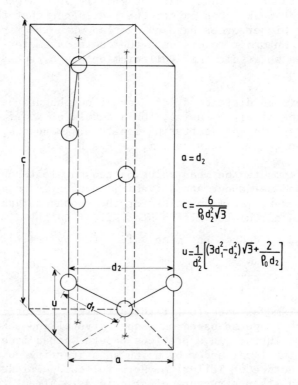

$$a = d_2$$

$$c = \frac{6}{\rho_0 d_2^2 \sqrt{3}}$$

$$u = \frac{1}{d_2^2} \left[(3d_1^2 - d_2^2) \sqrt{3} + \frac{2}{\rho_0 d_2} \right]$$

Fig.4 : Elementary cell of As lattice type used as a structural
        basis for the model Te.

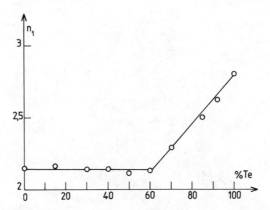

Fig.5 : Coordination number N_1 as a function of the concentration
        x in Tellurium.

        The parameters a, c, u of the crystalline Arsenic have
been changed to take into account the first and second neighbour
distances and the atomic density as shown in figure 4.

        For Te-rich melts we have assumed that the $Se_{1-x}Te_x$
system consists of a quasi-crystalline matrix of As-type Tellurium
in which are present Se-Te chains with the limit concentration of
about 60% Te.This assumption is based on the variation of the
coordination number versus concentration that we have plotted in
figure 5. Moreover this behaviour of the coordination number is
in fair agreement with previous results obtained from density[10]
and X-ray diffraction measurements[11].

4. DISCUSSION

        The experimental and simulated structure factors of
liquid Se Te melts have been represented in figures 6 and 7. The
observed fair agreement on the Selenium side is consistent with
the constant value of the coordination number in this range of
concentration. Moreover, the high values of the resistivity[12,13]
up to 60 % Te at temperature of the order of 500 C seems to be a
confirmation of this random chain structure of SeTe melts.

        The agreement is not so good in the Te-rich region but
it seems that the As structure can be a reasonable approximation
for local order in liquid Te.

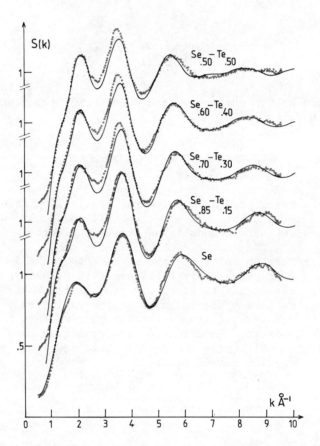

Fig.6 : Experimental (open circles) and simulated (full line)
structure factors of liquid $Se_{1-x}Te_x$ mixtures ($0 \leqslant x \leqslant 0.5$).

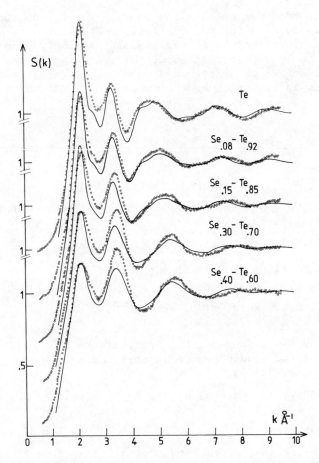

<u>Fig.7</u> : Experimental (open circles) and simulated (full line)
structure factors of liquid Se$_{1-x}$Te$_x$ mixtures ($0.6 \leqslant x \leqslant 1$).

CONCLUSION

      We have determined accurate structure factors for molten SeTe systems by neutron diffraction measurements using a multidetector.

      The 2 coordinations of liquid Se has been conserved up to 60 % Te which is very consistent with a chain structure of SeTe substitution real systems. In the Te-rich range, the coordination number grows up to 3 for pure Tellurium. We have therefore represented the structure factor assuming an As-type local order. This seems to be a confirmation of the weakening of the one dimensional behaviour of elements when one is going from Sulfur and Selenium to Tellurium and Polonium.

REFERENCES

1. Tourand, G., J. Physique 34 (1973) 937-942.

2. Gobrecht, H., Gawlik, D., Mahdjuri, F., Phys. kondens. Materie 13 (1971) 156-63.

3. Tourand, G., Phys. Lett. 54 A3 (1975) 209-210.

4. Perron, J.C., Adv. Phys. 16 (1967) 657-666.

5. Paalamn, H.H., Pings, C.J., J. Appl. Phys. 33 (1962) 2635.

6. Yarnell, J.L., Katz, M.J., Wenzel, R.G., Koenig, H.S., Phys. Rev. A7 (1973) 2130.

7. Bellissent, R., Tourand, G., Proc. of the 7th Int.Conf. on Amorphous and liquid semiconductors, Edinburgh June 1977, ed. Spear, W.E., 98-104.

8. Bellissent, R., Tourand, G., J. Non-Crystalline Solids 35-36 (1980) 1221-1226.

9. Cabane, B., Friedel, J., J. Physique 32 (1971) 73.

10. Thurn, H., Ruska, J., J. Non Crystalline Solids 22 (1976) 331-343.

11. Hoyer, W., Thomas, E., Wobst, M., Z. Naturforsch. 30a (1975) 1633-1639.

12. Perron, J.C., J. Non Crystalline Solids 8-10 (1972) 272-279.

13. Vollmann, J., Wobst, M., Fischer, B., Phys. Stat. Sol. (a) 17 (1973) K 165-K 168.

NUMERICAL INVESTIGATION OF THE DC-CONDUCTIVITY AND ELECTRON LOCALISATION IN ONE-DIMENSIONAL DISORDERED SYSTEMS

G. Czycholl
Institut für Physik, Universität Dortmund
D-4600 Dortmund 50, West-Germany

B. Kramer
Physikalisch-Technische Bundesanstalt
D-3300 Braunschweig, West-Germany

The Anderson model of a 1 d disordered system is investigated using two independent numerical methods. The dc-conductivity, the inverse participation number, and the mean square extension of some eigenstates are calculated. Some transition from a conducting to a non-conducting state is found at a finite disorder in contradiction to the conventional theory. The implications of the results for the theory of disordered systems are discussed.

1. INTRODUCTION

It was argued by Mott and Twose[1] that in 1 d disordered systems all states should be exponentially localised for any kind and degree of disorder. Consequently, such systems should be always insulating at absolute zero. Though a mathematically rigorous proof of this statement is still lacking[2], there is little doubt about its validity, and the localisation problem seems to be solved in one dimension. Therefore, it appears natural to use the 1 d case as a test for the numerical methods applied in higher dimensions to study localisation.

In order to study how natural limitations inherent in all numerical methods as, for instance, the finite size of a system, could affect the results about the localisation in a disordered system we investigate the dc-conductivity at zero temperature in 1 d with two different numerical procedures, both starting from Kubo's formula. Among other most unexpected results

159

we observe a strong increase of the conductivity with increasing
size of the system if the disorder is below a certain "critical
value". As previously[3] we are not able to offer any consistent
interpretation of our data. Instead, we shall present a collection
of possible conclusions ranging from the obvious "technical"
statement that the system sizes used ($\leqslant 10^4$) are still insufficient
to study localisation to a short discussion of the question whether
the Kubo-Greenwood conductivity is a useful concept in studying
localisation.

2. MODEL AND METHODS

We consider the Anderson model in 1 d

$$H = \sum_n [\varepsilon_n |n><n| + V(|n><n+1| + |n+1><n|)] \tag{1}$$

with fixed transfer elements V, and site energies ε_n distributed
independently at random over an energy interval of width W around
$\varepsilon = 0$.

The dc-conductivity within linear response theory

$$\sigma(E) = \frac{2\pi e^2}{m_e^2 a} \frac{1}{N} \text{Tr}[p\delta(E-H)p\delta(E-H)] \tag{2}$$

is determined by the same methods as described in refs. 3 and 4,
namely :

1. Calculating the matrix elements of the resolvent $G(E+i\delta) = (E+i\delta-H)^{-1}$ for "fixed" boundary conditions using continued
   fractions[5], and

2. Calculating the time dependence of a random state via
   Schrödinger's equation with periodic boundary conditions, and
   time averaging products of site amplitudes chosen to reproduce
   eq.2. The random initial state is energetically weighted in
   order to favour eigenstates with energies near the Fermi level
   E[6]. The width of the weighting function, $m^{-1/2}$ corresponds to
   δ in method 1, and may be regarded as simulating the coupling
   to a heat bath.

In order to check our results we determine two properties
of the eigenstates which are closely related to localisation. The
mean square extension of the eigenstates $|\alpha>$

$$\overline{\Delta x^2} = \frac{1}{N} \sum_\alpha (<\alpha|x^2|\alpha> - <\alpha|x|\alpha>^2) \tag{3}$$

is calculated by using method 1 [4]. $\overline{\Delta x^2}$ is expected to diverge in
the thermodynamic limit for extended states and to approach a

finite value in the limit $N \to \infty$ for localised states. The average inverse participation number of the eigenstates

$$P = \frac{1}{N} \sum_n (|<\alpha|n>|^4) \tag{4}$$

is calculated by using method 2 [7]. It should vanish for extended states.

3. RESULTS

Some results for the conductivity as a function of the disorder, W/V, with the Fermi energy in the center of the band, i.e. $E = 0$, and as a function of E with $W/V = 0.6$ are shown in figures 1 and 2, respectively.

The most important information one gets from Fig.1 is that for $W/V < 2$ the conductivity increases when increasing the size of the system and decreasing the "delta-function width" δ and $m^{-1/2}$, respectively. From the inserts in Fig.1 we learn that only for $W/V > 1$ there is a clear localisation of the eigenstates. For $W/V \leqslant 1$ the mean extension of the states seems to increase with the system size, though from the data in the insert of Fig.1a a saturation with $N \to \infty$ cannot completely be excluded. On the other hand, the inverse participation number in the insert of Fig.1b is vanishingly small for $W/V \leqslant 1.5$, and roughly proportional to $W-W_o$ for $W > W_o \sim 1.5$ V, thus indicating some sort of transition from extended to localised states at W_o.

Most peculiar, the conductivity as a function of the Fermi energy shown in Fig.2 for $W/V = 0.6$ is strongly fluctuating, this behaviour being most pronounced below some critical energy $E_c \simeq 1.5$ V. Again, for $E < E_c$ one observes an increase of σ with increasing system size and decreasing "width". For $E > E_c$ this trend seems to be reversed, at least for the data obtained by method 1 (Fig.2a), the data in Fig.2b being still incomplete. It is tempting to extrapolate the data for $E > E_c$ to zero with $N \to \infty$, $\delta \to 0$.

4. DISCUSSION

We are not yet in a position to offer any conclusive interpretation of our data. Let us add, instead, some preliminary remarks about possible explanations of the apparent discrepancies between our results and the conventional theory. A fuller discussion will be given elsewhere[4]. In our opinion, three

(a)

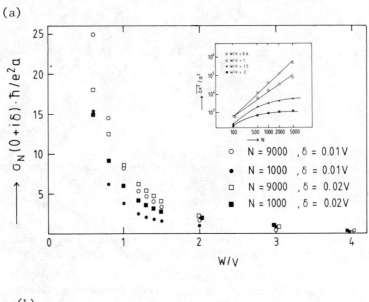

(b)

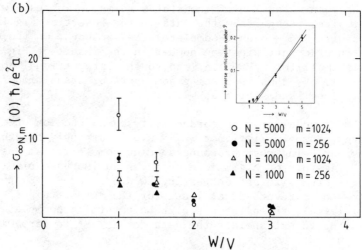

Fig.1 : Conductivity as a function of the disorder parameter W/V
for Fermi energy at the band center (E = 0) as obtained
a) by method 1, b) by method 2.
The inserts show a) the mean square extension $\overline{\Delta x^2}$ as a
function of the system size as obtained by method 1,
b) the inverse participation number for E = 0 as a function
of W/V as obtained by method 2.

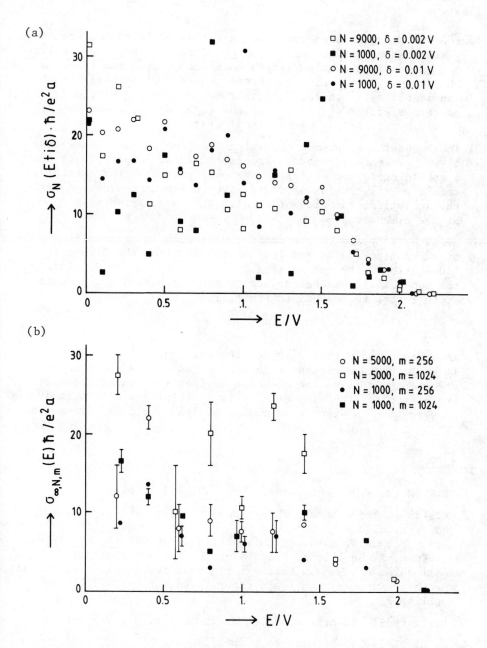

Fig.2 : Conductivity as a function of the Fermi energy for
W/V = 0.6 as obtained by a) method 1, b) method 2.

differing conclusions may be drawn from our numerical results,
namely :

1. The system sizes used are too small thus yielding the states
   in the center of the band as extended though they are localised
   but with a localisation length much larger than the length of
   the system. This would, of course, imply consequences for the [8]
   applicability of these, and other methods in higher dimensions,
   as larger system sizes than of the magnitude of about 10^4 are
   hardly available in numerical calculations.

2. The conventional wisdom about localisation in one dimension
   is wrong. Indeed, rigorous proofs for this, in particular for
   a vanishing dc-conductivity, are only available for special
   models[9]. This conclusion would have serious consequences for
   the analytical theories of localisation as applied also to
   higher dimensions.

3. From the increase in the conductivity with increasing length
   of the system and decreasing "width" and from the strong
   fluctuations with energy (Fig.2) it is tempting to conclude
   the conductivity to be subject to a distribution which remains
   broad in the thermodynamic limit, i.e.

$$\overline{\Delta\sigma^2} \geqslant N^2$$

This would imply that σ is not the right quantity to average.
It would be in agreement with most recent ideas of Anderson
et al.[10], and rises the question whether the linear response
expression for the conductivity is a useful concept to study
localisation in disordered media, at least in lower dimensions.

REFERENCES

1. N.F. Mott, W.D. Twose, Adv. Phys. 10, 107 (1961).

2. K. Ishii, Prog. Theor. Phys. Suppl. 53, 77 (1973), see,
   however, also
   S.A. Molchanov, Math. USSR Isvestija (Transl.) 12, 69 (1978).
   H. Kunz, B. Souillard, Proceedings of the International
   Colloquium on "Random Fields : rigorous results in statistical
   mechanics and quantum field theory", Esztergom 1979, to be
   published.

3. G. Czycholl, B. Kramer, Solid State Commun. 32, 945 (1979).

4. G. Czycholl, B. Kramer, to be published.

5. G. Czycholl, Z. Physik B 30, 383 (1978).

6. B. Kramer, D. Wearie, J. Phys. C 11, L 5 (1978).
   D. Wearie, B. Kramer, J. Non-Cryst. Solids 32, 131 (1979).

7. D. Wearie, A.R. Williams, J. Phys. C 10, 1239 (1977).
   D. Wearie, V. Srivasta, J. Phys. C 10, 4309 (1977).

8. D. Wearie, B. Kramer, Proceedings of 8th International
   Conference on Amorphous and Liquid Semiconductors, Cambridge
   Mass. 1979, J. Non-Cryst. Solids 35/36, 9 (1980).

9. A.A. Abrikosov, I.A. Ryzhkin, Adv. Phys. 27, 147 (1978).

10. P.W. Anderson, D.J. Thouless, E. Abrahams, D.S. Fisher,
    preprint (1980), to be published.

1D DISORDERED SYSTEM WITH PERIODIC BOUNDARY CONDITIONS:

DENSITY OF STATES AND SHAPE OF WAVEFUNCTIONS

J. Decker and H. Hahn

Institut A für Theoretische Physik der
Technischen Universität Braunschweig

3300 Braunschweig, Fed. Rep. Germany

By example of a one-dimensional relativistic random electron
problem, we show that the states of 1d disordered systems are
strongly influenced by the kind of boundary conditions imposed.
Treating the spatial propagation of the wavefunctions within the
framework of the transfer matrix method, we find that periodic
boundary conditions only become possible if we do not fix the
value of the wavefunctions at the edges of the system : Periodic
boundary conditions induce a certain kind of correlation between
the "random sequence" characterizing the disordered system and
the space dependence of the wavefunctions. Since this correlation
is absent for fixed boundary conditions we are led to an argument
making the "localisation of almost all states" in the usual sense
highly unplausible for periodic boundary conditions.

THE QUESTION

The most exciting statement in the theory of disordered
systems in one space dimension is that only localized (i.e. square
integrable) states should exist[1,2,3,4,5]. As a consequence, the
corresponding Hamiltonian should have a pure point spectrum[6,7].
In the framework of the transfer-matrix method, this may be
qualitatively explained as follows: For the wavefunctions of
an electron in a random potential, for the amplitudes of a
disordered tight-binding electron system, or for the vibrational
modes of a disordered harmonic chain, the spatial propagation is
mediated by successive operation of corresponding transfer
matrices, which in turn, because of the disorder, depend on

167

certain random numbers. For this product of random matrices, we can apply a theorem by Fürstenberg[8],[4] and find that, if we fix the amplitude, e.g., at the left-hand edge of the 1d system, then the absolute value of the amplitude shows a net increase when going trough the disordered material from the left-hand to the right-hand side. Thus, this "Fürstenberg type" of state is "localized" just at the right-hand "surface". Equally well, one can fix the amplitude at the other edge and let the transfer matrices act in the opposite direction. By this, we get states "localized" at the left boundary.

Since the states just mentioned do distinguish the "surfaces", in order to obtain a more realistic physical behaviour, we may fix the amplitudes at both boundaries, make them propagate simultaneously from both edges towards a given site x_o inside the material, and try to "match together" the two branches there. In the spirit of the Fürstenberg theorem, used for each branch separately, we here select states "localized" around x_o. However, none of the "matching points" is physically distinguished, and we thus have to apply the above procedure for all possible values of x_o in order to approach a really realistic picture[9].

In any way, we see that the shape of the wavefunctions strongly depends on the way we build them up. Therefore, the question of how the real states of 1d disordered systems do look, has not been answered quite satisfactorily till now. Since by choosing a certain "matching point" one actually violates the symmetry of the system (and can cure it only "on an average" within an ensemble of solutions taking into account all the "matching points"), it is our goal to attack this question by considering boundary conditions which allow us to study wave-functions each of them containing the full symmetry of the system, viz., periodic boundary conditions.

THE MODEL

The concrete example studied here is a system of non-interacting relativistic Fermi particles with a "rest mass" depending statistically on the position in the one-dimensional space:

$$^1\hat{H} = -i \begin{pmatrix} 0 & -i \\ i & 0 \end{pmatrix} \nabla_x + m(x) \begin{pmatrix} 1 & 0 \\ 0 & -1 \end{pmatrix} \tag{1}$$

This model system (arising in the treatment of 2d Ising models with spatially inhomogeneous exchange coupling[10]) has the advantage

i) that, in the spatially homogeneous case $(m(x) \equiv m_o)$, it possesses the simplest "band gap" imaginable : The forbidden region of the one-particle energy is solely given by $-m_o < \varepsilon < +m_o$

ii) that its transfer matrix directly has the physical meaning of a translation operator for the two-component spinor wave-function,

$$\Psi(x+\xi) = T(x,\xi) \, \Psi(x) \; , \tag{2}$$

and iii) that the transfer matrix itself shows a particularly simple structure : Since $T(x,\xi)$ contains only real matrix elements we can confine ourselves to real valued spinor functions $\Psi(x)$ visualizing them by a vector function in a plane. $\Psi(x)$ then favourably is parametrized by its orientation $\Phi(x)$ and its length $|\Psi(x)|$.

In the case of the "rest mass" being a piecewise constant function of x with values $m_{(\nu)}$ statistically distributed according to

$$\rho(m) = \frac{1}{2} \, \delta(m-m_1) + \frac{1}{2} \, \delta(m-m_2) \; , \tag{3}$$

for instance, in the regime $m_1 \ll \varepsilon \ll m_2$ (i.e. near the band edge $m_o = 1/2 \, (m_1 + m_2)$ of the corresponding homogeneous system), we can approximate[11] $T_{(\nu)}(\xi)$ by just a simple rotation matrix

$$T_{(\nu)}(\xi) = \begin{pmatrix} \cos q\xi & -\sin q\xi \\ \sin q\xi & \cos q\xi \end{pmatrix} , \tag{4a}$$

with $q = -(\varepsilon^2 - m_1^2)^{1/2}$, for $m_{(\nu)} = m_1$, and by a "hyperbolic rotation"

$$T_{(\nu)}(\xi) = \begin{pmatrix} \cosh \kappa\xi & \sinh \kappa\xi \\ \sinh \kappa\xi & \cosh \kappa\xi \end{pmatrix} \tag{4b}$$

with $\kappa = (m_2^2 - \varepsilon^2)^{1/2}$, for $m_{(\nu)} = m_2$.

PERIODIC BOUNDARY CONDITIONS

Let us start with a "fixed" boundary condition by prescribing the spinor $\Psi(x)$ at the left-hand side, x = 0: e.g., $\Phi(0) = 0$ and $|\Psi(0)| = 1$. The spinor at the right-hand side, x = L $= \sum_\nu 1_\nu$, then is given by

$$\Psi(L) = (\prod_\nu T_{(\nu)}(1_\nu)) \, \Psi(0) \quad . \tag{5}$$

Since the total rotation angle of $\Psi(x)$ along the whole system, successively induced by all the transfer matrices $T_{(\nu)}(1_\nu)$ present, fulfills

$$\frac{\partial}{\partial \varepsilon} \Delta \phi_t < 0 \tag{6}$$

we are allowed to require also a fixed orientation of the spinor
at the right boundary, e.g., $\Phi(L) = \Phi(0)$: Adjusting the value of
$\Phi(L)$ required by varying the one-particle energy, we just determine
one kind of energy spectrum possible, $\varepsilon_n(\Phi(0))$. From the theorem
by Fürstenberg, however, we know that $|\Psi(L)|$ behaves as $\exp(\gamma L) \gg 1$.

Thus, in order to satisfy the periodic boundary conditions
$\Phi(0) = \Phi(L)$ and $|\Psi(0)| = |\Psi(L)|$, we may no longer prescribe a
fixed spinor orientation at the edges. Using the unimodularity of
each of the matrices $T_{(\nu)}(1_\nu)$, Det $T_{(\nu)}(1\nu) = 1$, we can show that
periodic solutions indeed do exist for certain values of $\varepsilon : \varepsilon_n$.
But each of these solutions is characterized by its own value
of $\Phi(0) = \Phi(L) = \Phi_B$ which (apart from ε) depends on the whole
sequence $m_{(\nu)}$ of the "rest mass":

$$\Phi_B = \Phi_B(m_{(1)}, m_{(2)}, \dots; \varepsilon_n) . \tag{7}$$

(For details see [11]).

Additionally we find each energy niveau belonging to
"fixed" boundary conditions to be surrounded by just two niveaux
of periodic solutions :

$$\dots \tilde{\varepsilon}_{n-1} < \varepsilon_n \leqslant \varepsilon_n(\Phi(0)) \leqslant \tilde{\varepsilon}_n < \varepsilon_{n+1} \dots . \tag{8}$$

This means, in the limit of an infinite system, $L \to \infty$ (where the
$\varepsilon_n(\Phi(0))$ and, hence, also the ε_n, $\tilde{\varepsilon}_n$ lie dense), the density of
states, apart from a factor of two, is the same for both boundary
conditions considered.

A CERTAIN KIND OF CORRELATION INDUCED

In the case of fixed initial orientation $\Phi(0)$, the
propagation of $\phi(x)$, and especially of the angular variable $\Psi(x)$,
along the disordered system forms a Markovian stochastic process
(i.e., we here interpret the x-coordinate as the time axis!). In
particular, there is no correlation between the actual value of
$\Phi(x)$ and the values of the "rest mass", $m_{(\nu)}$, in the "future",
$x' > x$. In the case of periodic boundary conditions, however,
because of eq.(7), $\Phi(x)$ is correlated also to the "future" part
of the $m_{(\nu)}$-sequence via its dependence on the initial orientation
Φ_B. Thus, there is some tricky "threading in" taking place that

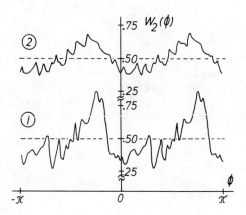

Numerical results for the correlation function $W_{"2"}(\Phi)$ defined
in equation (9) are shown for the two parameter sets
$q = \sqrt{\varepsilon^2 - m_1^2} = 0.5$, $\kappa = \sqrt{m_2^2 - \varepsilon^2} = 0.5$ (curve 1; this corresponds
to a value of ε near the "band-edge" of the related homogeneous
system, $\varepsilon \simeq m_o = \frac{1}{2} (m_1 + m_2)$.) and $q = 0.6$, $\kappa = 0.2$ (curve 2; here
ε is some distance above m_o). The peaky structure of the $W_{"2"}(\Phi)-$
graph stems from the fact that we did not analyse $\Phi(x)$ continuously
but only after each transfer step (of length $1_\nu \equiv 1$), i.e. we
took into account only the values of $\Phi(x)$ at the sites $x = \nu$,
$\nu = 1,2,3, \ldots$.

just prevents the wavefunctions from having a net increase of
its length in the "Fürstenberg sense". The $\Phi(x)$-process is no
longer an ordinary Markov process.

For the model system outlined, we can describe the main
effect of this correlation in an elementary fashion : Because of
the "hyperbolic rotation" induced by them, the transfer steps with
$m_{(\nu)} = m_2$ provide for the orientation of $\Psi(x)$ along the system
to become more or less concentrated in the regions around $\Phi_+ = \pi/4$
and $\Phi_- = -3\pi/4$. I.e., the frequency distribution of the values of
$\Phi(x)$, $g(\Phi)$, shows two maxima near Φ_+ and Φ_- respectively. On the
other hand, the transfer matrix (4b) just causes an increase of
$|\Psi(x)|$ whenever $\Psi(x)$ has an orientation near Φ_+ or Φ_-. For the
Markovian case this, together with the shape of $g(\Phi)$, results in
the "Fürstenberg behaviour". In order to prevent this net increase
of $|\Psi(x)|$, there must be a correlation in such a way that, whenever
$\Phi(x)$ is near Φ_+ or Φ_-, a subsequent transfer step with $m_{(\nu)} = m_2$
occurs only with a lowered probability. Defining

$$W_{"2"}(\Phi) = \left(\frac{\text{Number of events where } \Phi \leqslant \Phi_{(\nu)} < \Phi + \delta\Phi \text{ and } m_{(\nu+1)} = m_2}{\text{Total number of events where } \Phi \leqslant \Phi_{(\nu)} < \Phi + \delta\Phi} \right) , \qquad (9)$$

that correlation yields a Φ-dependence of this quantity just being
minimal in the regions around Φ_+ and Φ_-. The figure below shows
corresponding results obtained by direct numerical computations[11].
(For fixed boundary conditions, the correlation mentioned is
absent and, therefore, $W_{"2"}(\Phi)$ is a constant as indicated by the

dashed lines drawn in the figure).

THE PROBLEM OF LOCALISATION

Are the states belonging to periodic boundary conditions
"localized" or "extended"?
If there are no correlations within the $m_{(\nu)}$-sequence and if the
probability distribution for the random numbers $m_{(\nu)}$ is assumed
to be spatially homogeneous, then the $\Phi_{(\nu)} - m_{(\nu+1)}$-correlation
should maintain this symmetry, i.e. $W_{"2"}(\Phi)$ should not depend on
the position within the system. If this is really true then the
"Fürstenberg increase" of $|\Psi(x)|$ is suppressed homogeneously, and
the wavefunctions should represent "extended" states in the sense
that, in the limit of an infinite 1d disordered system, they
remain non-vanishing states only if we normalize them in the way

usual for extended states,

$$\lim_{L \to \infty} \frac{1}{2L} \int_{-L}^{+L} |\Psi(x)|^2 \, dx = 1 .$$ (10)

Of course, $|\Psi(x)|$ shows statistical fluctuations along the system which in fact can be very large in the energy regime near the band edge of the corresponding homogeneous system, so that locally the states possibly look like "localized" states. This conjecture is supported by direct numerical treatments[12,13] and recent calculations concerning the static conductivity[14]. (To be quite clear : By this plausibility argument, we cannot exclude that localized states obeying periodic boundary conditions also do exist. Since these states are characterized by a spatially inhomogeneous $\Phi_{(\nu)} - m_{(\nu+1)} -$ correlation and, hence, violate the translational symmetry of the probability distribution for the $m_{(\nu)}$, in the limit $L \to \infty$, they should have a vanishing statistical weight in comparison with the set of the "extended" states existing in the sense mentioned above).

REFERENCES

1. N.F. Mott and W.D. Twose, Advan. Phys. 10, 107 (1961).

2. R.E. Borland, Proc. Roy. Soc. A274, 529 (1963).

3. B.I. Halperin, Phys. Rev. 139A, 104 (1965).

4. K. Ishii, Progr. Theor. Phys. Suppl. 53, 77 (1973).

5. M. Goda, Progr. Theor. Phys. 62, 608 (1979).

6. I.Ya. Gol'dshtein, S.A. Molchanov, and L.A. Pastur, Functional Analysis and its Applications 11, 1 (1978).

7. S.A. Molchanov, Math. USSR Izvestija 12, 69 (1978).

8. H. Fürstenberg, Trans. Amer. Math. Soc. 108, 377 (1963).

9. F.J. Wegner, Z. Physik B22, 273 (1975).

10. I. Decker and H. Hahn, Physica 93A, 215 (1978).

11. I. Decker and H. Hahn, to appear in Z. Physik B.

12. R.D. Painter and W.M. Hartmann, Phys. Rev. B13, 479 (1976).

13. C. Papatriantafillou, E.N. Economou, and T.P. Eggarter, Phys. Rev. B13, 910 and 920 (1976).

14. G. Czycholl and B. Kramer, Solid State Commun. 32, 945 (1979).

NUMERICAL CALCULATION OF CONDUCTIVITY OF DISORDERED SYSTEMS

B. Kramer and A. MacKinnon

Physikalisch-Technische Bundesanstalt
Bundesallee 100

3300 Braunschweig, Federal Republic of Germany

The dc-conductivity of the two and three dimensional Anderson
model is studied numerically using the equation of motion method.
From the disorder dependence of the conductivity in the center of
the band a minimum metallic conductivity is estimated. Its numerical
value is 0.2 ± 0.1 $e^2/\hbar$ for the square lattice and 0.05 ± 0.02 $e^2/\hbar a$
for the simple cubic lattice.

1. INTRODUCTION

Increased interest in the Anderson transition[1] was
generated by the continuing discussion about the critical behaviour
of the conductivity near the mobility edge (energy which divides
extended from localized states). Mott's suggestion[2] of a
discontinuous drop, though often criticized[3], has been very
successful in the interpretation of transport properties of
amorphous solids and inversion layers[4]. Theoretically the concept
of the minimum metallic conductivity is at least as controversial
as the Anderson transition[5-11]. Recently developed scaling
approaches[6,7] containing somewhat arbitrary assumptions contradict
the results of their numerical implementations[8] as is the case
also for other recent analytical and numerical work[5,9,10,11].
Particularly in 2d analytical theory predicts all states localized,
whereas numerical calculations indicate a transition from extended
to localized states at a finite strenght of the disorder.

In this paper we investigate the behaviour of the dc-
conductivity at the Anderson transition. We present numerical
results, which are in favour of Mott's picture.

175

2. MODEL AND METHOD

We start from the well known Anderson-Hamiltonian

$$H = \sum_{j=1}^{N} \varepsilon_j |j><j| + V \sum_{j,j\Delta}^{N,Z} |j><j\Delta|$$

with constant nearest neighbour matrix elements, V, and uniformly distributed diagonal elements ε_j within $(-W/2, +W/2)$. The strength of the disorder is described by W/V, the width of the potential fluctuations divided by the kinetic energy. N is the size of the system.

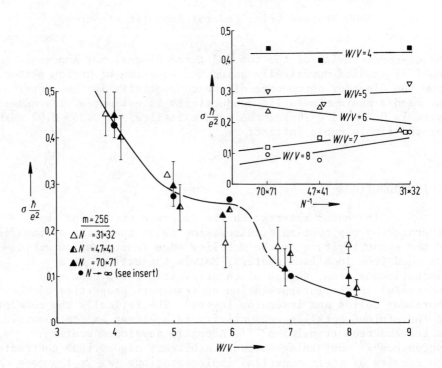

Fig.1 : The zero temperature conductivity in the center of the band of the two dimensional square lattice in units of $e^2/\hbar$ as a function of the disorder (●) as extrapolated from data obtained for systems containing N = 31 × 32 (Δ), N = 47 × 41 (▲), and N = 70 × 71 (▲) sites (see insert). The error bars are estimated from the time extrapolation. m = 256 corresponds to a width of the energy weighting function of about 5 % of the total band width.

The calculation of the dc-conductivity in the zero temperature limit using time dependent electron states is described elsewhere[12-14]. Roughly speaking combinations of the time dependent states are time averaged in order to represent, in the limit of infinite time, the Kubo formula. The time dependent states are obtained by integrating the time dependent Schrödinger equation numerically starting from a randomly defined initial state, which is energetically weighted in order to favour eigenstates near the Fermi energy. The width of the weighting function, $m^{-1/2}$, determines how many eigenstates contribute to the conductivity. Being interested in the zero temperature limit, we have to investigate the conductivity as a function of N and m.

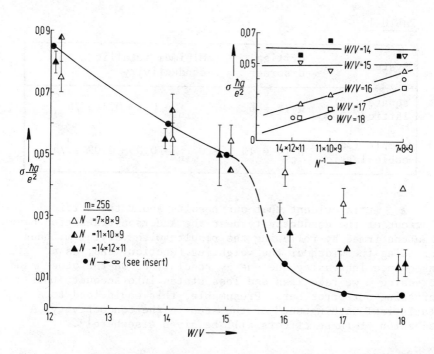

Fig.2 : The zero temperature conductivity in the center of the band of the three dimensional simple cubic lattice in units of $e^2/\hbar a$ as a function of the disorder (•) as extrapolated from the data obtained for various systems containing $N = 7 \times 8 \times 9$ ($\triangle$), $N = 11 \times 10 \times 9$ ($\blacktriangle$), and $N = 14 \times 12 \times 11$ ($\blacktriangle$) sites (see insert). The error bars are estimated from the time extrapolation. m = 256 corresponds to a width of the energy weighting function of about 5 % of the total band width.

3. RESULTS

The critical behaviour of the conductivity can be obtained by studying its disorder dependence for a given Fermi energy ($E = 0$, the center of the band, in our calculation). Following the data for a given value of m with varying system size (Figs.1 and 2), we observed in 3d as well as in 2d a decrease in the conductivity with increasing N for values of the disorder above W/V 15(6) in 3d(2d). Such a behaviour is expected for localized states. For lower disorder the conductivity is independent of the system size within the accuracy of our procedure. This indicates extended states. From the data extrapolated for $N \rightarrow \infty$ we estimate an Anderson transition to occur in both dimensions. We observe also a step-like behaviour of the conductivity. Identifying the magnitude of the step with the minimum metallic conductivity we estimate values as given in table 1.

Table 1

System	Critical disorder	Minimum metallic conductivity
square (2d) lattice	$W_c/V = 6$	$\sigma_{min} = 0.2 \pm 0.1 \ e^2/\hbar$
simple (3d) cubic lattice	$W_c/V = 15$	$\sigma_{min} = 0.05 \pm 0.02 \ e^2/\hbar a$

It is evident that our results about the critical behaviour of the conductivity near the Anderson transition have to be confirmed by following the results with varying m, thus decreasing the width of the weighting function, and, as a consequence increasing the energy resolution. On the other hand, increasing m we take less and less states into account, thus increasing our error bars. Presumable, this would lead to a "statistical washing out" of the step in the conductivity. A discussion of these effects will be given elsewhere[14].

4. CONCLUSION

Though being in agreement with experimental data[4] and conclusions from other numerical work[8,10] our results are certainly in contradiction with analytical ideas[6,7,9]. It remains to be seen whether the reasons for this are weak approximations in analytical theory or short comings of the numerical calculations (insufficient system sizes!).

REFERENCES

1. P.W. Anderson, 1958, Physical Review 109, 1492.

2. N.F. Mott, 1972, Philosophical Magazine 26, 1015.

3. M.H. Cohen and J. Jortner, 1973, Physical Review Letters 30, 699, 1974, Physical Review A 10, 978.

4. M. Pepper, 1977, Contemporary Physics 18, 423.

5. D. Weaire and B. Kramer, 1980, Journal of Non-Crystalline Solids 35 & 36, 9.

6. F. Wegner, preprint Nr. 30 of SFB 123, Universität Heidelberg.

7. E. Abrahams, P.W. Anderson, D.C. Licciardello and T.V. Ramakrishnan, 1979, Physical Review Letters 42, 673.

8. P. Lee, 1979, Physical Review Letters 42, 1492.

9. W. Götze, 1978, Solid State Communications 27, 1393, 1979, Journal of Physics C 12, 1279.

10. J. Stein and U. Krey, 1980, Zeitschrift für Physik B 37, 13.

11. P. Prelovsek, 1978, Physical Review B 18, 3657, 1979, Solid State Communications 31, 179.

12. B. Kramer and D. Weaire, 1978, Journal of Physics C 11, L 5.

13. D. Weaire and B. Kramer, 1979, Journal of Non-Crystalline Solids 32, 131.

14. B. Kramer, A. MacKinnon and D. Weaire, to be published.

TRANSPORT PROPERTIES OF THE BLUE BRONZE $K_{0.30}MoO_3$

R. Brusetti, B.K. Chakraverty, J. Devenyi,
J. Dumas, J. Marcus and C. Schlenker

Groupe des Transitions de Phases*, C.N.R.S.

B.P. 166, 38042 Grenoble Cedex, France

The nonmetal-metal transition in $K_{0.30}MoO_3$ has been studied by electrical conductivity, thermopower, magnetic susceptibility and calorimetric measurements. The conductivity and the thermopower are found to be strongly anisotropic in the high temperature phase. These results are discussed in terms of a one-dimensional band structure. At low temperatures, the transport properties are shown to be extrinsic, the carriers being possibly polarons. This could account for the sharp peak found in the thermopower at 38 K.

The "blue bronze" $K_{0.30}MoO_3$ belongs to the class of the ternary transition metal oxides, of general formula $M_x TO_m$, where TO_m is the highest binary oxide of the transition metal T and M usually an alkaline metal[1]. These materials are called bronzes as they sometimes have a metallic luster. In such compounds, the alkali metal usually donates its outer electron to the transition metal cation : this partly fills the d-states which would be otherwise empty. Among this class, the tungsten bronzes $M_x WO_3$ have been extensively studied; less information is available on the molybdenum bronzes.

The existence of two defined compounds of ideal formula $K_{0.33}MoO_3$ and $K_{0.30}MoO_3$ has been previously established[2]. While $K_{0.33}MoO_3$, the so-called "red bronze" is semiconducting at all temperatures, the "blue bronze" $K_{0.30}MoO_3$ shows a nonmetal-metal transition at 180 K [3]. In addition, its magnetic properties are unusual as it is quasi-diamagnetic in the semiconducting low

* Laboratory associated with Université Scientifique et Médicale de Grenoble.

temperature phase[4]. Up to now, no satisfactory model has been
proposed to account for these properties.

The crystal structure of the blue bronze at room
temperature is monoclinic (space group C 2/m)[5]. The unit cell
includes ten formula units. It is a layered structure in the sense
that it consists of infinite sheets of MoO_6 octahedra sharing edges
or corners, separated from each other by the K ions. These sheets
are parallel to the b-monoclinic axis and to the [102] direction;
they are built with clusters of 10 distorted MoO_6 octahedra sharing
edges. Each of these clusters shares corners in two directions
with other identical clusters. The structure may also be described
as containing infinite chains, parallel to $\vec{b}$, of MoO_6 octahedra
sharing corners only. There are three independent Mo sites, Mo(1),
Mo(2) and Mo(3); only the Mo(2) and Mo(3) are involved in the
infinite chains.

In this article, we report electrical resistivity data,
including anisotropy and the effect of a pressure of 6 kbar and
thermopower showing a strong anisotropy in the metallic phase.
These results suggest that the blue bronze may be considered as a
one-dimensional metal. We also report magnetic susceptibility and
calorimetric measurements. Preliminary studies of V doped samples
also indicate that the incorporation of V stabilises the semiconducting phase.

Single crystals of $K_{0.30}MoO_3$ have been grown by the
electrolytic reduction of a $K_2MoO_4-MoO_3$ melt, following the method
described in reference 2. The crystals are platelets, of typical
size $5 \times 2 \times 1$ mm^3, parallel to the monoclinic $\vec{b}$ axis and to the
[102] direction; the long dimension of the sample being parallel
to $\vec{b}$. They have been characterised by x-ray data. In the V-doped
samples, the vanadium concentration has been obtained by
microprobe analysis; its homogeneity through the crystal has been
checked by the same method.

The electrical conductivity has been measured between
4.2 K and 300 K by the four-point technique using indium contacts
soldered on freshly cleaved crystals; in some cases, the contacts
were evaporated on the samples. Thermopower data have been obtained
by the differential method using a dynamical technique[6]. Magnetic
susceptibility has been measured between 4.2 K and 300 K using a
vibrating sample magnetometer. Calorimetric data have been obtained
between 120 K and 300 K with a commercial scanning calorimeter
(Perkin Elmer DSC2).

The electrical resistivity ρ has been measured on more
than 10 samples, along the $\vec{b}$ axis and along the [102] direction.
Typical results for the logarithm of the resistivity vs reciprocal

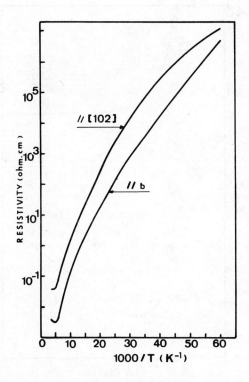

<u>Fig.1</u> : Electrical resistivity (logarithmic scale) of $K_{0.30}MoO_3$ vs reciprocal temperature, measured along $\vec{b}$ or along [102].

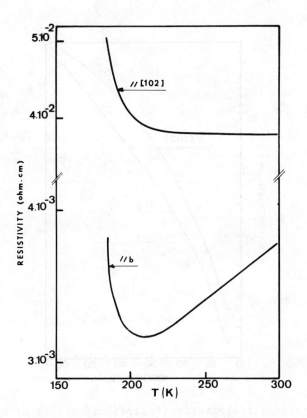

<u>Fig.2</u> : Resistivity (linear scale) of $K_{0.30}MoO_3$ vs temperature
showing the anisotropy in the high temperature phase.

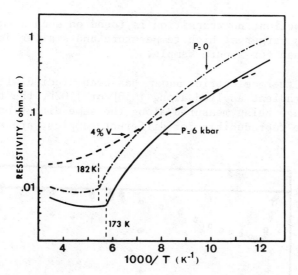

Fig.3 : Resistivity vs reciprocal temperature of $K_{0.30}MoO_3$ under
a pressure of 0 and 6 kbar, and of $K_{0.30}Mo_{1-x}V_xO_3$ (x = 4 %)
with P = 0.

temperature are shown on figure 1. For all our samples, the
transition temperature defined by the steep variation of ρ with T
is found to be 182 K; the absolute value of ρ may vary from one
sample to another by a factor of 10 to 50. Along the $\vec{b}$ axis, the
curve clearly shows a metallic behaviour above ∿220 K; at low
temperature (<50 K), the thermal dependence of ρ is well described
by an activation energy of 0.02 to 0.04 eV depending on the
samples; below ∿20 K, we have not been able to measure ρ due to
its high value. Along the [102] direction, the resistivity is
found to be 10 to 100 times larger than along $\vec{b}$ in all the
temperature range; at high temperature, the resistivity is
decreasing with increasing temperature; at low temperature ρ does
not obey a simple activated law and would be better fitted by a
plot of log vs $T^{-1/4}$. The anisotropy at high temperature is better
seen on a plot of ρ vs T (fig.2); the transition temperature is
not well-defined along [102] while it appears by a very steep
slope along [010] . At high temperature, the results may be
characteristic of a quasi one-dimensional metal. Figure 3 shows
the effect on the resistivity measured along $\vec{b}$ of a pressure of
6 kbar; the transition temperature is found to be displaced from
182 K down to 173 K; the resistivity is decreased by 25 % to 100 %
depending on the samples in the metallic phase and by 100 % in the
semiconducting phase. On the same figure are shown the results
obtained on a V doped sample $K_{0.30}Mo_{1-x}V_xO_3$, with x = 0.04, along

the [010] direction; no transition is found on such a sample; the resistivity is larger at high temperature and smaller at low temperature than for a pure sample.

The thermoelectric power has been studied with the temperature gradient applied along [010] or [102], the thermo-electric voltage being measured along the same direction. The results, quite reproducible for different samples, are shown

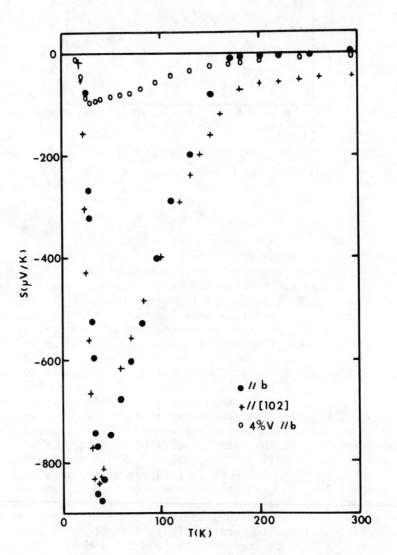

Fig.4 : Thermopower vs temperature for $K_{0.30}MoO_3$ measured either along $\vec{b}$ or [102], and for $K_{0.30}Mo_{1-x}V_xO_3$ (x = 4 %) along $\vec{b}$.

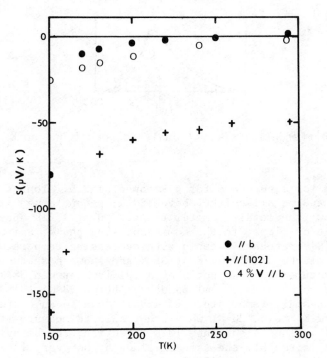

Fig.5 : Thermopower vs temperature above 150 K.

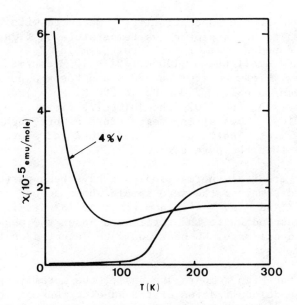

Fig.6 : Magnetic susceptibility vs temperature for K$_{0.30}$MoO$_3$ and for K$_{0.30}$Mo$_{1-x}$V$_x$O$_3$ (x = 4 %).

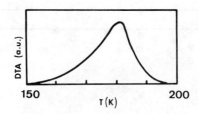

Fig.7 : Differential thermal analysis peak found on $K_{0.30}MoO_3$.

on figure 4 for a pure and for a V doped sample. Along the $\vec{b}$ axis, S is small and metallic-like above 182 K, p-type above ∿250 K and n-type below; the metal-nonmetal transition at 182 K appears as a sharp change of slope; in the semiconducting phase, S is always negative; $|S|$ increases steeply with decreasing temperature, reaches a value of ∿1000 μV/K at 38 K and shows a sharp peak at this temperature; it does not obey a 1/T law between 38 K and 182 K as expected for a simple semiconductor; below 38 K, $|S|$ decreases steeply in a small temperature interval, down to an extremely small value at ∿20 K; below 20 K, we are not able to measure it because of its small value. Along the [102] direction, the behaviour of S below 182 K is roughly the same as along $\vec{b}$, however, it is basically different above 182 K; this is shown on fig.2; S keeps a negative value of ∿60 μV/K up to room temperature.

        The magnetic susceptibility χ measured with the magnetic field parallel to $\vec{b}$, is plotted vs temperature on figure 6 for a pure and a V doped sample. For the pure sample, χ is temperature independent and small below ∿100 K ($|\chi| < 10^{-6}$ emu/mole); it increases slowly between 100 and 220 K and keeps a positive value of ∿2×10^{-5} emu/mole from 220 K up to 300 K. No anisotropy could be detected between the [102] and [010] directions. The V doped sample (x = 0.04) shows at low temperature a Curie behaviour, a slight increase of χ between 100 K and 220 K and above 220 K a smaller value than the pure sample.

        Differential thermal analyses have been performed on pure and doped samples; the pure sample shows a peak spread out between ∿155 K and ∿195 K with a maximum at 182 K (Fig.7); the enthalpy corresponding to the total area under the peak has been found to be ∿6 cal/mole. No peak could be detected on the V doped sample.

        In the high temperature phase, the metallic properties are due to the delocalisation of the 4d electrons of the Mo^{5+} ions. The low value of the conductivity has to be attributed to low mobility of the carriers in a narrow d band. The band structure of $K_{0.30}MoO_3$ may be similar to that of ReO_3 with an anionic valence band made primarily with p_π oxygen orbitals[7]; the conduction band would be built on hybridized cationic t_{2g} and anionic p_{π} states and would be partly filled; in the case of the blue bronze, the crystal structure with infinite chains of MoO$_6$ octahedra parallel to $\vec{b}$ may lead to the formation of a quasi one dimensional band with a very anisotropic Fermi surface allowing an easier conductivity along $\vec{b}$, as shown by our results. Optical reflectivity measurements have also established that K$_{0.30}$MoO$_3$ behaves optically like a

semiconductor for light polarized along [102] and like a metal along [010] [8]. The thermopower data showing a strong anisotropy are also consistent with a one-dimensional metal; the change of sign from n-type to p-type at 250 K indicates that the structure of the conduction band changes with temperature between 182 K and $\sim$250 K. This is in agreement with the susceptibility results which also show a temperature dependence above 182 K. If one assumes that the magnetic susceptibility is Pauli type, one obtains a density of states of $\sim$7 eV^{-1} per 4d electron; this value takes into account a diamagnetic contribution of $\sim -5 \times 10^{-5}$ emu/mole obtained for K$_{0.30}$MoO$_3$ with the values of reference 9. Such a density of states is consistent with a small bandwidth.

The metal-nonmetal transition is related to a localisation of the 4d electrons and therefore to the opening of a gap in the band structure. The magnetic susceptibility which does not show a Curie behaviour at low temperatures, indicates that the d electrons are engaged in diamagnetic Mo^{5+} - Mo^{5+} pairs and (or) participate to a filled bonding band, as in other transition metal oxides, for example Ti$_4$O$_7$ [10]. In K$_{0.30}$MoO$_3$ the

one-dimensional properties suggest the possibility of a Peierls type transition; crystallographic studies which could corroborate this model are presently on progress[11].

The absence of a well-defined activation energy in the electrical conductivity shows that the transport properties are rather extrinsic, related to some Mo^{5+} located near defects either due to non stoichiometry or impurities. The ionic nature of the compound suggests that the carriers may be polarons and that the conductivity takes place by a hopping process. In this process, the polaron energy depending on the electron-phonon interaction, as well as the Coulomb repulsion between neighbouring Mo^{5+} ions should be involved[12]. One may also notice that the data could be consistent with a variable range hopping of the carriers[13].

The sharp peak found in the thermopower at 38 K is similar to the results obtained on germanium at low temperature, which have been attributed to phonon drag[14]. However, we rather believe that this peak is due to the polaronic nature of the carriers; in this case and in a simple model, the thermopower would be expected to vary with temperature at low temperature as $T^{1/2}$ [15]; obviously such a law is not obeyed here. This could be due to an anomaly in the low energy part of the phonon spectrum; the layered structure of the blue bronze may be responsible for such an anomaly.

The incorporation of vanadium in $K_{0.30}MoO_3$ seems to stabilize the low temperature phase. The susceptibility data show that some of these V carry magnetic moments; EPR has also been performed and establishes the presence of V^{4+}; if one assumes an effective magnetic moment of 1.73 μ_B per V^{4+}, the Curie constant is consistent with a concentration of V^{4+} of 0.27 at % when the total concentration of V is 4 at %. Most of the V are then in non magnetic V^{5+} ($3d^0$) states. The smaller ionic size of V^{5+} compared to either Mo^{5+} or Mo^{6+} may then induce some structural distortion which could stabilize the semiconducting phase. One may also notice that the presence of V impurities substituted on Mo sites could shorten the chains of MoO_6 octahedra and may lead to a vanishing of the one-dimensional properties.

Acknowledgement - The authors wish to thank Sir Nevill Mott and P. Nozières for some helpful discussions.

REFERENCES

1. P. Hagenmuller, Progress in Sol. State Chem. 5, 71 (1972).

2. A. Wold, W. Kunnmann, R.J. Arnott and A. Ferretti, Inorg. Chem. 3, 545 (1964).

3. W. Fogle and J.H. Perlstein, Phys. Rev. B 6, 1402 (1972).

4. G. Bang and G. Sperlich, Z. Physik B 22, 1 (1975).

5. J. Graham and A.D. Wadsley, Acta Cryst. 20, 93 (1966).

6. P.M. Chaikin and J.F. Fwak, Rev. Sci. Inst. 46, 218 (1975).

7. J.B. Goodenough, Prog. Sol. St. Chem. 5, 145 (1972).

8. G. Travaglini, P. Wachter and C. Schlenker (to be published).

9. P.W. Selwood, Magnetochemistry (Interscience Publ., N.Y.), p.78 (1964).

10. C. Schlenker, S. Ahmed, R. Buder and M. Gourmala, J. Phys. C - Sol. State Phys. 12, 3503 (1979).

11. M. Marezio and M. Ghedira (to be published).

12. N.F. Mott, Festkörperprobleme XIX (J. Trusch Ed., Vieweg, Braunschweig), p.331 (1979).

13. N.F. Mott, Metal Insulator Transitions (Taylor and Francis, London), p.34 (1974).

14. C. Herring, Phys. Rev. 96, 1163 (1954).

15. D. Emin, Phys. Rev. Lett. 13, 882 (1975).

CRITICAL BEHAVIOUR NEAR THE CONDUCTOR-INSULATOR TRANSITION

IN DISORDERED SOLIDS: A MODE-COUPLING THEORY

P. Prelovšek

Department of Physics and J. Stefan Institute
E. Kardelj University of Ljubljana

61000 Ljubljana, Yugoslavia

The mode-coupling approach is employed for the investigation of the critical behaviour near the Anderson conductor-insulator transition. Critical exponents for the d.c. conductivity s as well as the localization length ν are determined for the model of noninteracting electrons in disordered lattice. Marginal dimensionalities appear to be d = 2 below which no transition exists and d = 4 above which s = 1 and ν = 1/2. For 2 < d < 4 we get s = 1 and ν = 1/(d-2). In addition, we show that for parabolic bands and d $\geqslant$ 6 the transition cannot always be reached by lowering the Fermi energy.

It is well established that in disordered systems the mobility edges (ME) separate extended and localized states in the electron energy spectrum[1,2]. It has become clear in recent years that the Anderson transition between a conductor and an insulator at ME has the characteristics of continuous phase transitions, although some of its properties are unusual. Continuous behaviour near ME of the localization length $r_{\perp}$, the participation ratio P and the d.c. conductivity σ_o has been confirmed by various numerical experiments[3-6]. In the theoretical investigations of the transition the renormalization group (RG) techniques have been applied to the problem[7-9] in order to obtain the critical behaviour of relevant quantities. They yield d = 2 as a lower critical dimensionality, i.e. for d $\leqslant$ 2 electron states are even for weak disorder always localized. The same result is reproduced by the theory introduced by Götze[9-12], which is based on an approximative treatment of electron kinetic equations, analogous to mode-coupling theories.

   This paper is devoted to the analysis of the critical behaviour within the mode-coupling theory approach[11,12]. It is known that mode-coupling theories yield for critical dynamics the scaling behaviour which agrees with RG approach, at least within the first order of the ε-expansion[13]. Although for the present problem a relation between these approaches is not yet established we expect in the same manner that our results for scaling laws could be improved but not invalidated by better theories.

   We shall consider the d-dimensional model of a gas of noninteracting electrons, scattered by random local potentials

$$H = \sum_k \frac{k^2}{2m^*} c_k^+ c_k + \sum_{k,q} u_q c_{k+q/2}^+ c_{k-q/2} , \tag{1}$$

where c_k^+ (c_k) denote creation (annihilation) operators for electron momentum eigenstates and u_q is the Fourier transform of the potential. We assume throughout the paper the temperature $T = 0$. Parameters of the model (1) are the Fermi energy E or the Fermi wavevector q_E and the disorder strength $W = \sqrt{<|u_q|^2>_c}$, which is due to the statistical independence of local potentials q independent. Here, $< >_c$ denotes the configurational averaging.

   In the approach employing kinetic equations[9-12] the time development of phase space density operators $\rho_k(q) = c_{k+q/2}^+ c_{k-q/2}$ is studied. Defining as usual the scalar products between operators, e.g. $(\rho_k(q)|\rho_p(q))$, in terms of static response functions, it is convenient to work with transformed variables (for details see[11])

$$A_n(q) = \sum_{q|k} a_n^q(k) \rho_k(q) , \tag{2a}$$

which are orthogonal with respect to the weighting factor $g_{kp}(q) = <(\rho_k(q)|\rho_p(q))>_c$

$$(A_n(q)|A_m(q)) = \sum_{k,p} a_n^q(k) a_m^q(p) g_{kp}(q) = \delta_{nm} . \tag{2b}$$

The lowest $a_n^q(k)$ are chosen so that they allow an obvious physical interpretation : $A_o(q) = \rho(q)/\sqrt{g(q)}$, where $g(q) = <(\rho(q)|\rho(q))>_c$ and $\rho(q) = \sum_k \rho_k(q)$ is the electron density operator, whereas $A_1(q) = j_L(q) \sqrt{m^*/n_E}$ is related to the longitudinal current density

$$j_L(q) = \frac{1}{m^*} \sum_k (\vec{k}.\vec{e}_q) \rho_k(q) , \tag{3}$$

and $n_E = <\rho(q=0)>_c$ is the total number of electrons. With the use of Eq.(3) and the Mori-Zwanzig formalism the following identity for relaxation functions

$$\Phi_{nm}(q,z) = <(A_n(q)|(z-L)^{-1}A_m(q))>_c \qquad (4a)$$

can be derived[11]

$$z\Phi_{nm}(q,z) = \delta_{nm} + \sum_1 [\Omega_{nl}(q) + C_{nl}(q,z)]\Phi_{\ell m}(q,z) , \qquad (4b)$$

with

$$\Omega_{nl}(q) \doteq \frac{q}{m^*} \sum_k (\vec{k}\cdot\vec{e}_q)\, a_n^q(k)\, a_l^q(p)\, g_{kk}(q) , \qquad (5)$$

$$C_{nl}(q,z) = \sum_{k,p} a_n^q(k)\, a_l^q(p) <(F_k(q)|(z-QLQ)^{-1}F_p(q))>_c , \qquad (6)$$

where $LA = [H,A]$, $F_k(q) = QL\,\rho_k(q)$ and Q denotes the projector on the operator subspace orthogonal to all $\rho_p(q)$ with fixed q. From the particle number conservation and Eq.(1) it follows generally $C_{on} = C_{no} = 0$. In addition, we get from Eqs.(3), (5) $\Omega_{on} = \Omega_{no} = \delta_{n1}$. Therefore, the density relaxation function $\Phi(q,z) = g(q)\,\Phi_{oo}(q,z)$ can be written as

$$\Phi(q,z) = g(q)/[z - \Omega_{o1}^2/[z - C_{11} - \Omega_{12}^2/[z - C_{22} - \ldots]]] . \qquad (7)$$

Eq.(7) reveals for $q \to 0$ the known diffusional form since $\Omega_{o1}, \Omega_{12} \propto q$.

It has been recognized by Götze[9], that C_{11} is the critical quantity which characterizes the conductor- insulator transition, i.e. it determines the conductivity on the one side $\sigma_o \propto C_{11}''(0,0)^{-1}$ and the localization length $r_L \propto \lim_{z\to 0} z\, C_{11}(0,z)$ etc. on the other side of the transition. Employing in Eq.(6) a decoupling approximation[9-12]

$$C_{11}(q,z) = \frac{1}{n_E\, m^*} \sum_q <|u_{q-q'}|^2>_c |(\vec{q}-\vec{q}')\cdot\vec{e}_q|^2 \Phi(q',z). \qquad (8)$$

C_{11} is related back to the density relaxation function, from Eq. (8) also a simple q dependence follows $C_{11}(q,z) = M(z) + q^2 N(z)$.

Let us first investigate the conducting side of the transition $W < W_c$ and determine the conductivity exponent s, $\sigma_o \propto 1/M''(0) \propto |W_c - W|^s$. We put $z = 0$ and expand Eqs.(7), (8) for $M''(0)$, $N''(0) \to \infty$

$$\left.\begin{array}{c} M''(0) \\ N''(0) \end{array}\right\} \doteq \zeta W^2 \int_0^{q_E} dq\, q^{d-1} \left[\frac{M''(0) + q^2 N''(0)}{q^2} - \frac{\eta}{C''_{22}} \right] \left\{ \begin{array}{c} q^2 \\ d \end{array} \right. \tag{9}$$

A transition where $M''(0)$, $N''(0) = \infty$ exists for $d > 2$ at $W_c > 0$, whereas for $d \leqslant 2$ $W_c = 0$. In the former nontrivial case two possibilities should be considered : a) $C''_{22}(0,0) \propto M''(0)$, i.e. $C''_{22}(0,0)$ is also a critical quantity, what yields from Eq.(9) $s = 1/2$. This result has been obtained in previous works[9,11,12] since there the specific approximation for C_{nl} implied $C_{22}(q,z) \propto M(z)$. b) It is more plausible that $C''_{22}(0,0)$ is a noncritical or at least less critical quantity. This could be realized from a decoupling approximation analogous to Eq.(8) which would relate C_{22} to current relaxation function, i.e. to the fluctuations of a nonconserved quantity in contrast to Eq.(8) where the coupling to diffusion processes is essential. Assuming $C''_{22}(0,0)$ finite at ME we get $s = 1$.

On the insulating side $W > W_c$ $M''(0)$, $N''(0)$ are infinite, meanwhile $\tau_m = zM(z)$ and $\tau_n = zN(z)$ remain finite for $z \to 0$. Consequently, $\Phi(q,z)$ in Eq.(7) can be written for $z \to 0$ as

$$\Phi(q,z) = \frac{1}{z}\, g(q) \left[1 + \frac{\Omega^2_{01}(q)}{\tau_m + q^2 \tau_n} \right]^{-1} \tag{10}$$

as far as $C_{22}(q,z)$ is nonvanishing. For τ_n, τ_m a set of equations is obtained

$$\left.\begin{array}{c} \tau_m \\ \tau_n \end{array}\right\} = \zeta W^2 \int_0^{q_E} dq\, q^{d-1} \frac{\tau_m + q^2 \tau_n}{\tau_m + \alpha q^2} \left\{ \begin{array}{c} q^2 \\ d \end{array} \right. \tag{11}$$

ME are determined from the condition τ_m, $\tau_n = 0$. In the critical regime we are interested in the localization length exponent ν, $r_L \propto 1/\sqrt{\tau_m} \propto |W - W_c|^{-\nu}$. In the case $d > 2$ where $W_c > 0$ it follows from Eq.(11) that $\nu = 1/(d-2)$ for $2 < d \leqslant 4$ while for $d \geqslant 4$ $\nu = 1/2$. The latter result is a consequence of the fact that for

$d \geqslant 4$ also derivatives of Eq.(11) with respect to τ_m exist at $\tau_m = \tau_n = 0$.

$\nu = 1/(d-2)$ and $s = 1$ satisfy the scaling relation $s = \nu(d-2)$ obtained by Wegner[6] and agree with the result of the $1/n$ expansion[7] at least within the calculated order of the expansion. $s\sim 1$ seems to be consistent also with the numerical studies[5]. However in the interpretation of real and numerical experiments a caution is needed, since the theory indicates[11,12] that the critical regime where the scaling laws are obeyed is possibly very narrow and thus not yet reached in experiments.

Besides the critical behaviour at ME also $W_c(E)$ dependence at small E is of interest since it should represent a universal feature depending only on the dimensionality. We expand Eqs.(8), (10) for τ_m, $\tau_n \to 0$ and take into account $g(q < q_E) \sim \rho_E \propto q_E^{d-2}$ and $n_E \propto q_E^d$, whereas $g(q > q_E) \propto n_E/q^2$

$$\tau_m = \gamma W^2 \left[\frac{\rho_E^2}{n_E^2} \int_0^{q_E} dq \, q^{d-1} \tau_q + \int_{q_E}^{q^*} dq \, q^{d-5} \tau_q \right] , \qquad (12a)$$

$$\tau_n = \gamma d W^2 \left[\frac{\rho_E^2}{n_E^2} \int_0^{q_E} dq \, q^{d-3} \tau_q + \int_{q_E}^{q^*} dq \, q^{d-7} \tau_q \right] , \qquad (12b)$$

where for convergence on upper cutoff q^* has been introduced and $\tau_q = \tau_m + q^2 \tau_n$. For $d > 2$ two regimes should be distinguished :
a) for $2 < d < 6$ we get from Eq.(12)

$$W_c \propto E^{(6-d)/8} , \qquad (13)$$

i.e. with increasing E also the critical disorder W_c increases.
b) At $d \geqslant 6$ we get $W_c \propto q^{*d-4}$. W_c is thus determined entirely by the upper cutoff q^*. Thus ME cannot be reached by lowering E and for $q^* \to \infty$ also $W_c \to \infty$. In this respect $d = 6$ represents the upper critical dimensionality of the problem, since in the parabolic bands electron states seem to be always extended for $d \geqslant 6$.

It should be noted that within our mode-coupling approximation the scaling behaviour of r_L and $W_c(E)$ has the same origin as the low-frequency (long-time) anomalies of $M(z)$, $N(z)$ and consequently $\sigma(\omega)$. The anomaly of the same type as found within the present approach has been confirmed also by a Green's function method[14]. There is however a disagreement in a prefactor by a power of W. In order to resolve this controversy one should note a nontrivial equivalence, which follows from the time-reversal symmetry[14]

$$(\rho_k(q) | (z-L)^{-1} \rho_p(q)) = (\rho_{\frac{k-p+q}{2}}(k+p) | (z-L)^{-1} \rho_{\frac{p-k+q}{2}}(k+p)) . \quad (14)$$

Relation (14) sets a condition on Φ_{nm} and C_{nm} which is not fulfilled in general within our approach. Nevertheless, we can insert in Eq. (8) instead of the left side the right hand side of Eq.(14) what yields a larger prefactor for the long-time anomaly[14]. It is easy to verify that the replacement does not influence the results for s and ν. Although the first term in Eq.(12a) is altered, there seems to be also no change in $W_c(E)$ and no lowering of $d_c = 6$.

REFERENCES

1. P.W. Anderson, Phys. Rev. 109, 1492 (1958).

2. N.F. Mott and E. Davis, Electronic processes in noncrystalline materials (Clarendon, Oxford, 1979).

3. S. Yoshino and M. Okazaki, J. Phys. Soc.Japan 43, 415 (1977).

4. D. Weaire and V. Srivastava, J. Phys. C 10, 4309 (1977).

5. P. Prelovšek, Phys.Rev. B 18, 3657 (1978); Solid State Commun. 30, 369 (1979).

6. F.J. Wegner, Z. Physik B 25, 327 (1976); Z. Physik B 35, 207 (1979).

7. R. Oppermann and F.J. Wegner, Z. Physik B 34, 327 (1979).

8. E. Abrahams, P.W. Anderson, D.C. Licciardello and T.V. Ramakrishnan, Phys. Rev. Lett. 42, 673 (1979).

9. W. Götze, J. Phys. C 12, 1279 (1979).

10. W. Götze, P. Prelovšek and P. Wölfle, Solid State Commun. 30, 369 (1979).

11. W. Götze, Proc. of the Int. Conf. on Impurity bands in semiconductors, Würzburg 1979, to be published.

12. P. Prelovšek, to be published.

13. J.D. Gunton, Proc. of the Int. Conf. on Dynamical critical phenomena and related topics, Geneve, ed. C.P. Enz, Springer-Verlag (Berlin 1979), p.1.

14. L.P. Gorkov, A.I. Larkin and D.E. Khmelnitzkii, Zh. Eksp. Teor. Fiz. Letters 30, 248 (1979).

CLASSIFICATION OF CHEMICAL ORDERING IN AMORPHOUS

AND LIQUID SEMICONDUCTORS

J. Robertson

Central Electricity Research Laboratories

Leatherhead, Surrey KT22 7SE, U.K.

Binary non-crystalline semiconductors are classified in terms of their bonding at and away from stoichiometry. Density of states models are developed for example systems : ℓ-CsAu, a-MgSb, a-GaAs, a-GeTe and ℓ-PbTe.

The composition dependence of the transport and optical properties of binary liquid or amorphous alloys[1] can be used to understand the variation of their bonding[2], especially when there is little structural data available. There are five basic composition dependences of the resistivity R and thermopower S in which the two constituents C and A behave symmetrically. Other, asymmetric, behaviours can be developed. Figs. 1 (a) and (b) show schematically R and S for good metallic alloys (e.g. ℓ-SnZn), and semiconductors with random covalent bonding (e.g. a-SeTe), respectively. These two behaviours occur if the electronegativity differences between C and A are small (C is taken as more electropositive hereafter). For larger differences, heteroatom bonding is maximised giving ordering at some stoichiometry x_o. If the covalent coordination obeys the 8-N rule, as is common in non-crystalline phases, compensated semiconduction occurs for all x (fig. 1 c), e.g. a-Ge_xTe_{1-x}. Large electronegativity differences between metals can also produce a semiconductor, and it is now generally acknowledged that ionic ordering has occurred[1,3] (e.g. ℓ-Li_xPb_{1-x}). For the compound C^+A^- the valence band is predominantly A-like and the conduction band C-like so that for $x \neq 0.5$ $C \to C^+ + e^-$ or $A \to A^- + e^+$, causing ε_F movements equivalent to the

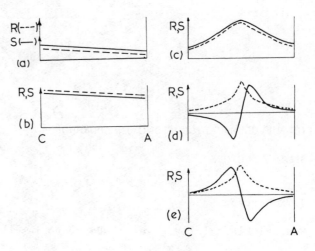

<u>Fig.1</u> : The five archetypal composition dependences of resistivity
and thermopower.

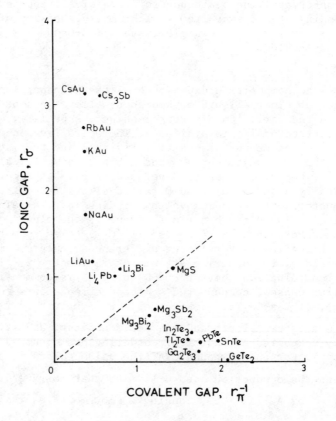

<u>Fig.2</u> : Quantal orbital coordinates for various systems.

doping of CA (fig. 1d). The fifth response has yet to be observed experimentally but is predicted here for a-Ga$_x$As$_{1-x}$. This paper seeks to classify the stoichiometric bonding configurations to identify the essential difference between behaviours (b)-(e) and to apply model electronic density of states (DOS) consistently over the complete range of x.

The principal bonding type at stoichiometry is found by comparing the various terms in the band-structure energy; the pseudopotential coefficients and the electronic kinetic energy. The kinetic energy dominates for simple metals and other terms appear to second order of perturbation. In the octet semi-conductors $C^N A^{8-N}$, the zincblendes, the V(111) pseudopotential term dominates, so V(111) is now first order and the others second order[4]. This generalises to non-crystalline with the necessary localized representations[4,2] of V(111) and V$_1$. Covalent octet semiconductors have a real V(111), polar covalent members have a complex V(111) and ionic members an imaginary V(111), the latter due to the symmetric sites in the rocksalt structure. This often applies more generally - most covalent non-crystalline structures have asymmetric sites and ionic systems more symmetric sites.

Bond ionicities are the major determinant of structure in a binary system. A recent quantal coordinate scheme[5] has proved a more sensitive separator of observed coordinations than earlier Mooser-Pearson plots or the dielectrically-determined ionicity. The coordinate r_σ acts as an ionic gap and r_π^{-1} as a covalent gap. These are plotted in fig.2 for various systems, although strictly each homologous series should be plotted separately. The alkali-gold systems appear highly ionic, as implied by their CsCl structure, and Cs$_3$Sb appears also to be highly ionic, contrary to some chemical arguments[6]. Li$_4$Pb, Li$_3$Bi and the magnesium pnictides are medium ionic, as are their crystals. The closed octet tellurides like Ga$_2$Te$_3$ are covalent.

To differentiate between the three behaviours of ordered non-crystalline semiconductors, the concept of formal charge is useful because it replaces the analogous Brillouin zone counting in crystalline systems in determining band occupancies. Formal charge is that integer charge associated with a site if covalent bonding is assumed symmetric and ionic charge transfer is assumed complete; therefore ionic and dative covalent bonding give formal charges. In order to show that the bonding configuration of the excess of each constituent beyond stoichiometry determines the composition-dependences, let us consider the possible configurations of excess Te added to a (hypothetical) chemically ordered a-MgTe. From its ionicity this compound could quite plausibly be either

ionic or tetrahedrally-covalent at stoichiometry. The formal
charge for the ionic model is $Mg^{2+}Te^{2-}$ and for the covalent
model $Mg^{2-}Te^{2+}$ because of the dative bonds. The stoichiometric
compound is an intrinsic semiconductor with a full valence band
within either model. If excess Te is added ionically (inter-
stitially), replacing a Mg atom to keep atom numbers constant,
4 holes are generated : $Mg^{2+} + Te \rightarrow Mg^{\uparrow} + Te^{2-} - 4e^{-}$ within either
model. Alternatively, in the perfectly ordered covalent model, Te
could substitute at an Mg site, giving 4 electrons :
$Mg^{2-} + Te \rightarrow Mg^{\uparrow} + Te^{2+} + 4e^{-}$. Finally, Te could enter in the
compensated bivalent mode with zero formal charge. Similar
possibilities are available for excess Mg (although in reality
bivalent Mg is unlikely). Since MgTe has a closed shell in either
model, an ionically bonded excess of either element gives the
behaviour of fig. 1(d). The behaviour for substitutional and
compensated excesses is that of 1(e) and 1(c), respectively.
Examples of this classification are shown in Table 1.

A-Ge_xTe_{1-x} is the simplest example of an ordered,
compensated semiconductor system. Ordering occurs at $GeTe_2$ (x_o =
0.33) with coordinations Ge:4, Te:2. The DOS at $GeTe_2$ (fig.3c)
shows Ge-Te σ and σ^* states and Te π states. For $x \neq 0.33$ homopolar
bonds are introduced because of coordination constancy. These have
smaller σ/σ^* splittings than Ge-Te bonds, causing the pseudogap to
narrow away from x_o, thereby reducing R and S since ε_F stays
midgap[7].

Table 1 : Bonding classification and examples : underlining means
that the property occurs only on that side of x_o

Bonding at Stoichiometry	Covalent	Ionic	Metallic
Bonding of excess			
Comp. covalent	a−GeTe	a−MgSb	
Doped covalent	a−GaAs		
Metallic	ℓ−T l Te	a−MgBi	ℓ−SnZn
Ionic		ℓ−CsAu	
		ℓ−PbTe	

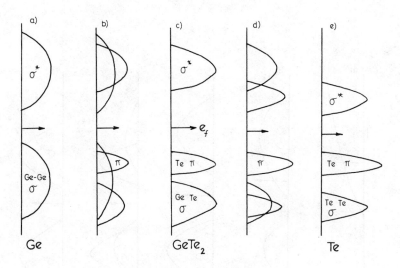

Fig.3 : Schematic DOS for a-Ge$_x$Te$_{1-x}$

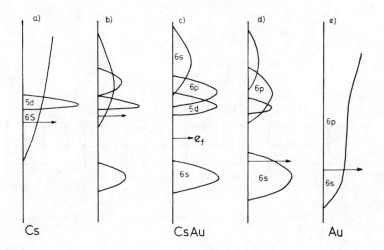

Fig.4 : Schematic DOS for ℓ-Cs$_x$Au$_{1-x}$.

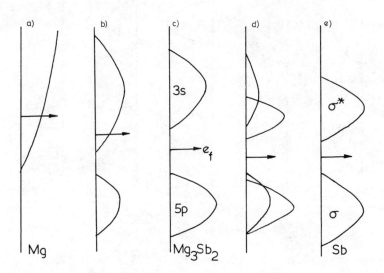

<u>Fig.5</u> : Schematic DOS for a-Mg$_x$Sb$_{1-x}$.

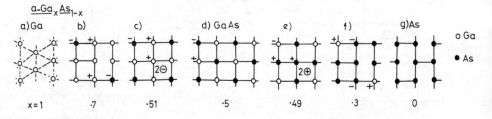

<u>Fig.6</u> : Schematic of local structures in a-Ga$_x$As$_{1-x}$.

Liquid Cs_xAu_{1-x} is probably the classic ionic ordered alloy with a model DOS shown in fig. 4, based on calculations on $CsAu$[8]. Gold exists as Au^- in CsAu, excess Cs as $Cs^+ + e^-$ and excess Au as $Au^- + e^+$. However, the 1.5 eV pseudogap allows ionic conductivity to dominate near $x = 0.5$ and doping behaviour is unfortunately not observed[3]. Liquid Li_4Pb and Li_3Bi appear to be ionic from their transport properties[9], but the large concentration of Li gives quite small cation-cation distances even at stoichiometry, increasing the complexity of structural models.

In these systems the anions Pb^{4-} and Bi^{3-} have p-orbital valence maxima. Liquid Pb and Bi have high coordinations. $A-Mg_xSb_{1-x}$ also has a p-electron anion Sb^{3-}. This, however, has a much lower self-coordination than Pb or Bi (3 in a-Sb), and hence homopolar Sb-Sb bonding can maintain a pseudogap[2] for $0 < x < 0.6$ (fig. 5). In essence the Mg-Sb system is ionic at stoichiometry with excess Mg entering metallically and excess Sb covalently. By comparison, $\ell-Tl_xTe_{1-x}$ has similar pseudogap behaviour, with metallic excess Tl and bivalent excess Te, but the ordering at Tl_2Te is covalent. This is an example of the bonding for $x \neq x_o$ rather than at x_o determining transport variations (although often atomic configurations are the same for all x so the distinction fades). $A-Ga_xAs_{1-x}$ and other III-V systems are the most likely examples of self-substitutional doping. If the stoichiometric a-GaAs is highly chemically ordered[10], then, by analogy with a-Si, excess of Ga or As should enter substitutionally giving shallow ionisable levels yielding the doping response (1e), if the pseudogap state density is sufficiently low. This doping regime only appears close to stoichiometry because, as in a-Si, larger excesses enter in a self-compensated mode (As) or in some auto-compensated mode (Ga), as in fig. 6.

The group IV chalcogenides are interesting. One member a-GeTe is a typical compensated system[7] with a stoichiometry at $x_o = 0.33$, whilst $\ell-PbTe$ has the characteristic ionic/interstitial excess response[11] with $x_o = 0.5$. This difference is intimately connected with bonding differences. The structures of C^NA^{10-N} crystals have been classified by the quantal coordinate method[12]. The orthorhombic phase is stable above a critical covalent gap and the rhombohedral form below a critical ionic gap (fig. 7). The covalency of 10-electron systems is that of resonant half filled p-bands rather than of paired electrons and becomes destabilised if the atomic positions in a non-crystalline phase prevent the alignment of these p orbitals. The ionic 6:6 coordinated structure transfers satisfactorily into a non-crystalline environment as in

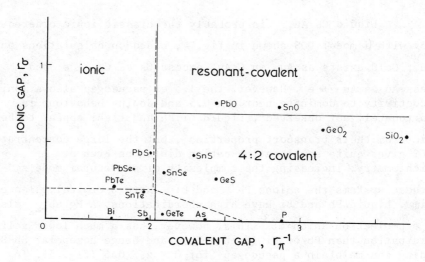

<u>Fig.7</u> : Quantal orbital coordinates for $C^N A^{10-N}$ systems;
— non-crystalline phase boundaries,
— — crystal phase boundaries.

ℓ-PbTe, but the more covalent alloys above the critical covalent
gap in fig. 7 revert to a more molecular form - the 4:2 coordinated
SiO_2-like structure with its compensated off-stoichiometric
response[2], as in a-GeTe.

REFERENCES

1. M. Cutler, 'Liquid Semiconductors', Academic Press (1977).
   J.E. Enderby, 'Metal-Non-metal Transition in Disordered
   Systems', Scottish Summer Schools, 19, (1978), p.361.
   W.W. Warren, 'Liquid Metals', Inst. of Physics, 30, 436, (1977).

2. J. Robertson, Phil. Mag., B39, 479 (1979); ibid., to be
   published.

3. F. Hensel, Adv. Phys., 28, 555 (1979).

4. W.A. Harrison, Phys. Rev., B14, 702 (1976), and 13th
   Phys. Semiconductors, Rome, ed. F.G. Fumi, p.111 (1976).

5. J.R. Chelikowsky, J.C. Philips, Phys. Rev. B, 17, 2453 (1978).

6. W. Freyland, G. Steinleitner, Ber Bunsenges Phys. Chem., 80,
   815 (1977).

7. H.K. Rockstad, J.P. De Neufville, 11th Phys. Semiconductors,
   Warsaw, ed. M. Miasek, Polish Scientific Publishers, (1972).

8. H. Overhof, J. Knecht, R. Fischer, F. Hensel, J. Phys. F, $\underline{8}$, 1607 (1978).

9. V.T. Nguyen, J.E. Enderby, Phil. Mag., $\underline{35}$, 1013 (1977).

10. N.J. Sevchik, W. Paul, J. Non Cryst. Solids, $\underline{13}$, 1 (1974).
    J.A. DelCueto, N.J. Shevchik, J. Phys. C, $\underline{11}$, L829 (1978).

11. J.C. Valiant, T.E. Faber, Phil. Mag., $\underline{29}$, 571 (1974).

12. P. Littlewood, J. Phys. C, to be published.

PROPERTIES OF A MODEL STRUCTURE FOR HYDROGENATED

AMORPHOUS SILICON

A. MacKinnon and B. Kramer

Physikalisch-Technische Bundesanstalt
Bundesallee 100
3300 Braunschweig, Federal Republic of Germany

Infrared absorption (IA), radial distribution function (RDF) and
proton magnetic resonance (PMR) spectra of a random network model
for a-Si:H have been calculated and compared with experiment.

1. INTRODUCTION

In the study of amorphous silicon and related tetra-
hedrally bonded amorphous semiconductors (e.g. Ge, GaAs) a number
of random network models have been proposed[1-3]. These were models
of the "ideal" pure amorphous silicon. Recently however interest
has been focussed on hydrogenated amorphous silicon, which is a
promising material for various electronic applications,
particularily in solar cells[4]. Weaire et al.[5] have therefore build
a model containing 314 Si and 83 H atoms. In building the model
they found that they were forced to introduce the hydrogen in
clusters in order to avoid leaving dangling bonds.

In this paper some calculated properties of the above
model will be discussed and after comparison with experiment some
conclusions drawn regarding the relationship between structural
details and observable properties of hydrogenated amorphous silicon.

2. RELAXATION OF THE MODEL

A table of nearest neighbours and approximate atomic
coordinates of the above model were stored on a computer and the

coordinates relaxed to the energy minimum of a modified Keating Hamiltonian[6].

The Keating expression for the elastic energy can be written as

$$V = \frac{3}{8} \left[\sum_{ij} \frac{\alpha_{ij}}{d_{ij}^2} [r_{ij}^2 - d_{ij}^2]^2 \right.$$

$$\left. + \sum_{ijk} \frac{\beta_{ijk}}{d_{ij}d_{ik}} [\vec{r}_{ij} \cdot \vec{r}_{ik} - \vec{d}_{ij} \cdot \vec{d}_{ik}]^2 \right] \qquad (1)$$

where $\vec{r}_{ij}$ is the vector from atom i to atom j and $\vec{d}_{ij}$ is its ideal value. The first sum is over all pairs of nearest neighbours and the second sum over all triplets of an atom and two of its nearest neighbours. The above expression is properly invariant under translation and rotation of the whole system. In particular the terms d_{ij}^2 and $\vec{d}_{ij} \cdot \vec{d}_{ik}$ simply define the ideal bond lengths and bond angles.

In the case of a-Si:H there are two α's and three β's which are to be defined. In practice α_{SiSi} and β_{SiSiSi} are fixed from crystalline silicon.

In addition to the above two further terms were included. Firstly a simple repulsion term between non-bonded atoms

$$V = \sum_{ij} B_{ij} [\exp(-A_{ij} r_{ij}) + \exp(-A_{ij} r_{ij}^{max})(A_{ij} r_{ij})] \qquad (2)$$

which was found necessary to prevent the hydrogen atoms from moving too close to one another.

r^{max} is the maximum atomic distance for which the above term was included. It was chosen as the second nearest neighbour distance. The values for the constants A and B were taken from published data for crystals of small hydrocarbon molecules[7]. In practice the values $A = 37.4$ nm^{-1} and $B = 2$ kNm^{-1}, appropriate for H-H, were used for all combinations of atoms since the results were very insensitive to the Si-H and Si-Si terms.

The second modification to the Keating Hamiltonian was the introduction of Coulomb forces

$$V = \sum_{ij} \frac{1}{4\pi\varepsilon_o} \frac{q_i q_j}{r_{ij}} \qquad (3)$$

where q_i is an effective charge on atom i to simulate the polarisation of the Si-H bond.

It is necessary however to introduce additional terms (between lst and 2nd nearest neighbours) in order to prevent the Coulomb forces distorting the bond lengths and angles. This was accomplished with the terms

$$V = \sum_{ij} - \frac{1}{4\pi\varepsilon_o} \frac{q_i q_j}{d_{ij}^2} r_{ij} \qquad \text{and} \qquad (4)$$

$$V = \sum_{ij} - \frac{1}{4\pi\varepsilon_o} \frac{q_i q_j d_{ij}^2}{3r_{ij}^3}$$

between like and unlike charges respectively, the difference being necessary to preserve the positive definite dynamical matrix.

In the presence of the Coulomb forces the exponential repulsion term was found to be inadequate between unlike charges and unnecessary between like charges. A term

$$V = \sum_{ij} - \frac{1}{4\pi\varepsilon_o} \frac{q_i q_j}{3r_{ij}^3} (r_{ij}^{max})^2 \qquad (5)$$

was therefore substituted in this case.

Using the relaxed coordinates the vibrational spectrum and radial distribution function were calculated together with the dipole–dipole broadening of the proton magnetic resonance.

3. RADIAL DISTRIBUTION FUNCTION

The radial distribution function (RDF) of amorphous Silicon is generally similar to that of the crystalline phase with the exception that the peak by 0.45 nm is almost totally absent in the amorphous case[8]. This is generally interpreted as implying that the dihedral angle is randomly distributed. In many samples however a small peak is observed in this region which is sometimes interpreted as being due to trapped SiH_4 molecules[9].
The calculated RDF of the model Fig. 1 was found to have such a small peak for some combinations of force constants. In general this region of the RDF was found to be sensitive to the force constants. Although this was certainly partly due to the small size of the cluster the effect does appear to be significant. An attempt to strengthen the peak by introducing a small attractive force between all Silicon atoms further than second nearest neighbours failed due to numerical instability.

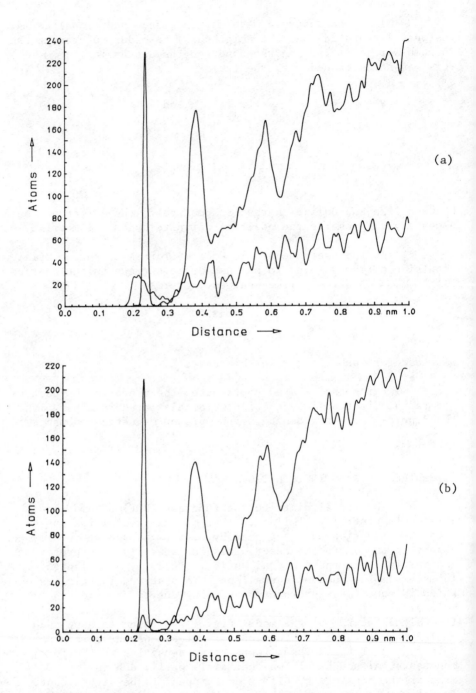

<u>Fig. 1:</u> Calculated radial distribution functions for (a) unpolarised
and (b) polarised bonds. The upper curve is the Si-Si
distribution, the lower curve H-H.

The widths of the other stronger peaks are consistent with experiment indicating a bond length distortion of about 1 % and bond angle distortion of about 7°.
The H-H radial distribution function shows fewer significant features. This is due to the much smaller number of H-H pairs (3403 as compared with 49141 Si-Si pairs) which results in a much noisier distribution. The shoulder at about 0.2 nm in the absence of Coulomb forces (Fig. 1a) is due to the mutual repulsion of H atoms that were too close together in the original model.
With Coulomb forces (Fig. 1b) it was found possible to remove this feature without disturbing the other properties of the cluster, whereas very large exponential forces led to large tails in the infrared spectrum.
A peak is expected about 0.24 nm due to = SiH_2. This peak does not show strongly in the calculation due to the presence of only five such "doubly bonded" hydrogen pairs (i.e. 10 atoms) and the proximity of the feature at 0.2 nm.

There is some evidence of a feature between 0.38 nm and 0.44 nm which can be related to the staggered configuration of H-Si-Si-H since it becomes larger in the presence of stronger H-H repulsion.

4. PROTON MAGNETIC RESONANCE

Recently Reimer et al.[10] carried out proton magnetic resonance (PMR) measurements on various a-Si:H films. They obtained resonance lines whose shape could be interpreted as a superposition of two gaussians with widths (FWHM) 22-27 kHz and 3.4-4.6 kHz respectively. The most significant contribution to the width of PMR lines is the dipole-dipole broadening. The second moment of a gaussian line is given by[11]

$$M = \frac{3}{5} \left(\frac{\mu_o}{4\pi}\right)^2 \gamma_p{}^4 \hbar^2 \ell(\ell+1) \sum_j r_{ij}^{-6}$$

where 1 is the nuclear spin, 1/2 for protons, γ_p is the nuclear gyromagnetic ratio and r_{ij} the spacing between spins. This formula was applied to the model for each hydrogen atom separately, giving a sum of 83 gaussians.

Without Coulomb forces the resulting curve had a value of between 60 kHz and 80 kHz, whereas a value of about 25 kHz was possible in the presence of polarized bonds. This is clearly related to the shoulder at 0.2 nm in the H-H RDF.

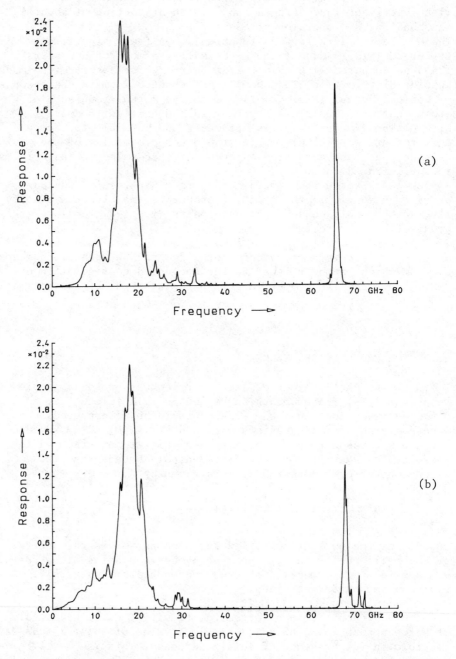

<u>Fig. 2:</u>   Calculated infrared response function for (a) unpolarised
          and (b) polarised bonds.

5. VIBRATIONAL PROPERTIES

There have been a number of measurements of the infrared absorption etc. of a-Si: H[12-15]. The last two groups have also carried out such measurements before and after thermal dehydrogenation of their samples.

The results from different groups and different samples differ in detail but there are a number of consistent features : peaks at 60 THz (2000 cm^{-1}) and 63 THz (2090 cm^{-1}), the Si-H stretching modes, between 25 THz (850 cm^{-1}) and 27 THz (900 cm^{-1}), the =SiH$_2$ and -SiH$_3$ bending modes, and between 12 THz (400 cm^{-1}) and 23 THz (750 cm^{-1}), the wagging, rocking etc. modes.

The diagonalisation of the full dynamical matrix for the model (1191 × 1191) was impractical so only the submatrix of the hydrogen atoms (249 × 249) was diagonalised. This is equivalent to the limit $\frac{m_H}{m_{Si}} \to 0$ where m_H and m_{Si} are the atomic masses of hydrogen and silicon respectively. The approximation is good for those eigen-frequencies which are well above the Si-Si stretching frequency 15 THz (500 cm^{-1}), i.e. the stretching and bending modes.

Initially the calculations were carried out without Coulomb forces. In this case (Fig.2a) it was found impossible to obtain the splitting of the stretching modes at the same time as the bending mode. In particular the symmetric and antisymmetric =SiH$_2$ stretching modes appear above and below the ≡SiH peak. In addition a tail appeared on the upper side of the wagging modes when the exponential repulsion forces were included. This tail made the bending modes hard to resolve. This is yet another undesirable result of the exponential forces.
The model contained only five =SiH$_2$ configurations and no -SiH$_3$'s, so no conclusions can be drawn about features due to -SiH$_3$, and =SiH$_2$ features appear much weaker than in the experiment. Lucovsky[16] has related the two Si-H stretching peaks to the electronegativities of the other neighbours of the silicon in hydro-silicon molecules (e.g. SiH$_4$, SiHF$_3$, SiH$_2$Cl$_2$) where the Si-H bond becomes shorter in =SiH$_2$ than in SiH and the stretching frequency higher.

In an attempt to simulate such an effect, Coulomb forces were built into the model. This creates on additional parameter q, the charge transferred to a hydrogen atom from the silicon atom to which it is bonded. In the ≡SiH configuration each hydrogen is

bonded to a silicon with charge $-q$, whereas in the $=SiH_2$
configuration the silicon charge is $-2q$. This reults in a shift
in both $=SiH_2$ stretching modes to higher frequency. In this
connection it is worth mentioning that it is essential that
vibrational calculations are carried out only after the model
has been fully relaxed with the appropriate forces.
Several spectral features appear in an unrelaxed system which
disappear after relaxation.

        The calculated spectrum with Coulomb forces is shown
in Fig.2b. The stretching and bending modes are well reproduced.
The tail on the upper side of the wagging mode peak in Fig.2a has
disappeared. This tail was due to the interaction of non-bonded
hydrogens and is another feature caused by the exponential repulsion
forces. The longer range Coulomb forces manage however to keep the
hydrogens apart without generating such features. The effective
charge used is equivalent to about 1/3 of an electronic charge.

6. CONCLUSIONS

        The model studied in this work provides a reasonable
description of hydrogenated amorphous silicon. In future attempts
to build such models, however, care must be taken that the
hydrogens do not get too close to one another since this tends
to dominate certain properties, e.g. PMR, calculated in this work
and could possibly yield unrealistic electronic properties. It has
been found possible to provide an adequate description of the
forces present in such a system and therefore of the vibrational
properties, only by the inclusion of forces describing the
polarisation of the Si-H bonds. The results confirm the usual
interpretation of the spectral features.

REFERENCES

1. D.E. Polk, 1971, Journal of Non-Crystalline Solids, 5, 365.

2. P. Steinhardt, R. Alben, D. Weaire, 1974, Journal of Non-
   Crystalline Solids, 15, 215.

3. G.A.N. Connell and R.I. Tempkin, 1974, Physical Review B9, 5323.

4. J.I.B. Wilson and D. Weaire, 1978, Nature, Lond. 275, 93.

5. D. Weaire, N. Higgins, P. Moore, I. Marshall, 1979,
   Philosophical Magazine B 40, 243-245.

6. P.N. Keating, 1966, Physical Review 145, 637-645.

7. Donald E. Williams, 1967, Journal of Chemical Physics <u>47</u>, 4680-4684.

8. R. Mosseri, C. Sella and J. Dixmier, 1979, Physica Status Solidi <u>52</u>, 475.

9. D. Weaire and F. Wooten, 1979, 8th International Conference on Amorphous and Liquid Semiconductors, Cambridge, Mass.

10. J.A. Reimer, R.W. Vaughan and J.C. Knights, 1980, Physical Review Letter (to be published).

11. A. Abragam, 1961, The Principles of Nuclear Magnetism, Oxford, p.112.

12. M.H. Brodsky, Manuel Cardona and J.J. Cuomo, 1977, Physical Review B <u>16</u>, 3556 - 3571.

13. J.C. Knights, G. Lucovsky and R.J. Nemenich, 1978, Philosophical Magazine B <u>37</u>, 467-475.

14. S. Oguz, R.W. Collins, A.M. Paesler and William Paul, 1979, 8th International Conference on Amorphous and Liquid Semiconductors, Cambridge, Mass. (to be published).

15. P. John, M. Odeh, M.J.K. Thomas, M.I. Tricker, M. McGill, A. Wallace and J.I.B. Wilson, 1979, 8th International Conference on Amorphous and Liquid Semiconductors, Cambridge, Mass. (to be published).

16. G. Lucovsky, 1979, Solid State Commun· <u>29</u>, 571-576.

COMPARISON OF IRRADIATION EFFECTS IN VITREOUS SILICA

BOMBARDED BY PARTICLES OF DIFFERENT MASS AND ENERGY

M. Antonini*, P. Camagni, and A. Manara

Joint Research Centre

21020 Ispra, Varese (Italy)

Amorphous silica samples have been bombarded with Ni^{+6} ions of 46.5 MeV to produce a large number of atomic displacements. Other samples with the same impurity content have been irradiated with electrons of 1.5 MeV. Isochronal annealings at various temperatures up to 600°C were made on the irradiated samples. Optical absorption spectra, induced with these treatments, in the UV region up to 190 nm have been analyzed and the relevant parameters of B_2, E_1', D_o and E centers have been evaluated. The comparison between results obtained from heavy ion and electron irradiated samples seems to indicate that the B_2 and E bands are related to displacive efficiency of impinging particles while the E_1' and D_o are related to the total energy deposition, irrespective of dpa values.

INTRODUCTION

        The nature of defects in irradiated vitreous Silica is not yet fully understood, in spite of a large number of investigations[1,2,3]. In fact the mixed character of bonding and its relatively open structure represent a source of complexity, which prevents in many cases a clear attribution of the observed effects in terms of electronic or atomistic modifications. This applies in particular to colour centres as observed by optical absorption and ESR. Even the extensively studied E' centre at 5.8 eV, though well characterized as an electron on a dangling orbital of a threefold coordinated Silicon, seems to be compatible with a

* Gruppo Nazionale di Struttura della Materia and Istituto di
  Fisica dell' Universita , 41100 Modena, Italy.

full oxygen vacancy or with relaxation of a broken bond[2,3,4].
Greater uncertainty exists as to other typical centres produced
by irradiation, such as the so called D_o (~ 4,8 eV)[5,6],
B_2 (5.06 eV)[7] and E (~ 7.6 eV)[8,10].

A good approach towards clarification of defects
structures in SiO_2 might be to compare the type and behaviour
of colour centres obtained in irradiations with particles of
widely different mass and energy, in such a way as to afford
a controlled separation between purely electronic (ionization)
and displacement processes. A program of this type is under way
in our laboratories,[11]; recent results are discussed in the
folowing.

RESULTS AND DISCUSSION

Pure Silica samples are obtained from Quartz et Silice
(France) in the form of synthetic-based Tetrasil or natural
Puropsil, differing in total OH-radical content (200 or 10 p.p.m.,
respectively). Both materials were irradiated with 46.5 MeV, Ni^{6+}
ions in the Variable Energy Cyclotron of A.E.R.E., Harwell (England)
at various doses between 0.01 and 1 dpa, corresponding to ~10^{14} -
10^{16} ions/cm^2. Details concerning the samples and the heavy-ion
irradiations can be found in a previous paper[12]. Synthetic samples
have also been irradiated in our laboratory with 1 MeV electrons
from a Van de Graaf accelerator. Optical absorption was measured
in the range 190 to 1500 nanometers, as a means to monitor defect
production in the irradiated material.

Following post-irradiation measurements, isochronal
annealings at various temperatures up to 600°C were made, and the
samples were re-examined. A representative picture of colour centres
produced by heavy-ion irradiation and of their successive
annealings is given in fig.1). Similar data were collected for
electron-irradiated samples.

All the observed spectra were analyzed so as to obtain
best-fits of line width, peak position and strength of distinct
colour centre components. In this way quantitative monitoring of
concentrations was performed. Figs. 2) and 3) show typical results
of this analysis. We start by summarizing the main features thus
obtained.

As shown in the previous figures, major differences
are found between heavy-ion and electron irradiation effects.
The former are characterized principally by the appearance of
the B_2 band and of a strong u.v. absorption tail (E-band) which
are entirely absent after electron bombardment. The E_1' band is

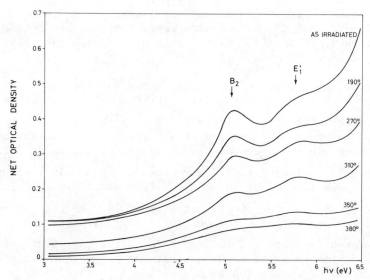

<u>Fig. 1:</u> Absorption spectra of heavy-ion bombarded silica as
irradiated at a dose of $1.7 \; 10^{15}$ particles/cm^2 (0.2 dpa)
and after isochronal annealings at the indicated
temperatures.

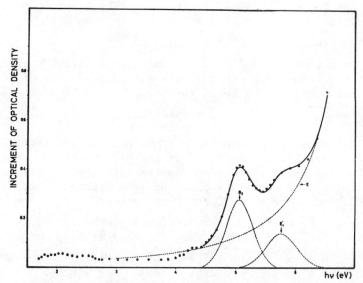

<u>Fig. 2:</u> Analysis of a typical coloration spectrum of heavy-ion
irradiated SiO_2 in terms of B_2 and E'_1 bands, superimposed
on the tail of the E band located at 7.6 eV. The full line
shows the obtained best fit. Note the absence of the D_0
band present in the electron irradiated sample (see
Fig. 3).

common to both treatments, while a subsidiary band D_o is
specifically related to the electron case. Analytic separation
shows that the FWHM of E_1' is markedly larger in this case (1.0 eV
as against 0.7 eV with heavy-ions).

The study of dose dependence shows saturation of colour
centres for both types of irradiation. With heavy ions, saturation
is attained when the dose exceeds 10^{14} Ni^{+6} particles/cm^2. This
corresponds to about 0.01 dpa and to a total energy dissipation
of ~60 eV/atom. In the electron case, saturation begins at doses
greater than 10^{17} particles/cm^2, corresponding to ~5.10^{-6} dpa and
to an energy dissipation of ~6 eV/atom. From spectral analysis,
assuming f = 0.14 for the oscillator strength[13], the saturation
level of E_1' concentration turns out to be approximative
10^{-4} atom for heavy ions and 10^{-5} atom for electrons. Confronting
these sets of figures, one observes that the production yield of
E_1' at the onset of saturation is proportional to total energy
dissipated per-atom, irrespective of particle type and dpa values.
On the other hand, the much greater dpa estimated for the heavy-ion
case seems to have its counterpart in the appearance of new intense
components, i.e. the B_2 band and the ultraviolet tail. We interpret
the former as the electron centre attributed by Arnold[7] to an
oxygen vacancy; the latter as the tail of the E-band at ~7.6 eV
produced by displacing radiation[8] and attributed to an oxygen hole
centre[10]. From spectral analysis, assuming a tentative oscillator
strength f = 0.5 for both centres, the saturation level of
concentration turns out to be of the order of 5.10^{-5}/atom and
10^{-3}/atom for B_2^- and E-centres, respectively.

Evidence for a connection between these two centres is
obtained from preliminary analysis of the experiments on thermal
annealing of heavy-ion-irradiated samples (see fig.1). The results
are given in fig. 4, clearly indicating a fair proportionality
between relative numbers of B_2^- and E-centres present in the sample
prior to or in the course of annealing. No such correlation was
found between the above centres, and E_1', thus marking a further
point of distinction for the latter. Thermal bleaching for the E_1'
band is in fact a complex process, depending on whether heavy-
ion or electron irradiated samples are involved. In the last case
the rate of annealing appears substantially faster; concurrently
D_o is also seen to decrease with a rough proportion. In conclusion,
the type and behaviour of electron and hole centres produced in
vitreous SiO_2 may be classified in two broad categories :

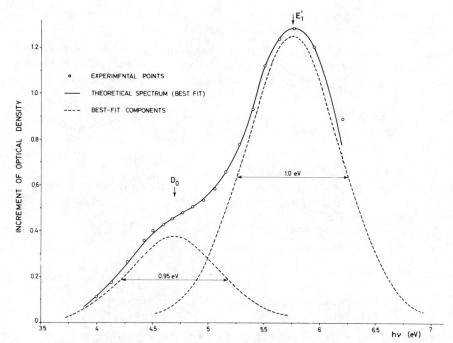

Fig. 3: Analysis of a typical coloration spectrum of electron irradiated SiO_2 in terms of D_0 and E_1' bands. The full line indicates the obtained best fit. Note the absence of the B_2 and E bands, both present in the heavy-ion irradiated sample (see Fig. 2).

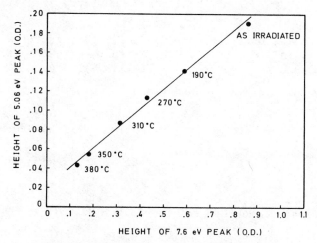

Fig. 4: Correlated annealing of B_2 and E bands produced during heavy-ion irradiation; points indicate the results of the best fit analysis.

i) centres whose production can be related to total energy deposition, i.e. essentially electronic excitation. E'_1 and D_o are entities of this type. Without prejudice as to the possible nature of the underlying defects, it must be accepted that these cannot be a primary product of displacement processes; ii) centres related more directly to the displacive efficiency of the impinging particles, i.e. B_2 and E, for which a specific precursor defect could be envisaged in the light of their correlated behaviour.

Acknowledgements — We wish to thank Miss T. Cavioni and Mr. N. Gibson for their cooperation in the experimental work and in the analysis of data.

REFERENCES

1. E. Lell, N.J. Kreidl and R. Hensler : Progress in Ceramic Science, J.E. Burke, ed. 4 (Pergamon, Oxford, 1966) 1.

2. D.L. Griscom : Defects and Their Structure in Nonmetallic Solids, B. Henderson and A.E. Hughes, ed. (Plenum Press, New York, 1975), 373.

3. G.N. Greaves : Philosophical Magazine, 37, 447 (1978).

4. F.J. Feigl, W. Beall Fowler and K.L. Vip : Solid State Communications, 14, 225 (1974).

5. P.W. Levy : Journal of Physics and Chemistry of Solids, 13, 287 (1960).

6. E.J. Friebele, D.L. Griscom and G.H. Sigel, jr.: The Physics of Non-Crystalline Solids, G.H. Frishat, ed., (Trans.Techn. Publications, Aedermansdorf, Switz. 1977), 154.

7. G.W. Arnold, I.E.E.E. Transactions : Nuclear Science, NS-20, 220 (1973).

8. E.W. Mitchell and E.G.S. Paige, Philosophical Magazine 1, 1085 (1965).

9. C.M. Nelson and R.A. Weeks, Journal of Applied Physics, 32, 883 (1961).

10. M. Stapelbrack, D.L. Griscom, E.J. Friebele and G.H. Sigel, Jr.: Journal of Non-Crystalline Solids 32, 313 (1979).

11. M. Antonini, P. Camagni, A. Manara and L. Moro : Journal of Non-Crystalline Solids, to be published.

12. M. Antonini, A. Manara and P. Lensi : The Physics of SiO_2 and its Interfaces, S.T. Pantelides ed. (Pergamon, New York, 1978) 232.

13. R.A. Weeks and E. Sonder : Paramagnetic Resonance II, W. Low ed. (Academic Press, New York, 1963) 869.

EXAFS OF SILICA AND SILICATE GLASSES

G.N. Greaves[+], A. Fontaine[++], P. Lagarde[++],
D. Raoux[++], S.J. Gurman[+++], and S. Parke[++++]

+   SRC, Daresbury Laboratory, Warrington WA4 4AD,
    England.

++ LURE, CNRS, Laboratoire associé à l'Université
    de Paris-Sud, Bâtiment 209C, 91405 Orsay, France.

+++ Department of Physics, University of Leicester,
    Leicester, England.

++++ Department of Ceramics, Glasses and Polymers,
    University of Sheffield, Sheffield, England.

EXAFS of Si K-edge of α-quartz, silica and several silicate
glasses is reported. The atomic distributions obtained using
calculated phase shifts are discussed comparatively. The
crystalline Si-O bond length which is common to silica is also
common to the silicate glasses. The crystalline Si-O-Si bond
angle is also retained in all the glasses but with some distortion.

We report the first use of EXAFS as a local structural
probe in the study of silica and multicomponent silicate glasses.
Measurements at the K edge of silicon (1.82 KeV) were made using
the soft x-ray crystal monochromator[1] at the 540 MeV storage ring
ACO at LURE. Three silicate glasses were chosen : silica (SiO_2),
sodium disilicate ($Na_2Si_2O_5$) and sodium calcium disilicate
($Na_2CaSi_5O_{12}$) - the last being a close approximation to the
commercial 'universal' glass composition. The glasses were
melted by conventional techniques and specimens approximately 1μ

thick were made by blowing films from the remelted glasses. These
were studied in conjunction with α-quartz. A suitable specimen
of α-quartz was fabricated by casting a suspension of the powder
in collodion on a clean water surface. Details of transmission
measurements of thin samples at these wavelengths have already
been given[2]. For these measurements a single ionization chamber
was used and spectra were normalised by a retrospective run without
the sample.

Normalised fine structure spectra for α-quartz and the
glasses were obtained from the transmission spectra by conventional
techniques and are presented in Fig.1. They are typical of light
atom systems – the EXAFS structure being spent within 500 eV. The
white lines of the spectra were used for alignment. All the
spectra are broadly similar and reflect the approximate invariance
of the SiO_4 tetrahedral unit. There are, however, significant
differences beyond the first shell of 4 oxygen atoms and these
give rise to detail in the structure below 150 eV. In α-quartz
there is a feature in the first minimum close to 70 eV which is
absent in the glasses. By curve fitting we can assign this to the
second shell of 4 silicon atoms.

To look more closely at the distribution of atoms
around silicon atoms the spectra shown in Fig. 1 were Fourier
transformed. A Gaussian window function was used to reduce
termination effects[3]. In order to make a quantitative appraisal
the $\ell = 1$ phase shifts for an excited silicon atom and for
surrounding oxygen atoms were computed using a muffin-tin
calculation due to Pendry[4]. This was performed for the α-quartz
structure incorporating corrections for the partial ionicity of
the Si-O bond. The Fourier transform approach implies that the
EXAFS is suitably represented by the plane wave approximation[5]
over the Fourier window. This becomes least reliable close to
the edge. The limited energy extent of the silicate spectra
necessitated centring the Fourier window such as to embrace
structure fairly close to the edge. Nevertheless by adjusting
the window position and the kinetic energy zero we have found it
possible to convincingly fit the α-quartz data. The window, which
was heavily damped, was centred at 122 eV and the energy zero was
taken 13 eV below the inflection point of the absorption edge.

The atomic distribution about silicon in α-quartz is
displayed in Fig.2. It runs out to 5 Å beyond which structure
due to the white line becomes significant. The histogram below
the experimental distribution is derived from the crystallographic
data[6]. The EXAFS nearest neighbour peak comes at 1.55 Å, 0.06 Å
below the crystallographic value and we take this as the systematic
error. The remaining two major peaks come at 2.94 Å and 4.37 Å close
to the first and second Si shells at 3.06 Å and 4.36 Å - A and B.

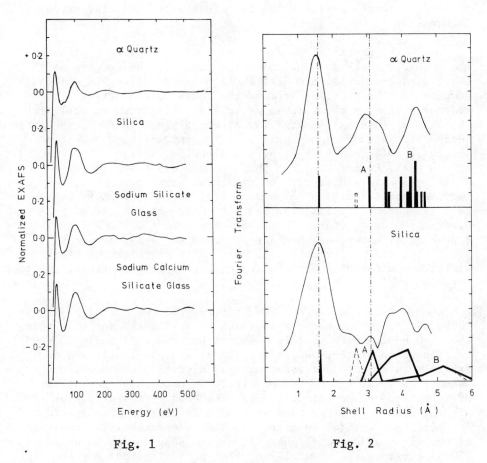

Fig. 1   Fig. 2

Fig. 1: Normalised EXAFS spectra for α-quartz, silica and the two
         silicate glasses.

Fig. 2: Fourier transform atomic distributions for α-quartz and
         silica.

The error for these shells evidently is ~0.1 Å. The experimental
positions of the Si-O (1.55 Å) and Si-Si (2.94 Å) peaks imply a
Si-O-Si bond angle of 143° compared to the crystallographic value
of 144°. Between the two Si-Si peaks comes some structure due to
oxygen shells - a shoulder at 3.4 Å and a minimum at 3.7 Å which
reasonably match the histogram. The prominence of Si-Si compared
to Si-O distances reflects the fact that silicon is the stronger
backscatterer.

        The identical Fourier window and energy zero used for
the α-quartz EXAFS in Fig.2 was also used for the spectra of the
three glasses. The atomic distribution about silicon in silica is
compared to that for α-quartz in the lower half of Fig.2. The
nearest neighbour shell comes at 1.57 Å. This compares well with
the 1.58(5) Å value reported for electron yield EXAFS at the oxygen
K-edge of silica grown on Si(111)[7]. The initial Si-O peak for
silica is followed in Fig.2 by a diminutive version of the α-quartz
2.94 Å peak. Beyond this the correspondence between the two
distributions disappears : the remaining peaks come at 3.97 Å
(with a shoulder at 3.65 Å) and at 4.62 Å. It is in this region
(> 3 Å) that the radial distribution curve obtained from
diffraction data departs from that of quartz[8,9]. The RDF exhibits
a peak at 4.1 Å with a shoulder at 3.6 Å and a further peak at
5.2 Å with a shoulder at 4.6 Å. So as with α-quartz the EXAFS
distribution for silica is in reasonable agreement with diffraction
data.

        Mozzi and Warren[8] have interpreted the RDF of silica in
terms of the random network approach of Bell and Dean[10]. In this
the Si-O bond length is left constant but the bond angle (Si-O-Si)
is allowed to distort whilst the dihedral angle is allowed to
freely rotate. The calculated partial distributions are shown
schematically at the bottom of Fig.2. In particular the Si-Si
distribution centred at 3.12 Å is broadened by distortion of the
Si-O-Si bond angle. Our peak at 3.03 Å is clearly greatly reduced
in going from crystal to glass for the same reason. The position
of the first and second peaks for silica suggest a bond angle of
150° - close to the crystalline value but distorted. For the
oxygen atomic distribution Stohr et al.[7] find a weak second peak
at 2.57 Å close to the O-O distance shown by dotted lines in
Fig.2. This indicates the crystalline tetrahedral O-Si-O bond
angle is retained in silica. The mass of individual Si-O distances
beyond 3 Å in the structure of α-quartz however, are broadened in
the random network model by virtue of a variable dihedral angle
into an asymmetric distribution rising to a peak at 4.1 Å but with
a shoulder at 3.6 Å. This is replicated in our EXAFS distribution.
In the same vein the Si-Si(B) distances for α-quartz are broadened
into a distribution with a peak at 5.2 Å. The experimental peak at
4.6 Å in Fig.2 may belong to this distribution.

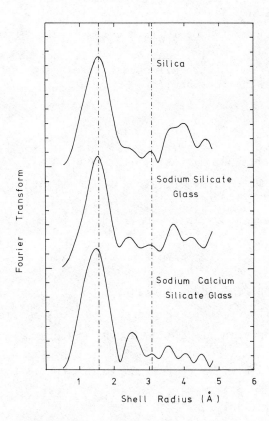

<u>Fig. 3</u>: Fourier transform atomic distributions for silica and
the two silicate glasses.

The EXAFS atomic distribution about silicon for the two silicate glasses are compared to the distribution for silica in Fig.3. The first and second peaks – close to 1.57 and 3.03 Å are common to all three distributions. This demonstrates for the first time the Si-O-Si bond angle of around 150° persists in the modified silicate structures. This is not a general property of crystalline silicates. The Si-O features between 3 and 4.5 Å are also common to all three distributions but the shoulder at 3.6 Å in silica becomes a peak in the silicate glasses. In crystalline sodium disilicate ($Na_2Si_2O_5$) the Si-Na distance comes at 3.35 Å; in sodium calcium orthosilicate (Na_2CaSiO_4) it comes at 3.15 Å with the Si-Ca distance at 3.42 Å[6]. The implication then is that the experimental peak around 3.6 Å in the two silicate glasses contains a contribution from modifier shells.

REFERENCES

1. M. Lemonier, O. Collet, C. Depautex, J.M. Esteva and D. Raoux, Nucl. Instrum. Meth. 152, 109 (1978).

2. A. Fontaine, P. Lagarde, D. Raoux and J.M. Esteva, J. Phys. F. 9, 2143 (1979).

3. S.J. Gurman and J.B. Pendry, Solid State Commun. 20, 287 (1976).

4. J.B. Pendry, Low Energy Electron Diffraction. Academic Press, London (1974).

5. P.A. Lee and J.B. Pendry, Phys. Rev. B11, 2795 (1975).

6. R.W.G. Wyckoff, Crystal Structures, New York : Wiley (1964).

7. Stohr, L. Johansson, I. Lindau and P. Pianetta, Phys. Rev. B20 664 (1979).

8. R.L. Mozzi and B.E. Warren, J. Appl. Cryst. 2, 164 (1969).

9. M. Misawa, D.L. Price and K. Suzuki, J. Non-Crystalline Solids 37, 85 (1980).

10. R.J. Bell and P. Dean, Nature, London 212, 1354 (1966).

DOPANT CONFIGURATIONS IN a-Se AND a-Si

J. Robertson

Central Electricity Research Laboratories

Leatherhead, Surrey KT22 7SE, U.K.

Impurity configurations in a-Se and a-Si are discussed in terms of the ionicity of the impurity-host bond and their quantal coordinates defined in terms of the Pauli-force model pseudo-potentials. Self-compensation and auto-compensation in a-Si is also discussed.

There are many diverse configurations for defects and impurities in crystalline semiconductors, due in part to the constraints of periodicity on the surrounding perfect crystal; ultimately the study of impurities in amorphous semiconductors may reveal a commensurate level of complexity, but presently the simplification that the lack of periodicity provides allows their configurations to be studied using a simpler, chemical bonding language[1,2]. In particular we show here the importance of ionicity of the impurity to host bond in determining the basic character of the impurity site. The cohesive energy of an element is maximised, giving a semiconductor, if a gap is opened up at the faces of a filled Jones zone, which is equivalent in chemical language to attempting to maximise the bonding character of the occupied orbitals. Thus originates the octet rule of covalent coordination, viz. N_c = 8-N where N is the valence electron number. It used to be believed that the coordinations of alloys and impurities were also determined by this rule. It is now realised that there are many other configurations available for a dilute impurity in an amorphous semiconducting host which are stable provided that the electronic structure of the host is sufficiently adaptable. In general, the covalent part of the cohesion is maximised if a tetrahedral impurity coordination is taken up, whereas an interstitial configuration tends to maximise any ionic contribution[3].

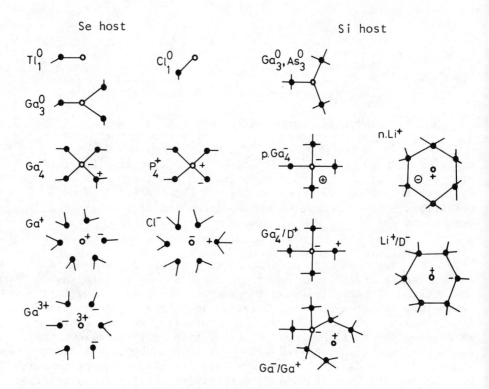

Fig.1 : Various impurity configurations in a-Se, (a) for electron-
        deficient elements, e.g. of group III, (b) for electron-
        rich elements; configurations in a-Si, (c) group III
        elements and (d) group I elements.

This is strictly true only for octahedral interstices for which
the principal interactions contributing to the cohesion are
antisymmetric in the lattice potential and hence solely ionic
- as in the rocksalt crystals[3]. This observation can be generalised
fairly easily to the hexagonal interstice which is also symmetric
but not so obviously to the 'back-bond' T_d interstice in Si. However, if this connection can be made, as it will also apply to non-
crystalline environments, there is then a general equivalence between
the covalent network-forming sites of glass science and the
substitutional sites in a-Si, and another between the ionic,
network-modifying sites and the interstitial sites in a-Si.

    Formal charge is now introduced in order to follow
changes of band occupancies caused by the non-octet configurations.
This idea is useful in non-crystalline systems and parallels the
Brillouin zone counting used in crystals. Formal charge is that
integer associated with an atomic site if covalent charge is
shared symmetrically between participating atoms and if atomic
charge transfer is complete, so that both dative covalent and
ionic bonding produce formal charges. If the covalent coordination
obeys the octet rule the site has zero formal charge and is self-
compensated, since it does not by itself change band occupancies.
Fig. 1 shows two examples of this for a group III element in a-Se
- univalent Tl p^1 and trivalent sp^2 - hybridized B. The tetravalent
Ga_4^- site bears a negative formal charge because one host electron
must be accepted. Conversely the ionic configurations Ga^+ and Ga^{3+}
donate electrons, but the former ion has retained its valence s^2
electrons. The electron-deficient elements with fewer than four
valence electrons are frequently found in these charged
configurations showing that they have greater stability than the
self-compensated alternatives. On the other hand the electron-rich
elements are usually found in self-compensated configurations (e.g.
As as trivalent p^3) but occasionally tetravalence is observed, as
in the P_4^+ sites in alumino-phosphate glasses. The Cl^- interstitial
would be an example of ionic bonding (fig.1). Formal charge shows
that tetravalent sites donate N-4 electrons to the host, while
ionic sites donate N or N-8 electrons (or N-2 electrons in e.g.Tl^+).
Therefore without compensation group V elements act as tetravalent,
substitutional donors, group III elements as substitutional
acceptors, group I elements as interstitial ionic donors and
group VII elements as interstitial acceptors. The chemical
periodicity of eight (of a host) has been translated into an
impurity periodicity of four, which is centred on columns IV and
VIII.

    The hosts a-Se and a-Si adapt to the non-octet impurity
configurations in two characteristically different ways, which
are actually quite similar from an energetic viewpoint. A chalcogen
host adjusts its coordination and the numbers of its lone pair p_π
electrons to retain overall compensation. Denoting coordinations

by subscripts[4] donated electrons are absorbed by transferring
shared σ electrons to non-bonded p_π states, giving $Se_2^o \rightarrow Se_1^-$.
Electrons withdrawn from the host results in the loss of p_π
electrons, thereby increasing the number of σ states and
coordination numbers, firstly to Se_3^+ and finally Se_4^{2+}. The
chalcogen acts as an ideal amphoteric Lewis host if only Se_3^+, Se_2^o
and Se_1^- sites are involved within a model in which the electrons
are transferred at the chalcogen p orbital energy ε_p (a constant
Lewis potential). Defining the impurity orbital as ε_i, the host-
impurity bond can be characterised by two parameters, its ionic
gap $V_3 = \frac{1}{2}(\varepsilon_i - \varepsilon_p)$ and its covalent gap $V_2 = <i|H|p>$. With the
total gap V written as $V = (V_2^2 + V_3^2)^{1/2}$ and the impurity formal
charge as Z then the total energy of the 8 electrons around a
tetrahedral impurity site is

$$E = 8V - 2ZV_3 - E_{pro} \tag{1}$$

where $Z = N-4$ and E_{pro} is promotion energy of the impurity atom
for $s^2p^{N-2} \rightarrow (sp^3)^N$. For example a group III impurity has N = 3,
$Z = -1$, $\varepsilon_i > \varepsilon_p$ usually, so $V_3 > 0$ and E is destabilized from the
simple 8V by the dative contribution $2ZV_3$. The total energy of the
electrons associated with an interstitial, ionic site is

$$E = -2NV_3 - E_{pro} \tag{2}$$

with N the formal site charge; e.g. a group I element has N = 1,
$V_3 > 0$ and a highly negative total energy. (These energies are
referred to ε_p).

     The tetrahedral-bonded semiconductors act as hosts in a
remarkably similar fashion, instead of accepting carriers at a
relatively constant Lewis energy ε_p as in a-Se; they cause a partial
band occupancy. However, the band-widths are sufficiently large
that the occupation of the band-edge states is much less
destabilizing than the occupation of the basic non-interacting σ^*
(or σ) levels. In this Weaire-Thorpe model, let us define extra,
intrahost interactions; the interaction $V_2^o = <h_1|H|h_2>$ and the
intra-atomic $V_1^o = <h_1|H|h_1>$. The host σ and σ^* states are at V_2^o
and $-V_2^o$, respectively, and the corresponding valence maximum and
conduction minimum are at $V_2^o - V_1^o$ and $-V_2^o + 3V_1^o$. The total energy

of the 8 electrons around the impurity together with that of the donated carriers can now be found, assuming that the latter occupy the band-edge states. For an n-type substitutional site with $N > 4$ and $Z > 0$:

$$E = 8V - Z(V_2^o + 2V_3 - 3V_1^o) - E_{pro} \tag{3}$$

and for a p-type substitutional site with $N < 4$:

$$E = 8V + Z(V_2^o - 2V_3 - V_1^o) - E_{pro} \tag{4}$$

For an n-type interstitial site with $N > 0$:

$$E = N(-2V_3 - V_2^o + 3V_1^o) - E_{pro} \tag{5}$$

where $V_3 > 0$ usually, and for a p-type interstitial $N < 0$ we find :

$$E = N(2V_3 - V_2^o + V_1^o) - E_{pro} \tag{6}$$

The similarity of doping of a-Se and a-Si is due to the extra carriers attempting to occupy levels as close to ε_h as possible. In a-Se the carriers are compensated at a relatively constant energy. On the other hand a-Si leaves the carriers free and is a less than ideal Lewis host ; in our model it donates electrons at $\varepsilon_h + V_2^o + V_1^o$ and accepts them at $\varepsilon_h - V_2^o + 3V_1^o$. The amphoteric character of the host increases with increasing metallicity $(2V_1^o/V_2^o)$.

Earlier it was noted that the sign of the carrier donated by a non-transition element impurity varied as -+-+ across the periodic table, with the elements of groups IV and III, neutral dopants, acting as the two +/- boundaries. In contrast the -/+ boundaries do not occur at fixed N but primarily at critical values of host to impurity ionicity. This could be determined by (1) and (2) for chalcogen hosts, but this turns out not be sufficiently accurate. For the present purposes of classifying coordination numbers these can be plotted for the ionic and covalent gaps V_3 and V_2 as in ref. 2 for electron-deficient elements in chalcogens. An improved classification results from a quantal coordinate plot[5], where r_σ plays the role of ionic gap and r_π^{-1} that of the covalent gap and fig. 2a shows this plot for these systems. A good separation is observed between those combinations known experimentally (by EXAFS or otherwise[2]) to be ionic, network-modifying and those known to be tetrahedrally-covalent. The combinations with even higher covalence are also separated out

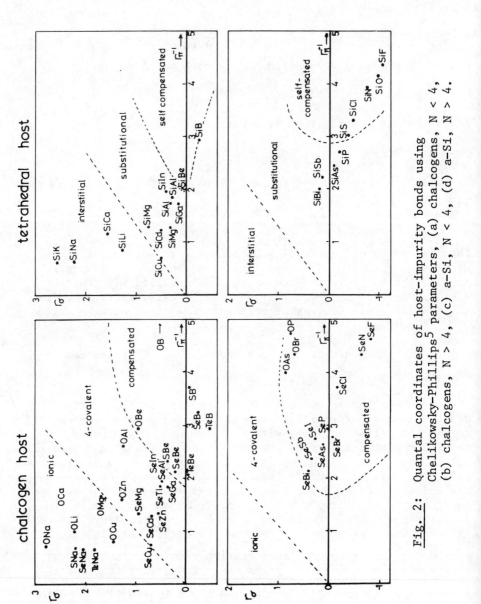

<u>Fig. 2:</u> Quantal coordinates of host-impurity bonds using
Chelikowsky-Phillips[5] parameters, (a) chalcogens, N < 4,
(b) chalcogens, N > 4, (c) a-Si, N < 4, (d) a-Si, N > 4.

into a self-compensated region, like Se-Be. This plot is similar
to that for the $A^N B^{8-N}$ series of crystals[5]. Note, however that
in the present case the electron per atom ratio varies via the p_π
number and this also reduces the bonding symmetry. Fig.2b is a
similar plot for electron-rich additives. If there is a
proportionality between V_3 and r_σ, and V_2 and r_π^{-1}, as there is for
the dielectrically-defined gaps[5], then the critical ionicity in
figs.2a and b should be similar, as equations (1) and (2) apply
equally to N < 4 and N > 4. The electron-rich additives possess
much higher r_π^{-1} values so that no interstitial combinations are
formed, and probably no tetrahedral combinations either (without
a higher host Lewis acidity from other mechanisms).

The impurity-tetrahedral host combinations can be treated
similarly. The exact relationship between r_π^{-1} and metallicity as
compared to the covalent gap has yet to be fully explored, but it
appears that the extra terms in (3)-(6) are handled by the quantal
coordinates and a similar critical ionicity is found to that for
chalcogen hosts. Again the electron-rich impurities are
associated with large covalent gaps and substitutional or self-
compensated sites. The latter separation is only achieved with
difficulty in fig.2c for such as Si:B.

Doping efficiencies[6] in a-Si are only of order 25 %,
indicating that both electrically active and inactive sites must
coexist in the same sample. The compensated (inactive) configurations for groups V-VII impurities are almost certainly the simple
self-compensated octet-rule sites. Similarly the inactive B site
is most likely the trivalent self-compensated site. However the
inorganic solid-state chemistry[7] of the other electron-deficient
elements indicates that they have little tendency to adopt formally
neutral sites - fig.2c shows that this is because only the host-
impurity combinations with high r_π^{-1} values can achieve this, e.g.
Na always appears as Na^+ and Al usually as Al^{3+} or Al_4^-. Although
self-compensation is prevented, experiment still requires some
form of electrically inactive form for these dopants, i.e. auto-
compensation. Fig.2c immediately suggests one type of auto-
compensation is to form bound complexes of substitutional and
interstitial sites, such as Mg^{2+}/Mg^{2-} or Ga^+/Ga^- as in fig. 1.
A second possibility is a dipole complex with a charged dangling
bond D, e.g. Li^+/D^- or Ga^-/D^+. As the majority of defects in a-Si
have a positive effective correlation energy[8], the excited
paramagnetic states of these complexes should be observable.
A final inactive site is an impurity configuration which is poorly

defined as substitutional or interstitial and which gives rise
to deep non-ionised states.

REFERENCES

1. J. Robertson, Phil. Mag. B, 40, 31 (1979).

2. J. Robertson, J. Non-Cryst. Solids, 35, 843 (1980), and
   Phil. Mag. B, 41, 177 (1980).

3. W.A. Harrison, 13th Int. Conf. Semiconductors, Rome, Ed.
   F.G. Fumi, North Holland.

4. M. Kastner, D. Adler, H. Fritzsche, Phys. Rev. Letts., 37,
   1504 (1976).

5. J.R. Chelikowsky, J.C. Phillips, Phys. Rev. B, 17, 2153 (1978).

6. J.C. Knights, T.M. Hayes, J.C. Mikkelsen, Phys. Rev. Letts.,
   39, 721 (1977).
   W.E. Spear, P.G. LeComber, Phil.Mag., 33, 935 (1976).
   W.E. Spear, P.G. LeComber, S. Kalbitzer, G. Muller, Phil.Mag.,
   B39, 159 (1979).
   P.G. LeComber, W.E. Spear, G. Muller, S. Kalbitzer, J. Non-Cryst.
   Solids, 35, 327 (1980).
   W. Beyer, B. Stritzker, H. Wagner, J. Non-Cryst. Solids, 35,
   321 (1980).

7. A.F. Wells, 'Structural Inorganic Chemistry', Clarendon Press
   (1962).

8. R.A. Street, D.K. Biegelsen, J. Non-Cryst. Solids, 35, 651
   (1980).

THE ELECTRONIC STRUCTURE OF SILICON NITRIDE

J. Robertson

Central Electricity Research Laboratories

Leatherhead, Surrey KT22 7SE, U.K.

Silicon Nitride is found to have a valence band maximum of nitrogen lone pair p electrons because of the planar nitrogen site. This contrasts with the usual lone pair semiconductors, such as SiO_2, caused by a p^4 valence configuration. Consequently although the valence band density of electron states shows a lone pair band and a deeper bonding band as usual, impurities have a greater effect in the nitride than in conventional lone pair semiconductors. Hole transport is also discussed.

Silicon nitride is an important ceramic and also used as an insulating and masking thin-film material in the solid state electronics industry. Chemical vapour deposited (CVD) nitride films are used in non-volatile MNOS memories in which the memory charge is stored in traps in the nitride and at the nitride/silicon dioxide interface[1]. The write/erase characteristics of the memories are controlled by the retention of the traps and by the tunnelling properties of carriers through the oxide barrier. Recently some interest has been focussed on hole transport in silicon nitride and its possible relationship to its electronic structure[2,3]. This note discusses the electronic structure and its relationship to the local atomic order in various phases of silicon nitride – the crystalline α and β phases[4] and an idealised amorphous phase from which it is possible to draw some conclusions for the CVD phase. Particular attention is paid to the nature of the band edge states and it is found that the geometry of the N site and not just its valence configuration determines the character of the valence band maximum – silicon nitride should be a lone pair semiconductor if the nitrogen site remains planar.

In the α and β phases of Si_3N_4 the silicon atoms occupy slightly distorted tetrahedral sites and the nitrogens occupy a mixture of almost planar and perfectly planar triply-coordinated sites. The local order is consistent with single covalent Si-N σ bonds between four Si sp^3 hybrids and three N p orbitals. However the nitrogen bond angle of around $120°$ implies that its three p orbitals cannot be assigned independently to each of these bonds as would be possible at a nitrogen with $90°$ bond angles. In order to discuss the orbital interactions in Si_3N_4 algebraically, it is helpful to define symmetry-adapted combinations of the Si hybrids which point from the three different Si atoms towards a single N site and interact separately with each N p orbital. (This is a similar construction to the use of super-bond orbitals[5] to allow for the expanded oxygen bond angle in SiO_2). For a general N site the Si-N-Si angle is given by : $\theta = 2 \sin^{-1} (\sqrt{3} \sin \phi/2)$ in terms of the pyramidal angle ϕ. Labelling the three hybrids on the three adjacent Si atoms as $|1>$, $|2>$, $|3>$, we define three new orbitals :

$$|s> = \{|1> + |2> + |3>\}/\sqrt{3}$$

$$|a> = \{|2> - |3>\}/\sqrt{2}$$

$$|b> = \{ 2|1> - |2> - |3>\}/\sqrt{6}$$

so that the Si-N two-orbital (covalent) interactions are now given by :

$$V_{2a} = <s|H|z> = \sqrt{3} \cos\phi \, V_2$$

$$V_{2e} = <a|H|y> = \sqrt{2} \sin \frac{\theta}{2} V_2 = <b|H|x>$$

in terms of the N 2p orbitals $|x>$, $|y>$ and $|z>$ (Fig.1a) and the basic two-centre Si-N interaction V_2. (The N 2s orbitals are omitted as core-like.) The ionic interaction V_3 is also included, $V_3 = \frac{1}{2} (\varepsilon_h - \varepsilon_N)$, where ε_h and ε_N are the Si hybrid and N p energies respectively. Each Si/N orbital pair, s and z etc., give rise to polar σ (S_+, A_+, B_+) and σ^* (S_-, A_-, B_-) states, one per N centre. The A_+ and B_+ states are deep valence states (Fig.1b). Above them lies the S_+ state, which is a pure N p_z lone-pair state for $\beta = 90°$ because $V_{2a} = 0$. This forms the valence band maximum and hence Si_3N_4 is a lone pair semiconductor for planar N sites.

The lowest state in the conduction band is S_- which is purely silicon s-like for $\phi = 90°$. Higher up are found A_- and B_-. If $\phi \neq 90°$ the S_+ and S_- states repel each other as $\bar{V}_{2a} \neq 0$ and assume

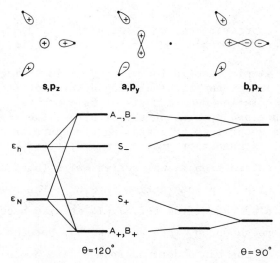

Fig. 1: (a) symmetrised Si orbitals around a N site and bonding
and antibonding states for (b) $\phi = 90°$ (c) $\phi < 90°$.

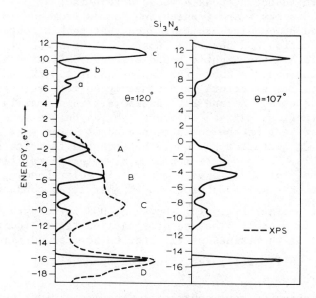

Fig. 2: (a) DOS for $\phi = 90°$ (b) DOS for $\phi = 67°$
with $\varepsilon_s^{Si} = -8.4$ eV, $\varepsilon_p^{Si} = -2.4$ eV, $\varepsilon_s^{N} = -24.06$ eV;
$\varepsilon_p^{N} = -11.5$ eV; first neighbour Si-N interactions
$P_\sigma = -4.76$ eV, sp $= -4.76$ eV, ps $= -2.73$ eV,
s $= -2.28$ eV, p $= -1.4$ eV, giving $V_2 = -6.5$ eV; and
a N-N interaction $V_p = -0.75$ eV.

a mixed atomic character once more. At $\sin\phi = (2/3)^{\frac{1}{2}}$ i.e. $\theta = 90°$, S_+, A_+ and B_+ are the three degenerate σ states (Fig. 1c).

This simple description of bonding in silicon nitride is now used to interpret the tight-binding density of states (DOS) in Fig.2. This calculation used a basis of 2s,p orbitals on the N atoms and valence sp^3 hybrids on the Si sites. The tight-binding parameters were determined by a scaling procedure from universal parameters as in a number of previous calculations on Si and SiO_2[5]. The two-centre interaction V_2 was scaled[6] with (bond length)$^{-2}$ from that of the sp^3 - p interaction in SiO_2[7]. The silicon energies ε_s and ε_p were also taken from ref. 7. It is also necessary to include second neighbour p_σ interactions, V_p, between nitrogens, otherwise some of the bands derived from S_+ are non-dispersive and the valence band width is under-estimated. Fig. 2a shows the calculated DOS for a planar N site with $\phi = 90°$. The valence band has four main features, peak A whose origin is the lone pair S_+ orbitals, and peaks B and C which are formed from combinations of the A_+ and B_+ bonding orbitals such that peak B contains a Si p-like combination and peak C a substantial Si s contribution. Finally there are the tightly bound nitrogen 2s states at 14 eV. The conduction band is above the gap of about 4.3 eV. Unlike the valence band, its structure is not nearly so well related to the parent σ^* states S_-, A_- and B_-. Instead, it is primarily determined by the interactions between adjacent hybrids on the same Si site, V_1. The lower part of the conduction band is largely Si s-like with some N 2s character, as would be expected from its parentage of S_- orbitals. The conduction band minimum has a relatively low effective mass for a solid with moderately polar bonding and in this respect is similar to SiO_2. The higher lying conduction band peaks are Si p-like. Fig. 2(b) shows the DOS for a pyramidal N site with $\theta=107°$ ($\phi = 68°$). (The use of localized tight-binding basis orbitals allows the values of V_{2a} etc. to be changed simulating this change in ϕ, whilst still retaining the connectivity of β-Si_3N_4.)

The lone-pair peak has now broadened towards peak B and both valence and conduction bands have contracted, leaving a wider optical gap of about 6 eV.

This is fair agreement between our calculations for β-Si_3N_4 with $\beta = 90°$ (fig. 2a) and experiment. The upper valence is found by XPS to consist of one large peak with three features[3], which line up well with the three peaks A, B and C in the calculations. The large difference between the XPS matrix elements[8] for 2s, 3s and 2p electrons is responsible for the large differences between the observed peak size of A-D and their calculated values.

The minimum gap is 5.3 eV optically[9] and 5.1 eV from internal photoemission[10], compared to our 4.3 eV. The peak in the optical ε_2 spectrum at 9 eV corresponds to the excitation of N lone pair electrons to Si s states. Much of the disagreement in details may be related to the nature of the CVD phase on which all these measurements were taken, and this will now be discussed.

CVD silicon nitride is an amorphous phase with a similar coordination to its crystalline phases[11]. Its electronic structure should therefore be related to that calculated for the crystalline phases having the same disposition of bonding and lone-pair states, but with some features of the DOS smoothed and with possible band tailing into the gap - for instance the two peaks of region C, having local s symmetry about the Si site, will be smoothed into one peak for the topologically disordered phase, as in a-SiO$_2$[5] and a-Si. CVD silicon nitride also contains appreciable impurities, notably chemically bonded hydrogen. 7 % of N-H sites and 0.5 % of Si-H sites were found for a 1000:1 NH$_3$ to SiH$_3$ flow ratio and 750°C deposition temperature[12]. These impurities may alter the valence band DOS both by introducing new bonding states and by reducing the average connectivity of the N sites thereby allowing the pyramidal angle at the N site to reduce[13]. They may also control the conduction properties of CVD nitride. The dominance of hole over electron conduction in silicon nitride contrasts with silicon dioxide in which the electrons have a remarkably high mobility[14,15] for a disordered and rather ionic material, but the holes form self-trapped polarons[15]. The holes in silicon dioxide have a high effective mass which aid polaron formation because they occur in an oxygen p$_\pi$ lone pair band. It was then a natural suggestion that the higher hole mobility in silicon nitride arose because it is not a lone pair semiconductor - the p^4 valence configuration of oxides gives a p$_\pi$ valence maximum whereas the p^3 electrons of triply bonded group V elements are all usually associated with σ bonding orbitals and no p$_\pi$ lone pair band is formed. As bands from σ bonds have a lower effective mass this could have accounted for the differences in conduction properties. The photoemission spectra[3] appeared to support this conclusion finding separated p$_\pi$ and σ peaks in SiO$_2$ but broadened peaks in silicon nitride. However, we showed that bot β-Si$_3$N$_4$ and CVD nitride possess p$_\pi$ valence maxima by virtue of the local atomic geometry, not their p^3 anion configurations.

A different interpretation is now required. CVD nitride is more typical of other amorphous semiconductors of groups IV and V in having a high average coordination number and relatively high impurity content. Group VI semiconductors are less sensitive to impurities because of their p_π valence maxima[16] and their amphoteric defect states[17], but the geometry-induced p_π valence maxima of silicon nitride is much more sensitive to impurities and other perturbations than in the lower coordinated p^4 semiconductors. SiO_2 may be anomalous in that there is little evidence of a mobility edge in its conduction band[15]. This remarkable property, which implies a nett cancellation of disorder potentials is unexplained within current band structure calculations. Electron transport is band-like, unlike in most other amorphous semiconductors in which it is within states highly influenced by disorder or defects. The electron mobility in $a-SiO_2$ is determined by the low effective mass and by scattering at dipole centres due to charged gap states[18]. Hydrogen in CVD nitride passivates most of the bonding imperfections but, unlike in a-Si say, it is calculated to leave Si-H σ states in the lower part of the pseudogap[13]. Hopping in these localized states can now give rise to hole conduction if the Fermi level is closer to these states than the conduction band. It is also probable that the greater coordination and the higher impurity content in silicon nitride does not allow the disorder potential at the conduction band to be minimised as in SiO_2. Consequently a mobility edge is now present lowering the electron mobility.

Acknowledgement - The work was carried out at the Central Electricity Research Laboratories and is published by permission of the Central Electricity Generating Board.

REFERENCES

1. J.J. Chang, Trans.IEEE ED-24, 511 (1977).

2. P.C. Arnett, Z.A. Weinberg, Trans. IEEE ED-25, 1014 (1978).

3. Z.A. Weinberg, R.A. Pollak, Appl. Phys. Letts. 27, 254 (1975).

4. R.W.G. Wyckoff, Crystal Structures, Vol.2, p.159, Wiley, N.Y., (1964).
   R. Marchand, Y. Laurent, J. Lang, Act. Cryst. B25, 2157 (1969).

5. S.T. Pantelides, W.A. Harrison, Phys.Rev. B 13, 2667 (1976).

6. J. Robertson, J. Phys. C 12, 4753 (1979).

7. R.B. Laughlin, J.D. Joannopoulos, D.J. Chadi, Phys.Rev. B 20, 5228 (1979).

8. F.R. McFeely, S.P. Kowalcyzk, L. Ley, R.G. Cavell, R.A. Pollak, D.A. Shirley, Phys. Rev. B 9, 5268 (1974).

9. H.R. Philipp, J. Electrochem. Soc. 120, 296 (1973).

10. D.J. DiMaria, P.C. Arnett, Appl. Phys. Letts. 26, 711 (1975).

11. M.V. Coleman, D.J.D. Thomas, Phys.Stat.Solidi. 25, 241 (1968).
    J. Stohr, L. Johansson, I. Lindau, P. Pianetta, Phys.Rev. B 20,
    644 (1979).
    F. Galeener concludes that the nitrogen site in a-Si_3N_4 is
    planar from scattering experiments; quoted by
    G. Lucovsky, J. Non Cryst. Solids 35, 825 (1980).

12. H.J. Stein, J. Electronic Mats., 5, 161 (1976).
    H.J. Stein, S.T. Picraux, P.H. Holloway, Trans. IEEE ED-25,
    1008 (1978).

13. J. Robertson, Phil. Mag., to be published.
    This paper also discusses the defect model of C.T. Kirk,
    J. Appl. Phys. 50, 4190 (1979).

14. R.C. Hughes, Phys. Rev. Letts. 30, 1333 (1973).

15. N.F. Mott, Adv. Phys. 26, 363 (1977).

16. M. Kastner, Phys. Rev. Letts. 28, 355 (1972).

17. N.F. Mott, E.A. Davis, R.A. Sreet, Phil.Mag. 32, 961 (1975).

18. G. Lucovsky, Phil.Mag. B 39, 531 (1979).

INVESTIGATIONS OF HYDROGENATED AMORPHOUS SILICON BY HIGH RESOLUTION NMR OF ^{29}Si

B. Lamotte, A. Rousseau
Centre d'Etudes Nucléaires de Grenoble
DRF-G/Section de Résonance Magnétique
85 X - 38041 Grenoble Cedex (France)

A. Chenevas-Paule
Centre d'Etudes Nucléaires de Grenoble
LETI/Nouveaux Composants Electroniques
85 X - 38041 Grenoble Cedex (France)

The new techniques of high resolution NMR in solids on "rare" spins via cross polarization and proton spin decoupling introduced by A. Pines, M.G. Gibby and J. Waugh, combined with the rapid rotation of the sample at the magic angle have been applied to amorphous hydrogenated silicon. These experiments can give structural informations on these materials and on their amorphous nature. First results will be presented.

1. INTRODUCTION

It is recognized by workers studying amorphous hydrogenated silicon that, while this compound raises great hopes for its potential applications in photovoltaic solar cells, it is a complex material which is not very well defined and for which the structure is not well known. Therefore, the application of any new spectroscopic method giving the possibility of obtaining additional information on it has much interest.

The new techniques of high resolution NMR in solids on "rare" spins via cross polarization and proton spin decoupling, introduced by A. Pines, M.G. Gibby and J. Waugh [1] (The P.E.N.I.S. method), combined with rapid rotation of the sample at the magic angle have proved very successful in the recent years, mainly for ^{13}C in molecular solids and polymers. This method also has potential for other nuclei such as ^{15}N, ^{29}Si, ^{31}P and ^{77}Se. The ^{29}Si isotope is in particular a good candidate because of its spin 1/2, its natural abundance of 4,7 % and its magnetic moment only 20 % less than that of ^{13}C.

Our intention is to use these new methods to obtain informations about the electronic environment of the silicon atoms and, if possible, to characterize the different kinds of silicon and their various bonds with hydrogen atoms by *measurements of the chemical shifts of the ^{29}Si.*

2. THE PULSE SEQUENCE : THE PENIS EXPERIMENT

Figure 1 shows the pulse sequence used for these experiments. It is a double resonance method which enables the detection of an enhanced ^{29}Si signal by cross-polarization with the protons under conditions of high resolution, its free precession being taken under strong proton decoupling.

The sequence preceeds as follows : by a $\Pi/2$ pulse along the x axis (H_0 being along the z axis), immediately followed by a x → y phase shift of this signal at constant level, the protons are put into their rotating frame and strongly polarized. During time ②, the silicon 29 nuclei are simultaneously irradiated at their resonance frequency with a pulse of several milli-seconds length. If the amplitude H_1 of this pulse is such that :

$$H_{1(Si)} = H_{1(protons)} \times \frac{\gamma(protons)}{\gamma Si} \quad \text{(The Hartmann and Hahn condition)}$$

- where the γ are the magnetic moments of the two nuclei - a resonant transfer of the magnetization from the protons to the ^{29}Si will occur. This gives a free precession signal of the ^{29}Si which is greater by a factor of five (the γ ratio of the two nuclei), than the one which would have been obtained after a classical $\pi/2$ pulse on the ^{29}Si nuclei.

During time ③, the free precession of the ^{29}Si nuclei is registered under strong irradiation of the protons, thus cancelling the dipolar interactions between protons and ^{29}Si nuclei and giving the possibility of a high resolution spectrum where the chemical shifts are apparent.

This sequence of pulses is then repeated N times with a time interval of the order of the proton T_1, the free precessions

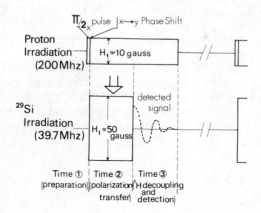

Fig.1 : The P.E.N.I.S. pulse sequence.

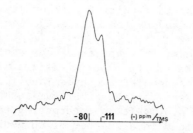

Fig.2 : High resolution NMR of ^{29}Si of a sample of amorphous hydrogenated silicon - Number of scans : 100; Time repetition delay : 10 sec.

are summed and Fourier transformed to give the spectrum. The
third advantage of this technique is the great saving of time and
thus the increase of sensitivity due to the fact that the repetition
time between two successive sequences of pulses is defined here
with reference to the proton T_1 instead to the ^{29}Si T_1, the first
being of order seconds whereas the second is about hours.

In addition, in order to simplify the spectrum and
smear out the anisotropic components of the chemical shift, the
sample is rotated at high speed (3000 – 4000 Hz) during the entire
experiment at the "magic angle" orientation with respect to the
magnetic field, which produces a spectrum containing only the
isotropic chemical shifts.

3. FIRST RESULTS

These experiments have been made on three different
samples of hydrogenated amorphous silicon prepared by reactive
sputtering. They are characterized by different proton NMR spectra
having linewidths (at half height) between 4.8 and 9.5 KHz.

The samples weights are around 350 mg. They are prepared
by deposition on thin aluminium sheets which are dissolved afterwards by attack with a diluted hydrochloric acid.

The experiments are done on a Bruker CXP 200 spectrometer working at 200 MHz for the proton and 39.7 MHz for the ^{29}Si.

In figure 2 we show a typical high resolution NMR spectrum
for ^{29}Si. Two main features appear on the spectrum of this sample.
First, in spite of the high resolution method used here, the lines
are broad in contrast to results for other crystalline and even
amorphous compounds used by the same method on ^{13}C.

Nevertheless, this spectrum does exhibit a certain
resolution: two lines can be clearly distinguished at −80 and −111
p.p.m. from the reference Tetramethylsilane (T.M.S.). The line at
−80 p.p.m. is also seen for the other samples but with an additional
shoulder around −28 p.p.m. It is interesting to point out that such
spectra can be obtained in a quarter of an hour, instead of several
days for a roughly equivalent spectrum detected by the classical way
after a $\Pi/2$ pulse on the ^{29}Si.

The broadness of these spectra is clearly due to the
amorphous state and it comes from a continuous distribution of
the chemical shifts which, in this case, is the intrinsic factor
limiting the resolution of the method.

A reliable attribution of the different peak observed to definite silicon sites, either silicon bonded to hydrogens or silicon close to silicon bonded to hydrogens, requires further work on the basis of an extended set of well defined sample. However some preliminary indications can be obtained by comparison with related situations :

1) we have measured the chemical shift of Si in *crystalline silicon* by a simple $\Pi/2$ experiment in a powdered sample (gamma irradiated to lower its T_1) and found -82 p.p.m. from TMS.

2) interesting comparisons can be made with chemical shifts δ measured in solutions in molecules having some relation with our problem :

$$
\begin{array}{l}
H_3 \quad Si \\
H_3 \quad Si \longrightarrow Si - H \qquad \delta = -96 \text{ p.p.m./T.M.S.} \\
H_3 \quad Si \qquad\qquad\qquad\qquad\qquad (2)
\end{array}
$$

$$
\begin{array}{l}
H_3 \quad Si \qquad\qquad H \\
\qquad\qquad\qquad Si \qquad\qquad \delta = -116 \text{ p.p.m./T.M.S.} \\
H_3 \quad Si \qquad\qquad H \qquad\qquad (2)
\end{array}
$$

Further investigations of this kind are continuing as well as studies designed to appreciate whether the linewidths we obtain can give information in relation with calculations on models about the amount of short-range disorder present in this compound.

Acknowledgement - We wish to thank Mr. Cuchet for the preparation of the samples. One of us (B.L.) has special thanks to adress to Prs Erwin Hahn and Alex Pines who introduced him to the wizardry of High Resolution Solid State N.M.R. during a stay at Berkeley in 1976-1977.

REFERENCES

[1] A. Pines, M.G. Gibby and J. Waugh, J. Chem. Phys., 59, 569 (1973).

[2] R. Lower, M. Vongehr and H.C. Harsmann, Chem. Z., 99, 33 (1957).

REMANENT MAGNETIZATION OF AN INSULATING SPIN GLASS: $Eu_{0.4}Sr_{0.6}S$

J. Rajchenbach and J. Ferré
Laboratoire de Physique des Solides associé au CNRS
Université Paris-Sud, Bâtiment 510
91405 Orsay (France)

H. Maletta
Institut für Festkörperforschung Kernforschungsanlage
5170 Julich (West Germany)

We report on the time $(10^{-6} < t < 10^3$ s$)$, temperature $(1.3 < T < 2K)$ and magnetic field $(1 < H < 1\ 000$ Oe$)$ dependence of the thermo-remanent magnetization (TRM) of the insulating spin-glass $Eu_{0.4}Sr_{0.6}S$ $(T_f = 1.55$ K$)$ deduced from Faraday rotation measurements. The TRM is fitted over several decades of time with a power law $M_R = At^{-a}$ which, however, fails near from T_f and it varies linearly with T for low fields $(H < 8$ Oe$)$. We have demonstrated that the unusual field dependence of the TRM always found in spin-glass systems, i.e. $M_R(H)$ has a maximum, is only due to the increase with H of the demagnetization rate which compensates, at high fields, the larger initial magnetization. Our data agree qualitatively with recent Monte Carlo simulation studies and cannot be analysed on the basis of independent two level systems. This suggests a cooperative behavior below T_f.

The dynamic properties of spin glass (SG) are far from well understood and need to be characterized in detail to clarify the nature of the spin glass state. Previous studies of the magnetic aftereffect in both metallic /1-5/ and insulating SG /6-8/ are generally too limited to provide a definitive test of the proposed models /2,9/.

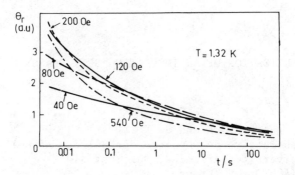

<u>Fig.1</u> : Time dependence of the TRM for different field values.

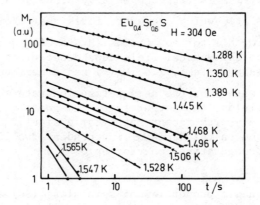

<u>Fig.2</u> : Log-Log plot of the saturated TRM versus t close to
            $T_f = 1.55°K$.

In order to go further in this area we have extensively studied the remanent magnetization M_r of the insulating SG : $Eu_xSr_{1-x}S$ by means of the Faraday rotation. Other aspects of this system have been extensively investigated by Maletta et al. /6,10/. The Faraday effect provides great sensitivity to the magnetization in this system : a small change induced by a field of 1 mOe can be detected. So we were able to perform experiments near the freezing temperature T_f, where M_r tends to vanish, even by using a weak initially applied field (H < 10 Oe). Another advantage is the rapid response of the detection after switching off the magnetic field. We have thus considerably extended the M_r time scale $(10^{-6} < t < 10^3$ s).

The principle of the used Faraday rotation set-up has been described elsewhere /11/ and it permits detection of rotations as small as 10^{-3} degree. The chosen wavelength of 8 000 Å ($\Delta\lambda$ = 50 Å) is located in a high transparent region of the crystal. In particular, we have determined T_f = 1.55 K from its static magnetization measured by Faraday rotation in a weak field of 1 Oe. In the reported experiments the magnetic field and the light beam were directed perpendicularly to the large (100) cleaved faces of our crystal plate ($4 \times 4 \times .83$ mm^3).

The investigated single crystal of $Eu_{0.4}Sr_{0.6}S$ showed all the characteristic properties of a SG /10/. After cooling the sample through T_f in an applied field H the thermoremanence (TRM) is obtained after switching off H at t = 0. In spite of the small H value and the weak remanence (M_r (t = 1 s)/M_o (t = 0) < 10^{-2}) measured between 1.3 and 1.55 K the TRM is not significantly affected by the earth's field : the results are non sensitive to its direction. To fully investigate the dynamics of such systems one has to measure the TRM dependence on time, temperature, and field values. Our data will be presented and discussed in the light of previously proposed models /2,9/.

Experimental results

After switching off a weak magnetic field (H $\simeq$ 4 Oe) we have found that the magnetization decreases very rapidly within less than 1 μs (M_r(t = 1 μs)/M_o (t = 0) < 0.1) for temperatures between 1.3 and 2.1 K, i.e., values which lie on both sides of T_f. This upper limit for the relaxation time after switching is due to the decay time of the current in our small superconducting coil. Afterwards the TRM relaxes very slowly to zero as reported earlier /1,5,6/ if T < T_f. Near above T_f, the TRM decreases rapidly to zero in a time which is consistent with the frequency dependence of T_f deduced from our dynamic susceptibility measurements.

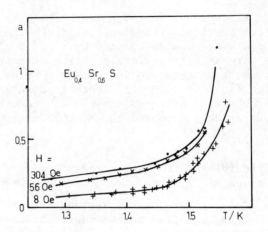

<u>Fig.3</u> : Temperature dependence of the <u>a</u> exponent for 3 fixed fields

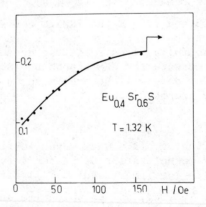

<u>Fig.4</u> : Field dependence of the <u>a</u> exponent (the arrow gives the
         saturation value).

Our analysis focuses on the time dependence of the TRM in the interval $5 \times 10^{-3} < t < 10^2$ s. As found previously for other SG /8,12/ M_r is no longer proportional to Log t over several decades of time if the temperature approaches T_f or if the applied field is strong enough (fig.1).

The power law $M = At^{-a}$ initially proposed by Binder /13/ to interpret his simulation data on a 3d-Ising SG seems to better appropriate our results over 2 or 3 decades of time (fig.2), but this law fails for more than 3 decades in the considered temperature range $(T/T_f > .84)$. A Log M_r - Log t plot shows a significant negative concavity. Note that the Monte Carlo simulations /13,14/ leading to this behavior are not precise enough to check this law over more than 2 decades of time. Consequently, even a power law cannot be considered as exact to describe the relaxation of the magnetization.

Since it appears to be justified to use the power law $M = At^{-a}$ between 1 and 30 s (fig.2) we have demonstrated, within this approximation, that a depends both on the temperature and on the initially applied field H. As shown on fig.3 the exponent a varies linearly with temperature up to $T/Tf = 0.94$ for all values of H in the range 80Oe $< H <$ 400 Oe. This result is consistent with recent Monte Carlo simulations for a 2d-Ising SG /14/ except for the experimentally found divergence of a near T_f (fig.3). The observed exponent a increases also with the applied field (fig.4) in a way that is also qualitatively in good agreement with the same simulation results. Note the similar field values found for M_r and a to reach the saturation /6/ (fig.5).

The dependence of M_r with temperature for fixed values of t and H is shown on fig.6. It looks very similar to that found by Tholence /1/ for CuMn alloys. Under weak fields (H < 10 Oe) M_r decreases linearly with temperature up to 1.53 K $(T/T_f = .985)$ as for CuMn /1/ and AuFe alloys /15/ but for higher fields it decreases more rapidly /1,3/. The change of the $M_r(T)$ behavior versus H is not revealed in the simulation results on Ising SG /13,14/ in spite of the rather large field values considered in such calculations. Note that just below T_f the thermoremanence in high fields is weaker than for small fields. This comes from the large increase of the demagnetization rate with H as reported above.

For all systems /1,6/, and in particular for $Eu_{0.4}Sr_{0.6}S$, one obtains an unusual field dependence of the TRM for given t and T values. $M_r(H)$ first increases with H up to a maximum value M_{rM} corresponding to a field H_M, and for stronger fields it decreases to reach finally a saturation at M_{rS} (fig.5). We have demonstrated here that this strange behavior is only due to a time effect. This results from an increase, with H, of the demagnetization rate $(dM/dt)_t$ which can compensate the larger initial t = 0 remanent magnetization value M_{r0}. Thus we are able to explain the increase of the ratio M_{rM}/M_{rS} = 1.57 + (0.091 + 0.009)Log t with the time of measurement $(5 \times 10^{-3} < t < 10^2 s)$ and the decrease of H_M are revealed on figure 5.

The most convincing information is brought forward on fig.1, i.e., the crossing of the $M_r(t)$ curves for different H values at experimental time t. As example, the thermoremanence for H = 540 Oe and measured at t = 1 s is smaller than that found for H = 40 Oe in spite of its larger initial value M_{r0}.

Discussion

Our short time measurements of the relaxation of the magnetization cannot yield a definitive answer on the existence of two processes as claimed earlier /2/ in spite of our possibility to extend experiments down to 1 µs.

Two different physical models, developed by Richter /16/, Street and Wooley /17/ and Neel /18/ have been used to interpret the time and temperature dependence of the slow part of TRM in SG. These theories apply similarly to non interacting single domain particles with uniaxial anisotropy as for independent multi-domain particles or domain walls. This comes from similar hypothesis for the TRM calculations.

α) The relaxation is no longer exponential because it is considered as a superposition of exponentials.
β) Individual contribution to the relaxation for a given activation energy E are described with Arrhenius laws $\tau = \tau_o \exp E/kT$.
γ) The distribution of energies is given by
$f(E)dE = 1/(E_2 - E_1)$ dE if $E_1 < E < E_2$
$\qquad\qquad = 0 \qquad\qquad$ if $E < E_1, E > E_2$
δ) The individual variations of the magnetization are not correlated to E.

So, the general equation for the TRM can be written as :

$$M_r = \frac{\overline{M}kT}{(E_2 - E_1)} \int_{\tau_1}^{\tau_2} \frac{e^{-t/\tau}}{\tau} d\tau \qquad\qquad (1)$$

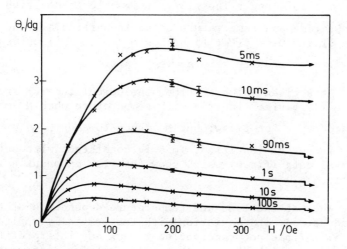

Fig.5 : Field dependence of the TRM at 1.32°K for several t values.

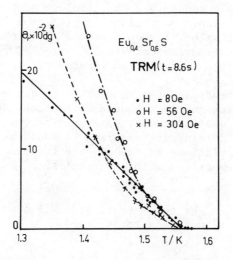

Fig.6 : Temperature dependence of the weak field (8 Oe), maximum
(56 Oe) and near saturated (304 Oe) thermoremanence.

This expression reduces in the blocking approximation $(\tau_1 < t < \tau_2)$ to a Log t law /18/. It is able to describe quite closely our time and temperature data, in particular, the observed curvature of our M_r = f(Log t) or Log M_r = f(Log t) curves (fig.1,2). When t > τ_2, M_r tends to zero as $(e^{-t/\tau_2})/t$.

At a given value of time, the calculated TRM from the expression (1) varies linearly with temperature when M_r may be approximated by a Log t law; but M_r tends towards $M_r = e^{-\alpha T}$, the power law is better justified. This is consistent with our experimental data (fig.6) which show that the exponent a increases with H (fig.4).

Whereas the time and temperature dependence of the TRM may be understood, its variation with the initially applied field H (fig.1,5) /6/ has not received up to now any interpretation. The crossing of the M_r(Log t) curves (fig.1) for different H cannot be explained with a scheme of independent two levels systems. Instead the crossing suggests a collective behavior as for ferromagnets for which the magnetization curves are controlled by irreversibility. Even more refined models /19/, which take into account the self-demagnetizing field, are not able to explain this behavior. Why at the crossing point are two states defined by the same magnetization not identical, as shown by the difference in $(dM_r/d\text{Log } t)_t$ for two H values (fig.1)? One possible explanation is to suppose that the topology of the spin system, for a given macroscopic magnetization, is strongly dependent of the initial field value. It would be useful to check this possibility by an experimental visualisation of the spin spatial distribution or from a simulation treatment.

Experiments at even shorter times are needed to analyse the processes which govern the demagnetization. Because our data verify, at least qualitatively, some predictions of Monte Carlo calculations on 2d-Ising systems /14/ it would be interesting to enlarge these simulations to the 3d case to permit a more quantitatively comparison with our results.

Acknowledgements - We thank Prof. S.J. Williamson for his critical reading of our manuscript.

REFERENCES

/1/ Tholence, J.L., Thesis, Grenoble (1973).

/2/ Tholence, J.L. and Tournier, R., J.Physique, 35, C4-229 (1974).

/3/ Tournier, R., Thesis, Grenoble (1965).

/4/ Lohneysen, H.V. and Tholence, J.L., Phys. Rev. B, 19, 5858 (1979).

/5/ Guy, C.N., J. Phys., F 8, 1309 (1978).

/6/ Maletta, H. and Felsch, W., Phys. Rev. B, 20, 1245 (1979).

/7/ Rechenberg, H.R., Bieman, L.H., Huang, F.S. and De Graaf, A.M., J. Appl. Phys. 49, 1638 (1978).

/8/ Rajchenbach, J., Ferré, J., Pommier, J., Knorr, K. and De Graaf, A.M., J. Magn. Mat., 15, 199 (1980).

/9/ Edwards, S. and Anderson, P.W., J. Phys., F 5, 965 (1975).

/10/ Maletta, H. and Felsch, W., Zeit. Phys., B 37, 55 (1980).

/11/ Badoz, J., Billardon, M., Canit, J.C. and Russel, M.F., J. Optics, 8, 373 (1977).

/12/ Prejean, J.J., J. Physique, 39, C6-907 (1978).

/13/ Binder, K., J. Physique, 39, C6-1527 (1978).

/14/ Kinzel, W., Phys. Rev. B, 19, 4595 (1979).

/15/ Guy, C.N., J. Phys., F 7, 1505 (1977).

/16/ Richter, G., Ann. Physik, 29, 605 (1937).

/17/ Street, R. and Wooley, J.C., Proc. Phys. Soc., A 62, 562 (1949).

/18/ Neel, L., Ann. Geophys., 5, 99 (1949).

/19/ Stacey, F.D., Advanc. Phys., 12, 45 (1963).

ELECTRICAL RESISTIVITIES OF $Fe_x Ni_{80-x} B_{18} Si_2$ ALLOYS

E. Babić and R. Krsnik
Institute of Physics of the University
Zagreb, Yugoslavia

H.H. Liebermann
General Electric Corporate Research and
Development Schenectady, N.Y. 12301, USA

ABSTRACT

Electrical resistivities (ρ) of several amorphous $Fe_x Ni_{80-x} B_{18} Si_2$ alloys have been measured from 1.5 – 450 K. The Curie (T_c) and crystallization temperatures (T_x) have also been determined from the thermogravimetric analysis and $d\rho/dT$ values. The magnetic properties of these alloys change from nonmagnetic ($x = 0$) to ferromagnetic ($x \geqslant 10$). In all the alloys a resistance minimum appears at low temperatures. Above the minimum resistivity varies as $T^{3/2}$. This variation agrees with the recent predictions for amorphous ferromagnets.

INTRODUCTION

There are two reasons for steadily increasing interest in metallic galsses. First, they are likely to have important technological applications and second the understanding of their physical properties is connected with some fundamental problems in solid state physics. A good example for that are the transition metal-metalloid alloys, some of which posses excelent solft magnetic properties which are highly desirable for the industrial applications. On the other hand these alloys exibit almost universally a small upturn (resistivity minimum) in the resistivity at the lowest temperatures which is still not properly understood. This little upturn with no practical importance clearly indicates the inadequacy of our knowledge of the electronic transport in metallic glasses.

In this paper we present the results of the systematic investigations of the electrical resistivities of $Fe_x Ni_{80-x} B_{18} Si_2$ alloys. In a way this is a continuation of our studies of the transport properties of Fe-Ni based metallic glasses. As before our aim is to contribute both to the understanding of the resistivity minimum and of the magnetic contribution to the resistivity of amorphous ferromagnets. This system was selected because it has some features similar to $Fe_x Ni_{80-x} B_{20}$ system[1] and in the same time the complete exchange of Fe with Ni is possible. The complete exchange of Fe with Ni leads to a loss of ferromagnetism as in $Fe_x Ni_{80-x} P_{14} B_6$[2] and $Fe_x Ni_{79-x} P_{13} B_8$[3] alloys.

EXPERIMENTAL

The samples were ribbons 1-2 mm wide and 20-30 μm thick. They were prepared by the special melt spining technique[4] in a vacuum from the master alloys of the predetermined concentrations. The ribbons were cut into pieces 3-5 cm long which were used in the actual resistivity measurements. The samples supplied with the potential and current thin platinum leads were mounted in a special cryostat suitable for the accurate resistivity measurements from 1.5 - 500 K. The details concerning the measurements technique and the temperature controll[5] were reported earlier and will not be repeated here. The relative accuracy of the resistivity measurements was few parts in 10^5 in a broad temperature interval (50 - 450 K) and 2.10^{-6} at the lowest temperatures. The temperature was measured by the calibrated Ge thermometer below 80 K and with the platinum resistance thermometer at higher temperatures.

In order to get some more direct information about the magnetic state and the stability of our samples the thermogravimetric analyses where performed on the samples taken from the same ribbons. These analyses have shown that all the samples have crystallization temperatures (T_x) in excess of 400°C (at the constant heating rate 20 degrees per minute) and they also gave the Curie temperatures (T_c) for the samples with $T_c > 300$ K.

RESULTS

The relative change in resistivity, $\Delta\rho/\rho_m$ (where $\Delta\rho = \rho(T) - \rho_m$ with ρ_m the resistivity value at the minimum) at the low temperatures (1.8 - 50 K) for five $Fe_xNi_{80-x}B_{18}Si_2$ alloys (with x = 0,20,40,60 and 80 respectively) is shown in figure 1. It can be seen that all these alloys exibit the resistance minima above 10 K and that the resistance varies in a logarithmic fashion below the minimum. The coefficients of these logarithmic terms $(\rho^{-1} d\rho/d \log T)$ vary smoothly with concentration exibiting a broad maximum centered probably around x = 50 as shown in the inset to figure 2. We note that a similar variation of these coefficients was also observed in $Fe_xNi_{80-x}B_{20}$ and $Fe_xNi_{80-x}P_{14}B_6$ alloys[7]. In the latter system however anomalously high coefficients were observed for x = 10 and 20 which were atributed to the mictomagnetic state of these alloys[7]. We also note that the resistance variations below the minimum in $Fe_xNi_{80-x}P_{14}B_6$ alloys with $10 \leqslant x \leqslant 30$ did not obey a unique logarithmic law[6].

Above the minimum the resistivity is dominated by the positive contribution. Our earlier results for $Fe_xNi_{80-x}B_{20}$ and $Fe_xNi_{80-x}P_{14}B_6$ alloys indicated that the dominant in the positive resistivity contribution of ferromagnetic alloys has a $T^{3/2}$ dependence. It was also found that the coefficient of this term decreases with increasing concentration.

In Fig.2 we plot the change in resistance above the minimum $(R(T) - R_{min})$ for two alloys ($Fe_{12}Ni_{78}B_{18}Si_2$ and $Fe_{40}Ni_{40}B_{18}Si_2$) vs $T^{3/2}$ in the temperature interval 20 - 100 K. The resistance variations of those two alloys represent also well those of other ferromagnetic alloys from the same system. It can be seen that in ferromagnetic $Fe_xNi_{80-x}B_{18}Si_2$ alloys the resistance varies above the minimum roughly as $T^{3/2}$ like in magnetic $Fe_xNi_{80-x}P_{14}B_6$ and $Fe_xNi_{80-x}B_{20}$ alloys. We note that the deviation from this dependence is usually observed at the lowest temperatures. In $Fe_{12}Ni_{68}B_{18}Si_2$ alloy the resistivity deviates from $T^{3/2}$ low also in the upper part of our temperature range. This seems to beconsistent with our earlier finding[7] that a $T^{3/2}$ variation usually does not extend above about $T_c/3$.

The resistances of our alloys from 50 up to 450 K are shown in Fig.3. We note that the resistance variation of $Ni_{80-}B_{18}Si_2$ alloy is different from those of other (ferromagnetic) alloys. While in magnetic alloys the resistivity curvature is positive up to T_c, $Ni_{80}B_{18}Si_2$ has a negative one in a most of our temperature range. From the selected alloys only $Fe_{18}Ni_{62}B_{18}Si_2$ has the Curie temperature within our temperature range (T_c = 370 K). In this alloy a clear change in slope of the resistivity is observed above T_c. This is again consistent with our earlier findings[8].

DISCUSSION

We start our discussion with the upturn in the resistivity at the lowest temperatures. This upturn which occurs almost universally in transition metal metalloid glasses is so poorly understood that even the physical interaction responsible for it is not agreed upon. Some of the models suggest the magnetic interaction to be responsible for the effect[9,10] while the others emphasize its structural origin[11,12]. The problem is that none of these theories can account properly for the experimental observations. While the structural models (which by the way have yet to be worked out satisfactorily) cannot account for the fact that no such (logarithmic) upturns are observed in the metallic glasses with no magnetic atoms present, the magnetic ones cannot explain a rather small magnetoresistances observed in these alloys.

As mentioned earlier the upturns in the low temperature resistivity of $Fe_xNi_{80-x}B_{18}Si_2$ alloys follow a $-\log T$ dependence down to the lowest temperature (Fig.1). In contrast to that in some $Fe_xNi_{80-x}P_{14}B_6$ alloys a change in slope was observed below about 5 K. This change of slope was ascribed to the additional contribution to low temperature resistivity upturn. Therefore we believe that the $-\log T$ terms in resistivity of $Fe_xNi_{80-x}B_{18}Si_2$ alloys are caused by the single mechanism. The concentration (x) dependence of the logarithmic slopes shown in the inset to Fig.2 may give us some indication of the interaction responsible for the upturns in the resistivity. We note that the concentration dependence of the Curie temperatures of these alloys (shown in the same inset) is remarkably similar to that of $\rho_m^{-1} \Delta\rho/\Delta \log T$. This correlation extends also to $Fe_xNi_{80-x}B_{20}$ alloys. Furthermore it is interesting to compare $\rho_m^{-1} \Delta\rho/\Delta \log T$ values for two limiting concentrations i.e. x = 0 and x = 80. A logarithmic slope of the alloy with x = 0 appears considerably lower than that for x = 80 as in $Fe_xNi_{80}P_{14}B_6$ alloys[7]. (We note that the quantitative

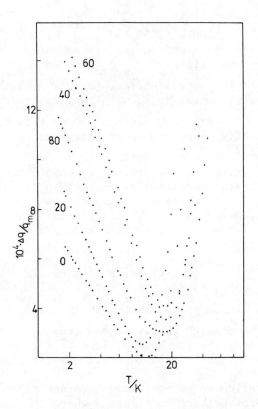

<u>Fig.1</u> : The relative changes in resistivity of $Fe_x Ni_{80-x} B_{18} Si_2$
alloys vs log T. The numbers denote Fe content (x). Note
that the curves are vertically displaced.

comparison of the logarithmic slopes in two limiting cases would
be useful only if the exact concentrations and purities of the
transition metal components are known). To conclude this part of
the discussion we note that the proper theory of the resistance
minimum should take into account the dependence of the upturn
on the particular transition metal (for a given metalloid content)
and its possible correlation with the Curie temperature (in the
intermediate concentration range).

A low temperature $T^{3/2}$ contribution to the resistivity
of ferromagnetic $Fe_x Ni_{80-x} B_{18} Si_2$ alloys (Fig.2) is rather
interesting. Since the same temperature dependence was also
observed in ferromagnetic $Fe_x Ni_{80-x} B_{20}$ and $Fe_x Ni_{80-x} P_{14} B_6$ alloys[7]
it may be typical for amorphous ferromagnets which contain
metalloids. It is interesting to note that at least two
calculations[13,14] predict a dominant $T^{3/2}$ contribution in the low
temperature resistivity of amorphous ferromagnets. In both
calculations a $T^{3/2}$ term originates from incoherent electron-
magnon scattering but its coefficient seem to differ for a factor
of about ten. For the quantitative comparison of the experimental
results with either of these theories one would need some
additional informations about the alloys (spin wave stiffness
constants etc.) which are not available for $Fe_x Ni_{80-x} B_{18} Si_2$ alloys
at present. We note however Richter et al.[14] give an estimate for
the coefficient of a $T^{3/2}$ term which compares favourably with those
of $Fe_x Ni_{80-x} P_{14} B_6$ and $Fe_x Ni_{80-x} B_{20}$ alloys[7]. Also the concentration
(x) dependences of the coefficients[7] in the above mentioned alloys
are not inconsistent with those deduced from both theories[13,14].
Namely a large increase in the coefficients for lower x values[7]
is probably due to decrease in the spin-wave stiffness constant.

The influence of the magnetic ordering on the resistivity
variation is seen the best in figure 3 where the resistances of
several alloys are shown in a broad temperature range. The change
in slope of the resistivity of $Fe_{18} Ni_{62} B_{18} Si_2$ alloy around T_c shows
clearly the importance of the spin disorder scattering in the
amorphous ferromagnets[8]. However the resistivity variation of
nonmagnetic $Ni_{80} B_{18} Si_2$ alloy is rather unusual. While the
resistivity of nonmagnetic $Ni_{80} P_{14} B_6$ alloy varied linearly with
temperature from about 60 to 400 K, $Ni_{80} B_{18} Si_2$ alloy has a negative
resistivity curvature in the same temperature range. At the moment
we cannot decide whether this is due to particular electronic
structure (like in pure paladium) or is the consequence of the
spin-glass character of this alloy. Magnetic measurements may help
to understand this unusual behaviour.

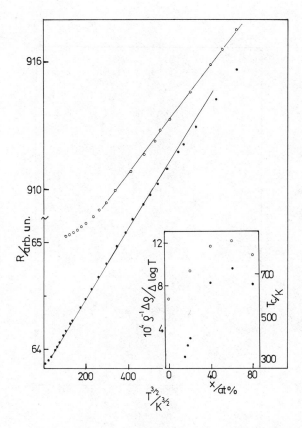

<u>Fig.2</u> : The resistance of $Fe_{12}Ni_{68}B_{18}Si_2$ (·) and $Fe_{40}Ni_{40}B_{18}Si_2$ (°)
alloy vs $T^{3/2}$. In the inset the dependence of the Curie
temperatures (·) and logarithmic slopes of resistivity at
the lowest temperatures (°) of $Fe_xNi_{80-x}B_{18}Si_2$ alloys on
Fe content (x).

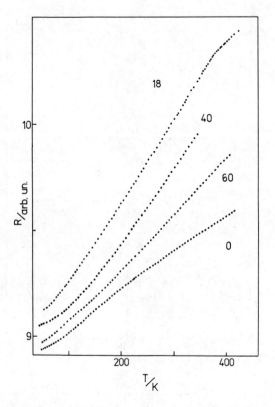

Fig.3 : The resistances of four $Fe_x Ni_{80-x} B_{18} Si_2$ alloys vs
temperature from 50 - 450 K. The numbers denote Fe content
(x).

To summarize the above, despite of the structural disorder strong effects of the magnetic ordering have been observed in resistivity of $Fe_xNi_{80-x}B_{18}Si_2$ alloy. While the origin of the resistance minimum in these alloys remains still unclear a low temperature resistivity variation may be associated with the incoherent electron-magnon scattering which is particularly important in disordered alloys.

REFERENCES

1. J.J. Becker, F.E. Luborsky and J.L. Walter, TEEE Trans. Magn. 13, 988 (1977).

2. C.L. Chien, D.P. Musser, F.E. Luborsky, J.J. Becker and J.L. Walter, Solid State Commun. 24, 231 (1977).

3. J. Durand, Amorphous Magnetism II (Edited by P.A. Levy and R. Hasegawa), p.305, Plenum Press, New York (1977).

4. H.H. Liebermann and C.D. Graham, IEEE Trans. Mag. 12, 921 (1976).

5. E. Babić, R. Krsnik and M. Očko, J. Phys. F : Metal Phys. 6, 73 (1976).

6. E. Babić, Ž. Marohnić and J. Ivkov, Solid State Commun. 27, 441 (1977).

7. E. Babić, Ž. Marohnić, M. Očko, A. Hamzić and B. Pivac, J.M.M.M., in press (1980).

8. Ž. Marohnić, K. Šaub, E. Babić and J. Ivkov, Solid State Commun. 30, 651 (1979).

9. R. Hasegawa, Phys. Letters, 36A, 175 (1971).

10. M.A. Continentino and N.Y. Rivier, J. Phys. F : Metal Phys. 8, 1187 (1978).

11. R.W. Cochrane, R. Harris, J.O. Ström-Olson and M.J. Zuckermann, Phys. Rev. Letters 35, 676 (1975).

12. F.J. Okhawa and K. Yoshida, J. Phys. Soc. Japan, 43, 1545 (1977).

13. G. Bergmann and P. Marquardt, Phys. Review B17, 1355 (1978).

14. R. Richter, M. Wolf and F. Goedsche, Phys. Status Solidi (b) 95, 473 (1979).

THERMOELECTRIC POWER OF AMORPHOUS $Fe_xNi_{80-x}B_{20}$ ALLOYS

B. Pivac and E. Babić*

Institute "Rudjer Bošković", Zagreb, Yugoslavia

*Institute of Physics of the University, Zagreb, Yugoslavia

The thermoelectric power of four ferromagnetic $Fe_xNi_{80-x}B_{20}$ alloys (where x varied in steps of 20) and of a nonmagnetic $Pd_{86}Si_{14}$ alloys has been measured in the temperature interval 50–470 K. The thermoelectric powers (S) of $Fe_xNi_{80-x}B_{20}$ alloys are negative and their temperature variation is similar to that in crystalline Fe-Ni alloys. However, their S values are for a factor of about ten smaller then those of crystalline alloys. In $Fe_{20}Ni_{60}B_{20}$ alloy the change in S above the Curie temperature (T_c = 426 K) is observed. An interpretation of our results in terms of strong ferromagnetism is offered.

1. INTRODUCTION

Metallic glasses form a new class of material which is likely to have significant industrial applications. Particularly promising are some transition metal-metalloid alloys which possess excellent soft magnetic properties. However, the effects of metalloid atoms and structural disorder on the electronic structure of these alloys are not well understood. Although it is well known that the thermoelectric power can give valuable information about the electronic structure there are only a few measurements of the thermoelectric power of metallic glasses. Moreover, most of these measurements were performed on nonmagnetic amorphous alloys. In this work we present to our knowledge the first systematic investigation of the thermoelectric power of ferromagnetic $Fe_xNi_{80-x}B_{20}$ alloys in a wide temperature interval.

273

This system was selected because the small size of the boron atom
seems to exclude the boron atoms as the nearest neighbours and
therefore the chemical short range order may be expected to be
similar to that in crystalline Fe-Ni alloys. In addition to the
thermoelectric of a nonmagnetic $Pd_{86}Si_{14}$ alloy has been also
measured.

2. EXPERIMENTAL

The alloys were prepared by the melt spinning technique
in a form of ribbons about 1 mm wide and 30 μm thick. The ribbons
were cut into pieces about two centimeters long which were measured
by the differential method as described earlier[1]. The only
difference was that in present measurements thermocouple (Au 0.03
at % Fe vs chrome[1]) and reference material (silver normal, Ag-0.37
at % Au) wires were directly spotwelded to the samples. No
difference was observed between the measurements obtained by
using either previous or a new method of attaching the wires. We
note that the advantage of using silver normal as a reference
arises from its rather small thermoelectric power which varies
linearly in a rather broad temperature interval (120-500 K). The
relative error of our measurements was a few percents and the
calibration of the apparatus was performed by measuring the
spectroscopically pure (annealed) lead sample of geometry similar
to that of our samples and the same temperature range. According
to this calibration the possible systematic errors in our absolute
thermoelectric power values (arising from errors in calibration of
the thermocouples) are estimated at about ± 5 %.

3. RESULTS AND ANALYSIS

Our experimental results for four $Fe_xNi_{80-x}B_{20}$ alloys
(where x varied in steps of twenty) and for nonmagnetic $Pd_{84}Si_{14}$
alloy are shown in figure 1 as plots of S against T. From figure 1
it can be seen that the thermoelectric power of $Pd_{86}Si_{14}$ alloy is
very different from those of $Fe_xNi_{80-x}Be_{20}$ alloys. In contrast to
those of $Fe_xNi_{80-x}20$ alloys the thermoelectric power of $Pd_{86}Si_{14}$
is positive and for a factor of more than ten smaller. A positive
thermoelectric power was also observed in some other nonmagnetic
metallic glasses[2]. We note however that in these alloys also linear
variation of the thermoelectric power with temperature was reported
and interpreted in terms of the Ziman's theory of transport
properties of liquid metals. It was also suggested[2] that a linear

and positive thermoelectric power is feature specific to metallic glasses. However the thermoelectric power variation in $Pd_{86} Si_{14}$ appears to be somewhat more complex. In that alloy a strictly nonlinear variation with a broad maximum centered around the room temperature is observed. Also the thermoelectric power tends to become negative both at high and low temperatures. These results indicate that the thermoelectric power of nonmagnetic metallic glasses may be neither linear nor positive and perhaps the only feature specific to these alloys is a relatively small thermoelectric power.

The thermoelectric powers of all $Fe_x Ni_{80-x} B_{20}$ alloys are negative throughout the explored temperature interval and for a factor of up to twenty bigger than that for $Pd_{86} Si_{14}$ alloy.

According to standard expression for the thermoelectric power :

$$S = \frac{\pi^2 k^2 T}{3e} \left(\frac{\partial \ln \sigma(E)}{\partial E} \right)_{E=E_F} \tag{1}$$

(where $\sigma(E)$ is the electrical conductivity, e is electronic charge, k is the Boltzmann constant and E_F is the Fermi energy) stronger energy dependence of the scattering would correspond to higher thermoelectric power values. Indeed in our Fe-Ni based ferromagnetic alloys one would expect a rather strong energy dependence of the conductivity. In order to proceed the analysis of our data in terms of eq.(1) we ought to make a more definite assumption about the dominant scattering mechanism. Following Mott[3] we assume that the s-d scattering dominates with the relaxation time inversely proportional to the d-component of the density of states. (This assumption may be supported by rather succesful explanation[4] of thermoelectric powers of liquid Fe, Ni and Co in terms of the same model). In that case the equation (1) can be written as :

$$S_d = - \frac{\pi^2 k^2 T}{3e} \left(\frac{2}{2E_F} - \frac{1}{N_d(E)} \frac{dN_d(E)}{dE} \right)_{E=E_F} \tag{2}$$

where S_d is the diffusion thermoelectric power and N_d is the density of d states.

Now the problem is that neither do our measurements extend to very low temperatures (where the diffusion thermoelectric power can be estimated with a less ambiguity) nor do we have any knowledge about the band structure of these alloys. Therefore we are limited to a qualitative discussion of our results. We start our discussion with the values of thermoelectric power at 60 K,S (60). Namely as all our alloys are ferromagnets with the Curie temperatures (T_c) well above room temperatures[5] we expect that any effects connected with the temperature dependence of the

magnetism (and thus of the density of states at the Fermi level)
would be less important. Also as we are dealing with concentrated
amorphous alloys (short electronic mean free path) any phonon or
magnon drag effects can be expected to be negligible. In Figs.1
and 2 it can be seen that $S(60)$ values increase monotonically with
Fe content. Since the contribution of the first term in eq.(2)
cannot account alone for the observed $S(60)$ values (for any
reasonable E_F values) we ascribe at least a part of this variation
to the second term in eq.(2). This term can increase with Fe
concentration either due to decrease in $N_d(E_F)$ or due to increase
in $(dN_d/dE)_{E_F}$. Furthermore since the $S(60)$ values are negative
(and yet larger than estimated from the first term of eq.2) N_d
must decrease with E. Although the actual $N_d(E_F)$ values for our
alloys are not known we note that a recent measurements of the
low temperature specific heat[6] of similar $Fe_{80-x}Ni_xP_{14}B_6$ alloys
indicate that the density of states at the Fermi level is increasing
with Fe content. (The actual increase in $N(E_F)$ in the same range
of Fe concentrations, 20-80 at % Fe was about 30 %). Therefore the
increase in $S(60)$ with increasing Fe content is probably due to
the faster decrease of $N_d(E)$ with the energy at the Fermi level.
This can be explained by taking into account the fact that in
amorphous alloys band edges are expected to be smeared due to
structural disorder. In a view of this result the electronic
structure of our alloys (including also $Fe_{80}B_{20}$) would appear to
be more similar to that of (crystalline) Ni than to that of Fe.

We note that similar conclusion has been reached by
Durand[7] on the basis of his magnetisation measurements in the
similar $Fe_xNi_{79-x}P_{13}B_8$ system. More precisely he concluded that
$Fe_{79}P_{13}B_8$ is a strong ferromagnet with the d band filled up so
that the rigid band model is roughly valid in these alloys[5]. The
magnetisation measurements on our $Fe_xNi_{80-x}B_{20}$ alloys also
indicate the approximate validity of rigid band model in the
concentration range $(x > 20)$ we explored.

The temperature dependence of the thermoelectric power
of our alloys is more difficult to explain. Qualitatively the
thermopower variation is similar to that in crystalline elemental
ferromagnets[8] and Fe-Ni alloys[9]. However, only in the case of
$Fe_{20}Ni_{60}B_{20}$ do our measurements extend over a temperature interval
broad enough to observe also the thermoelectric power variation
above T_c. From Fig.1 it can be seen that the magnitude of the
thermoelectric power of this alloy goes through a maximum around
$T_c/2$ and shows a small but definite change of slope above T_c.

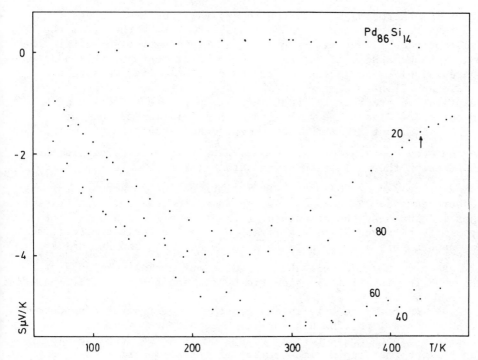

Fig.1 : Absolute thermoelectric power (S) of $Fe_xNi_{80-x}B_{20}$ alloys
vs temperature. Numbers denote x and arrow denotes the
Curie temperature. Curve at top represents S of the
nonmagnetic $Pd_{84}Si_{16}$ alloy.

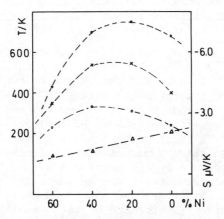

Fig.2 : Concentration dependence of : ● T_c, o T_{max} × S_{max} and
△ S(60). T_{max} denotes the temperature of the maximum in
thermoelectric power and S_{max} and S(60) denote the
thermoelectric power at T_{max} and 60 K respectively.

A similar thermopower variation in crystalline Fe-Ni alloys (in
the invar range of concentration) was described either from the
band approach[10] (but assuming a weak ferromagnetism) or within the
s-d exchange scheme[11]. Disregarding the fact that possibly none of
these calculations applies to our case we note that we have too
little additional information about our alloys to make any more
quantitative predictions about their temperature variation of S.
In our other alloys the variation of S is similar to the one
$Fe_{20}Ni_{60}B_{20}$, but its temperature dependence above the maximum is
less pronounced. In this sense our results seem to be consistent
with recent thermoelectric power data[12] for $Fe_{80}B_{20}$ alloy in the
temperature range 300-1000 K. It is interesting to note that these
measurements do not show any change in S around T_c. In our opinion
this is caused by the structural relaxation effects due to close
proximity between the crystallization (T_x) and Curie temperature.
To support this view we note that in $Fe_{20}Ni_{60}B_{20}$ and $Fe_{20}Ni_{60}B_{19}P_1$
alloys which have T_x well above T_c distinct anomalies around are
observed in resistivity[13] specific heat[13] and thermoelectric power[14].

Finally we briefly discuss the concentration dependence
of S in our alloys. The concentration dependence of various
quantities which characterize our system is shown in Fig.2. It
can be seen that the maximum values of thermoelectric power (S_{max})
vary smoothly with concentration exhibiting the maximum around
x = 50. The temperatures of S_{max}, T_{max}, show a similar x
dependence and T_{max} values are roughly around $T_c/2$. We note that
S_{max} values decrease more rapidly for lower x values (Ni rich
alloys) such a variation with x is similar to that of the T_c in
the same alloys, but it is different from that in crystalline
alloys. In the crystalline alloys S_{max} values for higher Fe
concentrations are lower than those on Ni rich side. This
difference may be due to the absence of invar effects in our
alloys. The absence of invar effects in our alloys is supported
by the following facts : i) smooth variation of the magnetic
moment per transition atom[5] with x; ii) smooth variation of
residual resistivities[14] (ρ_0) with x and iii) smooth variation
of the coefficients of the low temperature T^2 term[15] in resistivity
with x. According to that we may conclude that the structural
disorder and the presence of the metalloid atom (in spite of its
small size) have a strong influence on the electronic structure
of Fe-Ni alloys. In this sense the metalloid atoms seem to be
particularly effective as evidenced by the dramatic changes in
the thermoelectric powers of amorphous Fe-Ni alloys produced by
the exchange of one metalloid by the another[14] (B with P).

In conclusion we note that also the thermoelectric power of $Fe_xNi_{80-x}B_{20}$ alloys shows like the other transport properties strong effects of the magnetic ordering despite of the structural disorder. However, due to its relation to the energy dependence of the conductivity it may prove to be particularly useful in investigating the electronic structures of amorphous alloys.

Acknowledgement - We thank Drs. F.E. Luborsky and H.H. Liebermann for giving us the samples.

REFERENCES

1. J.R. Cooper, Z. Vučić and E. Babić, J. Phys. F. Metal. Phys. 4 (1974), 1489.

2. S.R. Nagel, Phys. Rev. Letters 41 (1978), 990.

3. N.F. Mott, Phil. Mag. 26 (1979), 1249.

4. J.E. Enderby and B.C. Dupree, Phil. Mag. 35 (1977), 791.

5. J.J. Becker, F.E. Luborsky and J.W. Walter, IEE Trans. Mag. 13 (1977), 988.

6. T.A. Donnely, T. Egami and David G. Onn, Phys. Rev. B 20 (1979), 1211.

7. J. Durand, Amorphous Magnetism II (R. Levy and R. Hasegawa eds.) (1976), p.305.

8. M.V. Vedernikov, Advances in Physics 18 (1969), 337.

9. O.V. Basargin, and A.I. Zakharov, Phys. Met. and Metallog. 37 (1974), 20.

10. B.F. Armatrong and R. Fletcher, Can. J. Phys. 50 (1972), 244.

11. T. Maeda and T. Somura, J. Phys. Soc. Japan 44 (1978), 148.

12. N. Teoh, W. Teoh, S. Arajs and C.A. Mayer, Phys. Rev. B 18 (1978), 2666.

13. E. Babić, B. Fogarassy, T. Kemeny and Ž. Marohnić, J.M.M.M. (1980) to be published.

14. E. Babić, Ž. Marohnić, M. Ocko, A. Hamzić and B. Pivac, J.M.M.M. (1980) to be published.

15. E. Babić, Ž. Marohnić, J. Ivkov and T. Ivezić, Fizika 10 Suppl. 2 (1978), 235.

LIGHT ABSORPTION AND PHOTOCURRENT OSCILLATIONS IN a-GeSe$_2$ FILMS

J. Hajto and M. Füstöss-Wégner
Central Research Institute for Physics of the
Hungarian Academy of Sciences
H-1525 Budapest, P.O.B. 49
Hungary

The steady state photocurrent and optical properties of amorphous GeSe$_2$ films has been investigated under the influence of continuous laser irradiation. A medium intensity (1.4 - 2.7 kW/cm^2) is found where a low frequency (3-50 Hz) periodic pulsation in the transmitted light and the photocurrent sets in. A model for light induced absorption oscillation is presented based on the bistable local bonding geometry of the amorphous network.

1. INTRODUCTION

The laser induced oscillation of the transmittance of vacuum evaporated a-GeSe$_2$ films has been described in a recent paper[1]. The oscillation of transmittance is accompanied by a similar oscillation of reflectance i.e. real absorption oscillation has been found[2]. A tentative explanation was suggested in terms of laser induced self-trapped exciton states of the material[3]. However the physical grounds of oscillation has not cleared yet. In this paper we present our primary results of the steady state photoconductivity measurements on a-GeSe$_2$ films under the influence of medium intensity laser irradiation. The paper is not concerned with a comprehensive study of the subject but aims at presenting a possible mechanism of the observations so far.

2. EXPERIMENTAL

GeSe$_2$ films were vacuum evaporated onto silica or mica substrates. Deposition rates usually ranged 20–40 Å/sec. The thicknesses were measured by quartz crystal monitor and later checked by Talystep. X-ray diffraction confirmed the layers to be amorphous. The continuous beam of 30 mW output power He-Ne laser (λ = 6328 Å) was focussed on the layers to a spot diameter of 20–350 μ. The incident light intensity could be reduced by a polarizer and the transmitted light signal was displayed on a storage oscilloscope. The photocurrent measurements were made on surface type samples prepared by vacuum evaporation of coplanar gold electrodes onto mica substrates (Fig.1). The steady state photocurrent (under continuous laser irradiation) was measured by an electrometer or amplified and displayed on a storage oscilloscope. The RC time constant of the electrical circuit was less than 1 msec. In this way simultaneous measurements of the transmittance and photocurrent pulsations could be performed during the light absorption oscillation.

3. RESULTS

3.1. Steady state photocurrent measurements

The typical plots of photocurrent versus 1/T for the GeSe$_2$ films are shown in Fig.2 with intensity as a parameter. The photoconductivity varies exponentially with an activation energy E_1 = 0.3–0.5 eV below 373°K depending on the incident photon flux and with an activation energy E_2 = 0.8 eV above 373°K which doesn't depend on the incident photon flux. In every cases the photocurrent increases linearly with the voltage indicating the lack of space charge limitation and electrode effects.

The photocurrent increases much more rapidly than proportional to the increase of absorption with temperature (Fig.3) indicating that other temperature dependent quantities (mobility, life-time) have to be taken into account.

3.2. Simultaneous transmittance and photocurrent oscillations

To produce pulsations[1] in the transmitted light intensity the photon flux should be increased to $4.5.10^{21}$ cm^{-2}.sec^{-1}. The oscillation of transmittance is accompanied with a simultaneous oscillation of the photocurrent signal (Fig.4). When the transmittance gets lower (higher absorption) the photocurrent rapidly increases to a higher level and when the transmittance gets higher (lower absorption) the photocurrent falls back at the same time to the lower value (Fig.5).

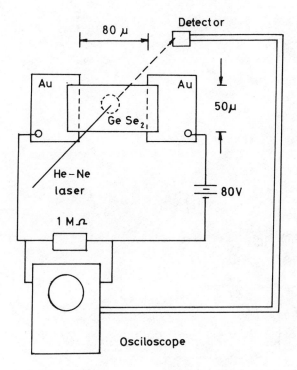

Fig.1 : Experimental arrangement.

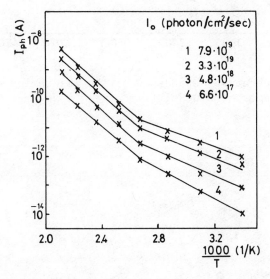

Fig.2 : Temperature dependence of photocurrent in amorphous
GeSe$_2$ films.

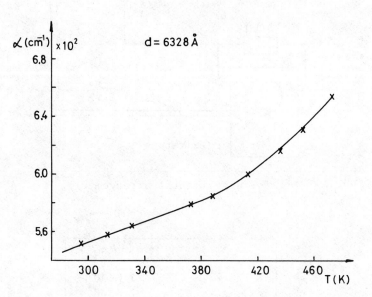

<u>Fig.3</u> : Temperature dependence of the absorption coefficient
in amorphous GeSe$_2$ film at λ = 6328 Å.

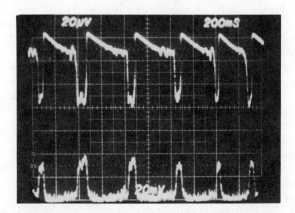

Fig.4 : Transmittance and photocurrent oscillations in 6 μ thick
a-GeSe$_2$ film under the influence of continuous laser
irradiation (I = $7.5.10^{21}$ photon cm^{-2} sec^{-1}).
Ordinate : Intensity of transmitted light and photocurrent.
Abscissa : time : 200 msec/div

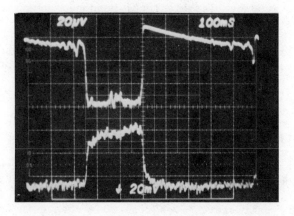

Fig.5 : Transmittance and photocurrent oscillations in the same
sample.
Abscissa : time : 100 msec/div.

4. DISCUSSION

The measured photocurrent data give direct information
on the electrical properties of GeSe$_2$ films because the measured
photocurrent was proportional to the voltage in agreement with
Ohm's law (no space charge limitation) and because the electrodes
were not illuminated (surface type cell). The increase of the
photocurrent with the decrease of the transmittance (higher
absorption) (Fig.4) is in accordance with the temperature
dependence of the photoconductivity (Fig.2). However the rapid
increase of the photocurrent during the oscillation can not be
explained by a pure thermal effect due to the higher light
absorption because the temperature increases exponentially and
slower as observed during the simultaneous transmittance oscillation
and temperature measurements[3]. The increased photocurrent signal is
not - in accordance with the proposed self trapped exciton model[2-3].
In this model the increasing absorption during the oscillation can
be attributed to the appearance of self trapped exciton states.
Because the self-trapped excitons (in form of D^+D^- pairs) are
localized and neutral formations therefore one should not observe
an increase in the photocurrent signal during the increasing
absorption period of the oscillation. The photocurrent
measurements show the opposite case therefore this model is not
consistent with the present data. Tanaka[4] has pointed out the
importance of the interaction between electron and lattice
vibrations producing local structural change in the amorphous
network which is more conceivable. The light induced increase
in fluctuation in Ge-Se bond angle has been directly observed in
amorphous sputtered Ge-Se film by Utsugi et al.[5]. Such a bending
flexibility may produce bistable local bonding geometries
responsible for the oscillation.

The bistable local bonding geometry represents local
bistable configurations which give two minima in the potential
energy of the system as a function of some appropriate local
atomic coordinates.

The energy of laser light used to induced the
oscillation is 1.96 eV (λ = 6328 Å), slightly less then the
value of the optical energy gap of the GeSe$_2$ films (E_q = 2.15 eV).
Therefore the assumption that the electron and holes are mostly
excited from the localized tail states of the valence band edge
(lone pair electrons) is plausible. Photoexcitation of a valence
electron in lone pair state gives rise to a change in interactions
between lone pair electrons or two neighbouring chalcogen atoms
and interactions with their local environment producing transition
from one to the other configuration (local phase changes). It
explains why one can observe a sudden change in the optical and
photocurrent signal during the oscillation. Because of the higher

absorptivity the temperature increases too. As the thermal
excitations overcomes the energy barriers between the two potential
minima the system turns back to the original state (local phase
change again). The absorptivity decreases therefore the temperature
falls down and the process should start again resulting in a
continuous oscillation of a light absorption and photocurrent.

REFERENCES

1. Hajto J., Zentai G., Kosa Somogyi I., Solid State Commun. 23,
   401 (1977).

2. Hajto J., Apai P., Journal of Non-Cryst. Solids 35-36, 1085
   (1980).

3. Hajto J., In Proc. of the European Physics Study Conf. on
   "Laser induced in Solids", Mons 1979, Belgium.

4. Tanaka K., In Proc. of the 8th Int. Conf. in Amorphous and
   Liquid Semiconductors. Cambridge Aug. 27-31, 1979 USA.

5. Utsugi Y. and Mizushima Y., J.Appl.Phys. 49, 3470 (1978).

SOME REFLECTIONS ON ENERGY GAPS IN AMORPHOUS As_2Se_3 AND As_2S_3

P. Van den Keybus, B. Vanhuyse and W. Grevendonk
Laboratorium voor Vaste Stof- en Hoge Drukfysica
Katholieke Universiteit Leuven
Celestijnenlaan 200 C
B - 3030 Leuven

Measurements of interband Faraday Rotation (F.R.) in amorphous
As_2Se_3 are reported. From theoretical formulae for F.R. as a
function of the photon energy an energy gap of 2.3 eV is determined.
This gap is compared with gaps obtained with other methods :
optical absorption usually gives a value of 1.8 eV but a formula
for "direct" transitions, applied to the region of high absorption,
confirms the F.R. result. Similar data are also found for As_2S_3
where the two gaps are about 2.4 and 3.1 eV.
Although it is usually accepted that there is no k-conservation
possible in amorphous semiconductors, it is suggested that in a
limited part of k-space the short range order allows to consider
direct transitions. The two determined gaps are then interpreted
as a kind of indirect resp. direct gap.

In a previous paper[1] we reported measurements of the
Faraday-rotation (FR) in amorphous As_2S_3. Fitting these results
to known dispersion formulae we found agreement with theories
based on direct transitions in crystalline material[2] and with
the theory of Mort and Scher[3] for amorphous semiconductors. All
these theories however revealed the same forbidden energy gap
in the vicinity of 3.1 eV whereas the usually accepted value,
obtained by other experiments, is 2.35 eV.
Since F.R. in crystals is a phenomenon, typically associated with
direct transitions, we interpreted the FR gap in amorphous As_2S_3
as a sort of 'direct' gap, in the sense that transitions in
these material can have a certain memory for k-values, within
a limited part in k-space. The procedure where results obtained
for amorphous semiconductors are interpreted in terms of well
established processes in crystals, is sometimes acceptable because

(short range) order is supposed to be responsible for observed semiconductor properties such as optical absorption edges and energy gaps.

In this paper we discuss further reflections about this hypothesis, and we argue that also from absorption measurements a similar direct gap in a-As_2Se_3 can be obtained, if we use absorption formulae for direct transitions. We also report new experimental F.R. results in a-As_2Se_3, confirming this gap value.

Using the technique, described in ref.1, we measured interband F.R. in a sample (thickness 0.33 mm) of amorphous As_2Se_3, obtained by quenching from the melt. The product nV (V is the Verdet-coefficient) is plotted as a function of $\hbar\omega$ in fig.1. The refractive index n was measured by Butterfield[4].
The curve, drawn through the experimental points, corresponds to the above mentioned formulae of KLN[2] or MS[3]. Both theories contain an energy gap E_G^{FR} as one of the parameters; the best fitting is obtained for E_G^{FR} = 2.35 eV. This gap value is quite different from most values, obtained from optical experiments as can be seen in table 1. The mean value of E_G^{OPT} is about 1.8 eV. The absorption curves $\alpha(\omega)$ are usually fitted with an expression of the form

$$\alpha\hbar\omega = Const \times (E_G - \hbar\omega)^2 \tag{1}$$

This formula for amorphous semiconductors is also valid for indirect transitions in crystalline material.

It is however possible to fit the absorption curves at higher ω and thus higher α, with an expression which, in crystals, is associated with indirect transitions :

$$\alpha = Const \times (E_g - \hbar\omega)^{1/2} \tag{2}$$

In fig.2 this fit is shown for data of Shaw et al.[12] and reveals a gap of 2.5 eV. In table 1 we also mention a value of 2.35 eV, obtained with dispersion data of Butterfield[4]; and applying an empirical relation of Wemple and DiDomenico[8]; in crystalline material the so determined gap should be direct.

Comparison with data for the corresponding crystalline material learns that c-As_2Se_3 is an indirect gap semiconductor with a direct gap E_G^D, and an indirect one E_G^I. In table 1 is mentioned which gaps were interpreted by the authors as direct or indirect. Also in a-As_2Se_3 the two different gaps (1.8 resp.

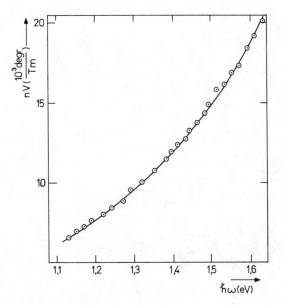

<u>Fig.1</u> : Verdet-Coefficient V, multiplied by the refractive index n,
as a function of photon energy $\hbar\omega$ in a-As₂Se₃.

     Circles : experimental points
     Curve   : best fit with theoretical expressions of either
                 KLN or MS.

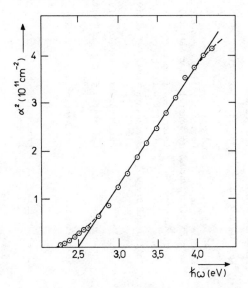

<u>Fig.2</u> : Optical absorption in a-As₂Se₃ as a function of photon
energy.

     Circles : experimental points
     Curve   : best fit of equation (2) to these points.

Table 1 : E_G (in eV), determined with different methods, for As_2Se_3 and As_2S_3

Authors	Method	As_2Se_3 c	As_2Se_3 a	As_2S_3 c	As_2S_3 a
Butterfield[4]	Optical Absorption		1.7-1.8		
Rockstad[5]	Photoconductivity		1.75		
Kolomiets et al.[6]	Optical Absorption (300 K)	2.0 (E_G^D) 1.87 (E_G^I)	1.8		
Kolomiets et al.[7] (+ several other auth.)	Opt. Absorption + Photocond. (+ several other methods)				2.4
These authors	Calculated from dispersion curves of ref.4 with theory of Wemple[8,9]		2.35		
Zallen and Drews[10]	Opt.absorptions + Reflectivity	2.0	1.73	2.9	2.34
Young[11]	Opt.Absorption (region of high α)				3.3 (E_G^D)
These authors	Calculated from ref.11 with theory of Wemple[8,9]			3.0	3.2
Shaw et al.[12]	Opt.Absorption + Photoconduction	2.0 (E_G^I) 2.2 (E_G^D)	1.8		
These authors	From ref.12 by fitting absorption curve at high α		2.5		
These authors[1]	Faraday Rotation				3.1
Evans and Young[13]	Opt.abs. + Refl. + Dispersion			2.8	
These authors	Faraday Rotation		2.35		
Althaus et al.[14]	Band Structure Calculation	1.85 (E_G^I) 2.15 (E_G^D)			
Gorban and Dashkovskaya[15]	Opt. Absorption			2.5 (E_G^I)	

2.3 eV) can be interpreted as "indirect" and "direct" as obtained by methods that, if applied to crystalline semiconductors, reveal such gaps. A similar hypothesis can be formulated for a-As$_2$S$_3$ (the summary of existing data is also shown in table 1). The value of 2.4 eV is usually accepted as the optical energy gap as determined with formula (1) from $\alpha(\omega)$ data. But also with formula (2), applicable in the region of large α, an energy gap (3.3 eV) can be determined[11]. Also with the theory of Wemple and with F.R. this value is obtained. Again the latter methods are typical for "direct" transitions. The data for c-As$_2$S$_3$ in table 1 are for direct transitions.

It is often believed that in glasses the uncertainty of k-values is of the order of the Brillouin-zone itself and that it does not make sense to use the terms direct and indirect; most experimental data are in agreement with this statement. To explain the results described above, we might suggest however that for some experiments these terms do make sense and that the short range order allows to consider a small region in k-space (much smaller than the Brillouin-zone in the corresponding c-material) where k may be conserved and where direct transitions, determining F.R. and optical absorption at high α, are possible.

Acknowledgement - This work was supported by the Belgian I.I.K.W. (Interuniversitair Instituut voor Kernwetenschappen).

REFERENCES

1. P. Van den Keybus, B. Vanhuyse and W. Grevendonk, To be published in Physica Status Solidi (b) 98 (1980).

2. J. Kolodziejczak, B. Lax and Y. Nishina, Physical Review 128, 2655 (1962).

3. J. Mort and H. Scher, Physical Review B 3, 334 (1971).

4. A.W. Butterfield, Thin Solid Films 21, 287 (1974).

5. H.K. Rockstad, Journal of Non-Crystalline Solids 2, 192 (1970).

6. B.T. Kolomiets, T.F. Mozets, SH.SH. Sarsembinov and SH.M. Efendiev , Journal of Non-Crystalline Solids 8-10, 1910 (1972).

7. B.T. Kolomiets, T.F. Mazets, SH.M. Efendiev, Journal of Non-Crystalline Solids 4, 45 (1970).

8. S.H. Wemple and M. DiDomenico Jr., Physical Review B 3, 1338 (1970).

9. S.H. Wemple, Physical Review B 7, 3767 (1972).

10. R. Zallen, R.E. Drews, R.L. Emerald and M.L. Slade, Physical
    Review Letters 26, 1564 (1971).

11. P.A. Young, Journal of Physics C 4, 93 (1971).

12. R.F. Shaw, W.Y. Liang and A.D. Yoffe, Journal of Non-Crystalline
    Solids 4, 29 (1970).

13. B.L. Evans and P.A. Young, Proceedings of the Royal Society
    A 297, 230 (1967).

14. H.L. Althaus, G. Weiser and S. Nagel, Phys.Status Solidi (b)
    81, 117 (1978).

15. I.S. Gorban and R.A. Dashkovskaya, Soviet Physics (Solid
    State) 6, 1895 (1965).

ELECTRICAL PROPERTIES OF SEMICONDUCTORS IN PRESENCE OF POTENTIAL FLUCTUATIONS
I: HIGH TEMPERATURE RANGE

J. M. Dusseau and J. L. Robert

Centre d-Etudes d'Electronique des Solides
associé au C.N.R.S.
Université des Sciences et Techniques du Languedoc
34060 Montpellier Cédex (France)

The localisation of a part of the band carriers in valleys of
potential is considered for the interpretation of the d.c.
electrical properties of highly compensated and disordered semiconductors. The fluctuations of potential are due to an inhomogeneous distribution of the impurities. Taking into account the
Boltzmann equation, the expressions of the conductivity and of
the thermoelectric power are derived. We propose an explanation
of the discrepancy observed between the slopes of the logarithm
of the conductivity and of the thermoelectric power versus 1/T
at high temperature. We apply this model to As_2Te_3, using numerous
results found in the literature.

The existence of spatial fluctuations of potential has
been put forward to interpret the electrical properties of highly
compensated[1] and disordered[2,3] systems.
The trapping of a part of the electrons in the potential valleys
has been considered to describe the experimental results, this
effect being the predominant factor that occurs in the variations
of the electrical properties with the temperature. So, most of
the experiments, like the temperature dependence of the conductivity of the drift mobility and of the photoconductivity can be
explained in this way.

295

However, this treatment seems to be rather rough since it is
concerned only with the change of the free carrier density with
temperature when it would be worthwhile to take into account the
effect of fluctuations on the transport coefficients. The aim of
this paper is to extend the previous idea starting from the
expression of the density of states in a disordered system, and
from the solution of the Boltzmann equation in the approximation
of the relaxation time.

In the first section, we give the method used to calculate the
various transport coefficients occuring in the d.c. experiments.
We show that our results agree with the previous ones in the case
of the d.c. conductivity and of the drift mobility. Concerning the
thermoelectric power, the present calculations show that its
activation energy is lower than the activation energy of the
conductivity. The physical meaning is that the mean energy of
transfer of the electrons is lower than in the case of an ideal
crystal, and varies with the temperature.

In the second section, the experimental results obtained on As_2Te_3
and published in the literature are discussed in the light of
this model.

1. TRANSPORT COEFFICIENTS

Density of States : The density of states in a disordered
system can be calculated assuming a Gaussian distribution of the
bottom of the conduction band around E_{co}, limited to a finite
interval to avoid considering degenerate valleys or probably
fluctuations without a physical meaning :

$$P(E_c) = \frac{1}{\gamma\sqrt{2\Pi}} \exp - [\frac{E_c - E_{co}}{\gamma\sqrt{2}}]^2$$

If we put the disorder parameter $\Gamma = \sqrt{2}\gamma$ we have $P(E_c) = 0$ when
$|E_c - E_{co}| > g\Gamma$. In the following calculations, g is chosen equal
to 3 because only deep levels are considered, the activation energy
of which is always larger than 3 .

The density of states can be formulated :

$$N_d(E) = \frac{4\Pi}{h^3}(2m)^{3/2}\frac{1}{\sqrt{\Pi}}\int_{-g}^{\min\,(g,\frac{E-E_{co}}{\Gamma})}(E-E_{co}-u)^{1/2}e^{-u^2}du$$

with

$$u = \frac{E_c - E_{co}}{\Gamma}$$

We assume that all the carriers above E_{co} are free to move through the sample. Their density n_F is given by

$$n_F = n_o\frac{2}{\Pi}\int_o^{\infty}e^{-\varepsilon}\int_{-g}^{\min\,(g,\frac{kT}{\Gamma}\varepsilon)}(\varepsilon-\frac{\Gamma u}{kT})^{1/2}e^{-u^2}\,du\,d\varepsilon \qquad (1)$$

with

$$\varepsilon = \frac{E - E_{co}}{kT}$$

The whole carrier concentration in the conduction band can be written :

$$n = n_o\frac{2}{\Pi}\int_{-g\frac{\Gamma}{kT}}^{\infty}e^{-\varepsilon}\int_{-g}^{\min\,(g,\frac{kT}{\Gamma}\varepsilon)}(\varepsilon-\frac{\Gamma u}{kT})^{1/2}e^{-u^2}\,du\,d\varepsilon \qquad (2)$$

in which we have put $n_o = N_c\,e^{\eta_o}$.

Current density : In presence of an external electric field E_X and of a temperature gradient ∇T the current J_x that flows through the sample is given by :

$$J_x = \sigma E_x + L_{ET}\,\nabla T$$

In order to avoid percolation paths, and using the Boltzmann equation[4], we calculate the contribution to the conductivity of the carriers within an energy E, E + dE :

$$d\sigma(E) \propto e^{-\varepsilon}[\int_{-g}^{\min\,(g,\frac{kT}{\Gamma}\varepsilon)}(\varepsilon-\frac{\Gamma u}{kT})^{\alpha+3/2}e^{-u^2}\,du]\,d\varepsilon$$

in which we get the classical energy dependence of the relaxation time $\tau \propto (E - E_c)^{\alpha}$.
From this expression, we can deduce the current density which would be obtained when all the carriers participate in the conduction :

$$\vec{J}_x = \frac{q}{m^*} A(kT)^\alpha \frac{2}{3} \frac{2}{\Pi} n_o \left[\int_{-g\frac{\Gamma}{kT}}^{\infty} e^{-\varepsilon} \int_{-g}^{\min\left(g, \frac{kT}{\Gamma}\varepsilon\right)} \left(\varepsilon - \frac{\Gamma u}{kT}\right)^{\alpha+2/3} e^{-u^2} du\, d\varepsilon [\vec{E}_x - \frac{1}{q}\vec{\nabla E}_F] \right.$$

$$\left. + \frac{k}{q} \int_{-g\frac{\Gamma}{kT}}^{\infty} (\varepsilon - \eta) e^{-\varepsilon} \int_{-g}^{\min\left(g, \frac{kT}{\Gamma}\varepsilon\right)} \left(\varepsilon - \frac{\Gamma u}{kT}\right)^{\alpha+3/2} e^{-u^2} du\, d\varepsilon\, \vec{\nabla T} \right] \qquad (3)$$

Our fundamental assumption is that the electrons below E_{co} are localised in the valleys of potential. In the absence of a gradient of temperature, the current due to these carriers is zero because a diffusion current is appearing in the valleys when an external electric field is applied.
So the current flow is actually equal to

$$J_x = n_F\, q\, \frac{q \langle \tau_o \rangle}{m^*} E_x \qquad (4)$$

with

$$\langle \tau_o \rangle = A(kT)^\alpha \frac{\displaystyle\int_{-g\Gamma/kT}^{\infty} e^{-\varepsilon} \int_{-g}^{\min\left(g, \frac{kT}{\Gamma}\varepsilon\right)} \left(\varepsilon - \frac{\Gamma u}{kT}\right)^{\alpha+3/2} e^{-u^2} du\, d\varepsilon}{\displaystyle\int_{-g\frac{\Gamma}{kT}}^{\infty} e^{-\varepsilon} \int_{-g}^{\min\left(g, \frac{kT}{\Gamma}\varepsilon\right)} \left(\varepsilon - \frac{\Gamma u}{kT}\right)^{\alpha+3/2} e^{-u^2} du\, d\varepsilon}$$

$\langle \tau_o \rangle = A(kT)^\alpha . F(\alpha)$, where $F(\alpha)$ depends only on the diffusion mode and is temperature independent.
In this expression, the mean mobility over energy does not depend on E_{co}. This calculation does not amount to the summation of $d\sigma(E)$ over energy above E_{co}, in which the exchange of carriers is not taken into account.

Because n_F can be represented by $n_F = n_o\, e^{\beta\Gamma/2kT}$ with $\beta = 0,08$, the conductivity can be approximated by :

$$\sigma \simeq N_c\, e^{\eta_o}\, q\, \mu_o \qquad (5)$$

since the value of the factor $e^{\beta \frac{\Gamma}{2kT}}$ is practically equal to one.
Now, introducing the impurity level and its compensation[4]

$$\left[\eta_o = -\varepsilon_d + Ln\, \beta\frac{(N_d - N_a)}{N_a} \right]$$

it is obvious that the activation energy of the conductivity is equal to ε_d.

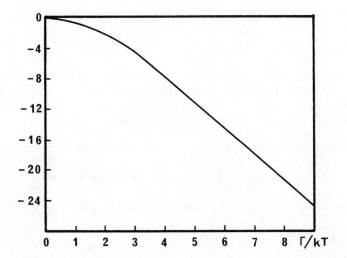

Fig.1 : Variation of the function which represents the influence
of the disorder on the mean kinetic energy of transfer of
the electrons.

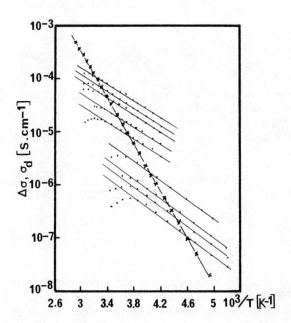

Fig.2 : Dark conductivity (X) and photoconductivity (.) in a-As$_2$Te$_3$
versus 1/T (Ref.8), —— fit according to eq.(5) and (8).

Concerning the drift mobility μ_D, we obtain

$$\sigma = n\, e\, \mu_D \quad (7) \qquad \text{and} \qquad \mu_D = \frac{n_F}{n}\, \mu_o \qquad (8)$$

It is clear that the temperature dependence of the photoconductivity is the same than the drift mobility one, provided that the photo-excited electron density remains constant with the temperature. This fact has been observed in several semiconductors[5].
Moreover, in presence of magnetic field, the measurement of the Hall coefficient leads directly to the determination of n_F.

The thermoelectric effect : When $E_x = 0$ and $J_x = 0$ in equation (3) and though the $n-n_F$ carriers are localised, we obtain the expression of the thermoelectric power

$$S = \frac{-\dfrac{1}{q}\,\nabla E_F}{\nabla T}$$

$$S = \frac{k}{q}\left[-\eta_o + \frac{\displaystyle\int_{-g}^{\infty} \frac{\Gamma}{kT}\,\varepsilon\, e^{-\varepsilon}\int_{-g}^{\min\left(g,\frac{kT}{\Gamma}\varepsilon\right)}\left(\varepsilon - \frac{\Gamma u}{kT}\right)^{\alpha+3/2} e^{-u^2}\,du\,d\varepsilon}{\displaystyle\int_{-g}^{\infty}\frac{\Gamma}{kT}\,e^{-\varepsilon}\int_{-g}^{\min\left(g,\frac{kT}{\Gamma}\varepsilon\right)}\left(\varepsilon - \frac{\Gamma u}{kT}\right)^{\alpha+3/2} e^{-u^2}\,du\,d\varepsilon}\right]$$

This is equivalent to

$$S = \frac{k}{q}\left[-\eta_o + \left(\frac{5}{2}+\alpha\right) + F\left(\frac{\Gamma}{kT}\right)\right] \qquad (9)$$

The difference between the above expression and the usual one obtained for an ideal semiconductor lies in the presence of the function $F(\Gamma/kT)$ which is plotted in Fig.1.
Because of the temperature dependence of this function, the activation energy of the thermoelectric power is lower than the activation energy of the conductivity.
The physical meaning is that the mean kinetic energy of transfer of the electrons is lowered in presence of fluctuations.

2. APPLICATION TO As$_2$Te$_3$

In this section, we use the proposed model to analyse the results obtained in the high temperature range on As_2Te_3, which are reported in the literature.

First we deduce the temperature dependence of μ_o from the Hall mobility measurements performed in the high temperature range (when $\mu_o \neq \mu_H$[6]). For example we use the results of A.J. Grant et al.[7]

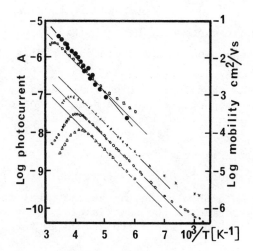

Fig.3 : Drift mobility (•) and photocurrent (0,X,□,Δ) in a-As$_2$Te$_3$ versus 1/T (Ref.9), —— fit according to equation 8.

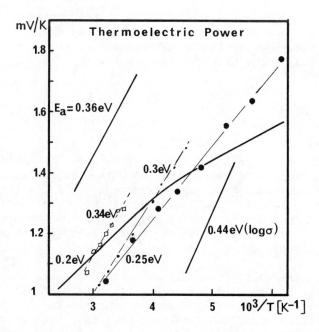

Fig.4 : Thermoelectric power in a-As$_2$Te$_3$ versus 1/T (Ref.8, □ ref.10, • ref.11), —— fit according to equation (9).

which lead to $\mu_o = 2.10^{-7} T^{2.2}$ and we fit the dark conductivity given by T.D. Moustakas and K. Weiser[8] (Fig.2) with E_a = 0.36 eV. The variations of μ_D and of the photoconductivity given by C. Main and A.E. Owen[9] are reproduced on Fig.3. For the temperature range in which the photoexcited electronic density remains constant (i.e. when i_p is proportional to the square root of the number of quanta absorbed per unit volume per unit time) we observe that the curves of μ_o and of i_p have same temperature dependence.

We calculate that Γ is equal to 0.085 eV from the curve μ_D versus $1/T$, while it is equal to 0.065 eV from the curve i_p versus $1/T$.

This difference is probably due to the fact that the measurements have not been performed on the same sample. The variations of i_p versus $1/T$ reported on figure 2 are accounted for by taking $\Gamma = 0.08$ eV.

Let us now consider the temperature dependence of the thermo-electric power. The results of T.D. Moustakas and K. Weiser[8], A. Van der Plas and H. Bube[10], and of C.H. Seager and R.K. Quinn are shown on the Fig.4. In order to fit these experimental values, we have used the expression (9) with E_a = 0,36 eV and Γ = 0,07 eV and the compensation ratio of the corresponding impurity level has been taken equal to 0.1. As expected, we found that the slope of the thermoelectric power is lower than the activation energy of the conductivity. ($E_s \simeq 0,2$ eV above 250 K when E = 0,44 eV).

CONCLUSION

The electrical properties of As_2Te_3 in the high temperature range have been explained without considering for example the existence of a hopping process or of a multitrapping effect on deep levels[12]. In the sight of the present analysis, a hopping process may be invoked only in the low temperature range when the contribution to the conductivity of the n_F carriers becomes negligible[6].

In any case, it is difficult to disregard the proposed effect of the fluctuations of potential, as a possible explanation of the behaviour of disordered semiconductors. In this hypothesis we found, for the samples of As_2Te_3 which have been studied, that the impurity level is 0,36 eV deep and the disorder parameter around 0.07 eV.

REFERENCES

1. J.L. Robert, B. Pistoulet, A. Raymond, J.M. Dusseau and G.M. Martin, J.Appl.Phys. 50, 349 (1979).

2. B. Pistoulet, J.L. Robert, J.M. Dusseau and L. Ensuque, J. Non-Cryst. Sol. 29, 29, 1978.

3. B. Pistoulet, J.L. Robert, J.M. Dusseau, Proc. 14th Int. Conf. Phys. Semic. (I.P.C.S. 43) Edinburgh, p.1009, 1978.

4. R.A. Smith, Semiconductors, Cambridge Univ. Press, 1968.

5. W.E. Spear, R.J. Loveland and A.Al. Sharbaty, J. Non-Cryst. Sol. 15, 410, 1974.

6. J.M. Dusseau and J.L. Robert, following paper.

7. A.J. Grant, T.D. Moustakas, T. Penney and K. Weiser, Proc. 5th Int. Conf. Am. Liquid Semicond. (Taylor and Francis) p.325, 1973.

8. T.D. Moustakas and K. Weiser, Phys. Rev. B 12, 6, 2448, 1975.

9. C. Main and A.E. Owen : Proc. 5th Int. Conf. Am. Liquid Semicond. (Taylor and Francis), p.783, 1973.

10. A. Vander Plast and H. Bube, J. Non-Cryst. Sol. 24, 377, 1977.

11. C.H. Seager and R.K. Quinn, J. Non-Cryst. Sol. 17, 386, 1975.

12. J.M. Marshall and A.E. Owen, Phil. Mag. 31, 1341, 1975.

ELECTRICAL PROPERTIES OF SEMICONDUCTORS IN PRESENCE OF POTENTIAL FLUCTUATIONS - II: HIGH ELECTRIC FIELD EFFECT AND LOW TEMPERATURE RANGE

J.M. Dusseau and J.L. Robert

Centre d'Etudes d'Electronique des Solides
associé au C.N.R.S.
Universités des Sciences et Techniques du Languedoc
34060 Montpellier Cedex (France)

The electrical properties of As_2Te_3 have been interpreted in the high temperature range on the basic of fluctuations of potential induced by inhomogeneities. The aim of this paper is to add to the preceeding one, by studying within the framework of the proposed model the d.c. conductivity and the drift mobility, first under high electric fields and then in the low temperature range.

      In the previous paper[1], we have shown that the fluctuations of potential due to inhomogeneities could be responsible for a localisation of a part of the band carriers in the associated potential valleys. The carriers above an energy E_{co} (the mean value of the conduction band) are free and their participation in the d.c. conductivity is dominant in the high temperature range. Considering the $n-n_F$ electrons, with an energy below E_{co}, their contribution to the d.c. conductivity cannot be neglected when high electric fields are applied on the one hand, and when a conduction process like a thermally assisted hopping is involved to the low temperature range on the other hand.

We apply these ideas on disordered As_2Te_3 and $As_{30}Te_{48}Si_{12}Ge_{10}$ compounds, using the results of J.M. Marshall and A.E. Owen[2].

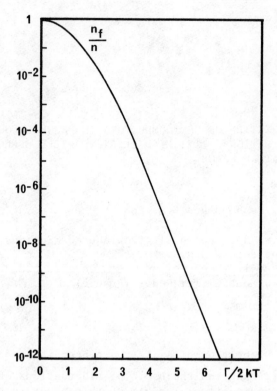

Fig.1 : Variations of the ratio n_F/n versus $\Gamma/2kT$.

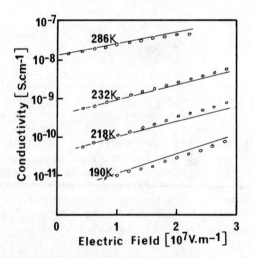

Fig.2 : Electric field dependences of the d.c. conductivity.
o : experimental points (Ref.2), — calculated values
according to the expression 1.

1. HIGH ELECTRIC FIELD EFFECT

The analysis of the high temperature range electrical properties of $As_{30} Te_{48} S_{12} Ge_{10}$[2], in the light of the proposed model[1,3] leads to a value of $E_a = 0,4$ eV (from the d.c. conductivity $\sigma = n_F q \mu_o$) and of $\Gamma = 0,135$ eV (from the drift mobility $\mu_n = \dfrac{n_F}{n} \mu_o$); in our computation we assume that the mobility μ_o varies as $T^{2.2}$ as in $As_2 Te_3$.

Due to the application of high electric fields the depht of the potential valleys is lowered and the conductivity is increased because of the increase of the free carrier density $n_F(E)$. (The electric fields dependence of the mobility μ_o can be neglected as for as the heating effects can be disregarded).

So we have $\quad \sigma(E) = n_F(E) q \mu_o = \dfrac{n_F(E)}{n(o)} \sigma(o) \qquad (1)$

The variations of n_F/n with $\Gamma/2kT$ are reported on fig.1. An approximation of these variations is given by $\dfrac{n_F}{n} = e^{-\left(\frac{\Gamma}{2kT}\right)\left(2 - \frac{\delta\Gamma}{2kT}\right)}$ where $\delta = 0.035$. We found that the lowering of the potential barriers as expressed by the simple expression $\Gamma(E) = \Gamma(o) - 7.10^{-10} E$ with E in $(V.m^{-1})$, allows us to account for the experimental variations of the conductivity, obtained by J.M. Marshall and A.E. Owen (fig.2).

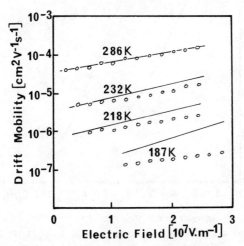

Fig.3 : Electric field dependences of the drift mobility.
o : experimental points (Ref.2), — calculated values according to the expression 2.

Let us now consider the behaviour of the drift mobility.
We have

$$\mu_D(E) = \frac{n_F(E)}{n(o)} \mu_o = \frac{n_F(E)}{n(o)} \frac{n(o)}{n_F(o)} \mu_D(o) \tag{2}$$

If we plot the calculated values on the experimental ones (fig.3),
a discrepancy appears particularly at low temperatures and in high
electric fields.

We have assumed in the interpretation of te d.c. conductivity
experiments and in the expression (2) that the whole band carrier
density remains constant when the electric field is increased.
Now, using the experimental results of $\sigma(E)$ and those of $\mu_D(E)$,
we find that the ratio $\sigma(E)/q \mu_D(E)$ is varying with the
the electric field and the temperature, except in the highest
temperature (T = 286 K) when extra photoinduced electrons can
be considered as thermalized.

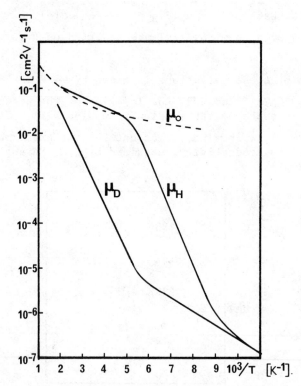

Fig.4 : Calculated variations versus temperature of the Hall,
drift and macroscopic mobilities according to the
equations (3) and (5) and with $\mu_o = 10^{-5} T^{1.5}$, H = 0.1 K
cm^2/Vs and $\Gamma = 0.1$ eV.

2. LOW TEMPERATURE RANGE

Provided that a thermally assisted hopping process were involved with an activation energy γ ($\gamma = \Gamma/\sqrt{2}$)[1], the contribution of the trapped electrons $n - n_F$ would be taken into account in the d.c. conductivity. The energy γ corresponds to the mean value of the hopping energy over the whole sample.

Of course, this possible process can only occur if the distances between the potential valleys are not too large and in the intermediate low temperature range before the classical hopping at the Fermi energy appears. Then, the mobility related to this mechanism is given by

$$\mu_h = \frac{H}{T} \exp \left(- \frac{\Gamma}{\sqrt{2}kT} \right) \tag{3}$$

where H is a constant.

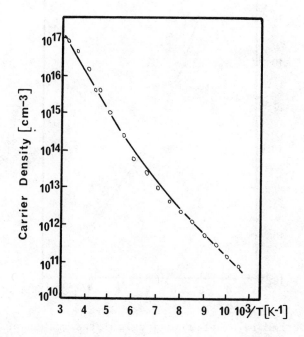

Fig.5 : Experimental (Ref.2) and calculated values of the carrier density in As_2Te_3.

In these conditions we obtain the drift mobility

$$\mu_D = \frac{n_F \mu_o + (n - n_F) \mu_h}{n} \tag{4}$$

while the Hall mobility is given by

$$\mu_H = \frac{n_F \mu_o^2 + (n - n_F) \mu_h^2}{n_F \mu_o + (n - n_F) \mu_h} \tag{5}$$

On figure 4 we have reported the drift and Hall theoretical dependence on the reverse of the temperature. We observe that the Hall mobility presents two activation energies, as it has been observed by C.H. Seager et al. in $As_{50}Te_{48}I_2$ glasses[4].

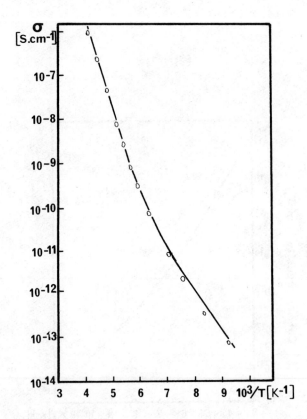

Fig.6 : Experimental (Ref.2) and calculated values of the d.c. conductivity in As_2Te_3.

We have reported the experimental variations versus temperature (between 300 K and 100 K) of the density of carriers (Fig.5), of the d.c. conductivity (Fig.6) and of the drift mobility (Fig.7) obtained on As_2Te_3 by J.M. Marshall and A.E. Owen[2]. Using the expression of n given by equation (2) in the preceeding paper[1], and the expression $\sigma = nq\mu_D$ in which μ_D is given by (4) it is worth while noticing that we obtain a good fit of these results (Fig.5,6,7) when $N_c T^{-3/2} = 7.25 \ 10^{17} \ cm^{-3}K^{-3/2}$, $H = 0.08 \ Kcm^2/Vs$, $\mu_0 = 10^{-7} T^{2.2}$ and $E_A = 0.36$ eV while $\Gamma = 0.082$ eV.

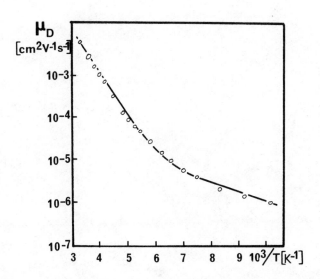

Fig.7 : Experimental (Ref.2) and calculated values of the drift mobility in As_2Te_3.

CONCLUSION

       We have obviously proposed a crude model of fluctuations of potential, that are probably due to inhomogeneities in compensation. As it can account in a self consistant way for most of the experimental results (such as those obtained in As_2Te_3 and related compound) it would be difficult to ignore this kind of fluctuations when one tries to explain the electrical properties of disordered semiconductors.

REFERENCES

1. J.M. Dusseau and J.L. Robert, preceeding paper.

2. J.M. Marshall and A.E. Owen, Phil. Mag. 31, 1341, 1975.

3. B. Pistoulet, J.L. Robert, J.M. Dusseau and L. Ensuque, S. Non Cryst. Sol. 29, 29, 1978.

4. C.H. Seager, D. Emin and R.K. Quinn, Phys. Rev. B 1, 4746, 1973.

ELECTRON RELAXATION IN Eu$_{3-x}$Gd$_x$S$_4$ BY MÖSSBAUER SPECTROSCOPY

W. Bedorf[+] and M. Rosenberg[++]

++ Institut für Experimentalphysik VI
Ruhr-Universität, NB+3/34
Postfach 102148, 4630 Bochum, BR Deutschland

+ Hoesch-Werke AG (ZDH) Dortmund

A Mössbauer study between 80 and 470 K of solid solutions of Eu$_{3-x}$Gd$_x$S$_4$ with x = 0; 0,3; 0,7; 1,0; 1,7; 2 and Th$_3$P$_4$ structure has been undertaken. As previously reported by Berkooz et al.[1] in the case of Eu$_3$S$_4$, the Mössbauer spectra provide strong evidence for thermally activated electronic relaxation. The low temperature spectra for the compositions with x > 2 show two well resolved absorption peaks with isomer shifts IS of −11,5 and +0,5 mm/s characteristic for the Eu^{2+}, respectively Eu^{3+} valency states with nearly the expected intensity ratio.

At higher temperatures a broadening and shifting of the both peaks occur so that at 300 K for instance in the case of x = 0,3 as for x = 0 they merge to a single relatively narrow resonance peak. With increasing x the activation energy of the electron relaxation shifts towards higher values so that at 470 K the relaxed lines for 1,3 > x > 0.7 are still quite broad. For compositions with x < 1,0 the relaxed spectra up to 470 K seems to be centred at about −3,5 mm/s, value of the isomer shift which corresponds to the narrow relaxation peak of Eu$_3$S$_4$.

The variation of the shape of the Mössbauer spectra with temperature given strong evidence for thermally activated electron hopping in this compounds with concentration dependent activation energies.

INTRODUCTION

Mössbauer spectroscopy was already used by Berkooz and al.[1] to find evidence for electron relaxation between the Eu^{2+} and the Eu^{3+} ions in the mixed valence system Eu_3S_4. An electron hopping model together with the exchange narrowing formula given by Wickman et al.[2] for magnetically splitted Mössbauer spectra involving relaxation processes yield a temperature dependent hopping time characteristic of thermally activated electron relaxation with an activation energy in acceptable agreement with the value obtained from electrical conductivity measurements[1,3].

Further electrical[3], differential thermal analysis and x-ray analysis experiments[4] have shown, that in Eu_3S_4 with a bcc Th_3P_4 structure a phase transformation occurs at 168 K with a lowering of the symmetry towards a tetragonal one due probably to a charge ordering[5].

For the electron transfer from Eu^{2+} to Eu^{3+} a theoretical model has been proposed by Mulak and Stevens[6] taking into account the intervening nonmagnetic S^{2-} ion and considering a simple $Eu^{2+} -S^{2-} -Eu^{3+}$ "molecule" in which the europium nuclei are regarded as fixed, whereas the sulphur ion can move. Thermally activated electron hopping involves a strong coupling to lattice vibrations. The phonon line at 425 cm^{-1} discovered by Vitins[7] in the inelastic light scattering spectrum of Eu_3S_4 has been ascribed to a vibration involving only the sulphur ions in the Th_3P_4 structure, describing a "breathing and twisting motion" which activates the electron transfer from Eu^{2+} to Eu^{3+}.

The purpose of the present paper has been to study with the Mössbauer spectroscopy the influence of the substitution of Gd for Eu in the $Eu_{3-x}Gd_xS_4$ solid solutions on the electron transfer $Eu^{2+} -Eu^{3+}$. Such a substitution has to make the electron transfer more and more difficult and a shift in the appearance of relaxation Mössbauer lines towards higher temperatures has to be expected.

EXPERIMENTAL AND RESULTS

Seven compositions in the system $Eu_{3-x}Gd_xS_4$ with x = 0; 0.3; 0.7; 1.0; 1.3; 1.7; 2.0 have been prepared using the solid state reaction in evacuated quartz ampoules between EuS, metallic Gd-sponge and sulphur in a sequence of 7 heat treatments at temperatures between 265 and 1000°C with duration of 24 up to 72 hours. The values of a for Eu_3S_4 (0,85322 $\pm$ 0,00018 nm) and $EuGd_2S_4$

0,84990 + 0,00011 nm) were in very good agreement with those given previously by other authors[1,3,8]. The magnetic susceptibility in the temperature range 20-300 K has been measured with a Brucker-Sartorius Faraday balance. The Mössbauer spectra in the temperature range 80-470 K were obtained with a standard spectrometer, a 50 mCl SmF_3 source (Amersham-Buchler) and a Marshaw GSHClM NaJ(Tl)-Detector for the 21,53 KeV Gamma line of ^{151}Eu.

The reciprocal magnetic susceptibility of the 7 compositions plotted against temperature is shown in Fig.1.

Both isoelectronic Eu^{2+} and Gd^{3+} have the same $4f^7$ electron configuration thus being in a $^8S_{7/2}$ ground state with an effective moment of 7,94 μ_B. Eu^{3+} with a $4f^6$ electron configuration ($S_7 = 3$ and $L = 3$) has to be in the nonmagnetic F_o ground state. The most recent results on the magnetic properties of Eu_3S_4 of Görlich et al.[9] show that the spins of Eu^{2+} order magnetically below $T_c = 3,8$ K. In the case of $EuGd_2S_4$ Lugscheider and al.[8] mentioned a paramagnetic Curie temperature of 1 K.

The substitution of Eu^{3+} with Gd^{3+} has to lead to a continuous increase in the Curie constant, a trend which is evident in the figure 1. But on the other side there is no linear dependence of χ^{-1} on T as would be expected in the case of magnetic ions in a $^8S_{7/2}$ state. The deviation from linearity is the strongest in the case of Eu_3S_4, whereas for $EuGd_2S_4$ the dependence is practically linear. The decreasing curvature of χ^{-1} with increasing x is due to the thermal mixing in the nonmagnetic ground states of the relatively low lying upper levels of Eu^{3+} with higher values of J.

The Mössbauer spectra of all the investigated compositions taken at 80, 300 and 470 K are shown in fig.2-4.

The most important trends appearing in the evolution of the Mössbauer spectra with temperature are as follows :
At 80 K the Eu^{2+} and Eu^{3+} Mössbauer lines with isomer shifts IS at -11,5 respectively +0,5 mm are clearly separated and the ratios of the intensities are in good agreement with the expected stoichiometry of the samples. Only in the case of $EuGd_2S_4$ the presence of a weak intensive Eu^{3+} line gives evidence for a small deviation from the right desired composition. At 300 K the Eu^{2+} and Eu^{3+} peaks of Eu_3S_4 and $Eu_{2,7}Gd_{0,3}S_4$ give rise to practically

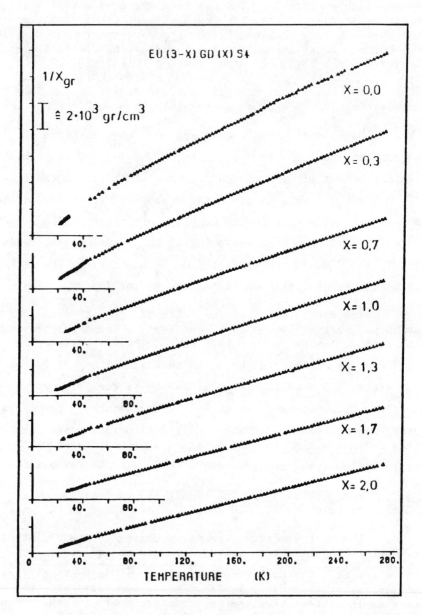

<u>Fig.1</u> : Reciprocal susceptibility versus temperature of $Eu_{3-x}Gd_xS_4$.

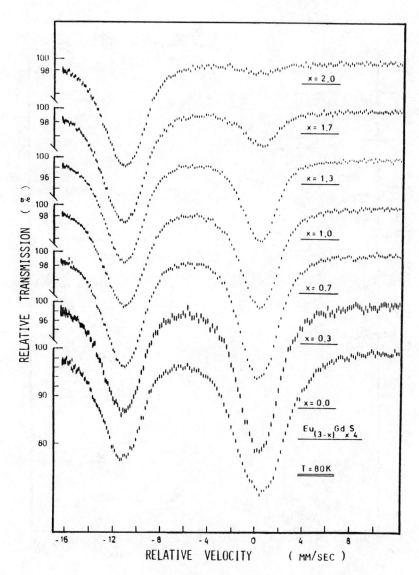

Fig.2 : Mössbauer spectra of $Eu_{3-x}Gd_xS_4$ at 80 K.

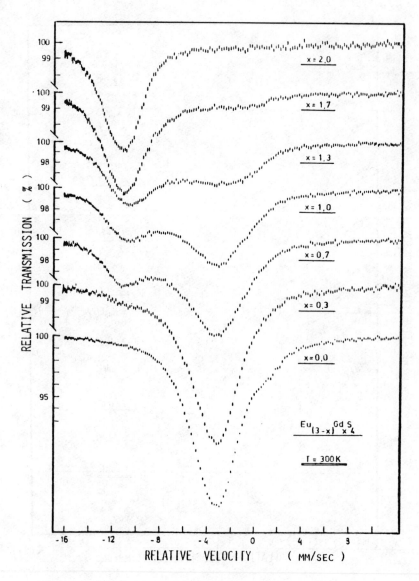

Fig.3 : Mössbauer spectra of $Eu_{3-x}Gd_xS_4$ at 300 K.

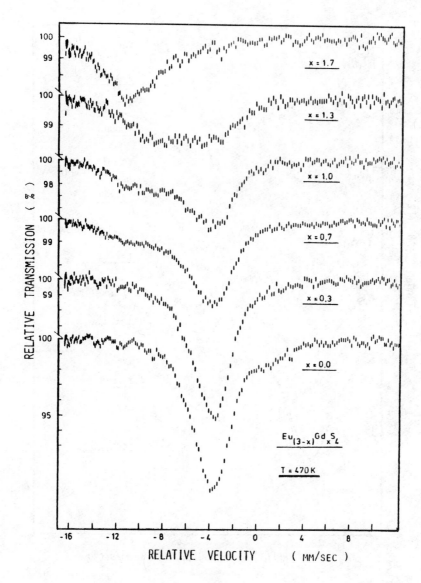

<u>Fig.4</u> : Mössbauer spectra of $Eu_{3-x}Gd_xS_4$ at 470 K.

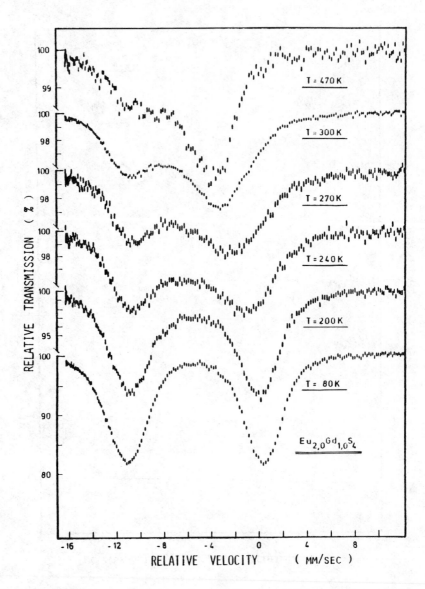

Fig.5 : Mössbauer spectra of Eu_2GdS_4 at different temperatures in the range 80-470 K.

single relaxation lines which become narrower at 470 K, centred at about -3,5 mm/s.

For the other compositions with $0.7 < x < 1.7$ two peaks spectra are still present but with broader overlapping lines centred around -3,5 and -11,5 mm/s. A peculiar feature is that the two shifted peaks in the case of $x = 1.0$ which at 80 K had the same intensity, have different intensities at 300 K, the one centred at -3.5 mm/s being more intensive (Fig.5).

No complete relaxation was reached at 470 K for $x > 0.7$.

DISCUSSION

Relaxation lines occur for those values of hopping frequencies ν_{hop} for which $\nu_{hop}^{-1} < \tau_M$, where τ_M is the relative time scale of the Mössbauer effect determined by the nuclear lifetime with values of about 10^{-9} s for Eu.

The hopping frequency can be expressed as

$$\nu_{hop} = A\nu_o \exp(- E_a/k_B T)$$

where ν_o is the frequency of the phonon mode coupled to the electron hopping and E_a an activation energy. The constant A contains geometrical factors and depends on the number of neighbouring places available for hoping. In the case of Eu_3S_4[1,3,4,7] $E_a = 0,21$ eV and $\nu_o = 4,10^{13}$ s^{-1} in good agreement with the frequency $\nu = 1,3.10^{13}$ s^{-1} of the so-called "breathing and twisting" mode of the sulphur ions[7]. Therefore the one-line absorption spectrum of Eu_3S_4 has to appear near $T = 230$ K in very good agreement with the experiment[1].

One can expect that the substitution with Gd will decrease the hopping frequency and shift the appearance of relaxation lines towards higher temperatures what in fact was observed. Most probably both the preexponential factor $A \nu_o$ and the activation energy will be influenced, the former one decreasing with x because of the decrease in the number of Eu^{3+} nearest neighbours whereas the latter one will increase because of the difficulty of the S^{2-}-ions in the presence of Gd to shift from one potential well to the other as a result of electron hopping between a pair $Eu^{2+}-Eu^{3+}$. A quasi-uniform distribution of Gd^{3+} could therefore only shift the merging of the Eu^{2+} and Eu^{3+} lines to a single relaxation line towards higher temperatures and

the shapes of the sequence of intermediate lines op to the single
relaxation one could be perhaps described with a model similar to
that of Wickman and al.[2] used by Berkooz and al.[1] in the case of
Eu_3S_4. But the evolution of the shape of the Mössbauer line in
the case of Eu_2GdS_4 doesn't agree with this expectation (Fig.5).
Because of the equal population of both states Eu^{2+} and Eu^{3+} the
shape of the deformed lines at intermediate temperatures up to
total relaxation has to be symmetrical according to Wickman and
al.[2]. That is but not the case.
Starting with 240 K the double-line becomes asymmetrical and at
470 K there is a strong peak centred at about -4 mm/s and a
broader one including a high amount of states with quasi-pure Eu^{2+}
character.

        This result can not be understood on the ground of only
one relaxation time for hopping, i.e. we have to assume that the
substitution with Gd gives rise to some degree of substitutional
disorder and to the occurrence of different types of local
surroundings of the Eu ions.

        In the Th_3P_4 structure an Eu-ion has 8 nearest metallic
neighbours and up to x = 2, two of them are always Eu^{2+} and the
other six Eu^{3+} and Gd^{3+} ions. Starting from a statistical
distribution of Gd^{3+} one can find 7 types of local surroundings
between those with 6 Eu^{3+} (Eu_3S_4) and 6 Gd^{3+} ($EuGd_2S_4$) and a
simple binomial distribution function will give the corresponding
7 probabilities $P_x(n)$ for every value of x so that n = 0,1,2,...,6
and $\sum_{n=0}^{6} P_x(n) = 1$. Assuming that in order to obtain a relaxation
line its intensity has to correspond at least to a $P_x(n) = 0,1$,
one has to expect spectra with up to 5 relaxation lines (for x = 1)
and isomer shifts between -3,5 and -11,5 mm/s. Assuming that the
relaxation occurs extremely fast and the influence of the
substitutional disorder is overcomed, one has to expect that a
single absorption line will result with an averaged shift given
by

$$IS = \frac{3\ IS(Eu^{2+}) + (6 - 3x)\ IS(Eu^{3+})}{9 - 3x}$$

with $IS(Eu^{2+})$ = -11,5 mm/s and $IS(Eu^{3+})$ = +0,5 mm/s. It can be seen
that at 470 K this is the case only for x = 0 and x = 0,3 and that
for higher Gd concentrations the full relaxation in the abovementioned sense has not yet been reached.

Unfortunately the simple model with a binomial distribution does not work satisfactorily. As we have mentioned above, the substitutional disorder can give rise for a given value of x to a distribution of relaxation times, so that at a given temperature the condition for the appearance of a single line can be reached for one type of surroundings, whereas for other ones not yet. On the other hand there is no spatial delimitation of the surroundings and we have to take into account the interconnections between them. It follows then that we have to observe deviations from the expectations from the binomial distribution model and the tails towards the Eu^{2+} states observed for $x > 0,3$ seem to give evidence for slower relaxation of the states with more Eu^{2+}-character, i.e. with more Gd^{3+} in the first metallic coordination sphere. For $0 < x < 1$ there is a strong tendency to relaxation in valence mixed states with IS in the range $-3,5$ ./. $-4,5$ mm/s. The reason could be either that the states with 4-6 Eu^{3+} ions between the nearest metallic neighbours have not very different hopping frequencies or that some degree of order occurs with a tendency towards the surroundings of Eu_3S_4 and $EuGd_2S_4$ type, both with different activation energies. For the time being there is no mean to distinguish between these two possibilities.

CONCLUSIONS

The temperature dependence of the Mössbauer spectra of $Eu_{3-x}Gd_xS_4$ ($0 < x < 2$) gives strong evidence for thermally activated electron hopping with temperature and concentration dependent frequencies. At 470 K only Eu_3S_4 and $Eu_{2,7}Gd_{0,3}S_4$ exhibit full relaxation.

Substitutional disorder of Gd^{3+} could give rise to a distribution of local surroundings and relaxation times. But for $x < 1$ the weight of the configuration with 6 Eu^{3+} which relaxes faster seems to be higher than expected from statistical disorder. This can be due to deviations from statistical disorder in favour of short range order configurations as in Eu_3S_4 and $EuGd_2S_4$.

REFERENCES

1. Berkooz O., Malamud M. and Shtrikman S., Solid State Communications, 6, 185 (1968).

2. Wickman H.H., Klein M.P. and Shirley D.A., Physical Review, 152, 345 (1966).

3. Bransky I., Tallan N.M. and Hed A.Z., Journal of Applied Physics, 41, 1787 (1970).

4. David H.H., Bransky I. and Tallan N.M., Journal of the Less-Common Metals, 22, 193 (1976).

5. Carter F.L., Journal of Solid State Chemistry, 5, 300 (1972).

6. Mulak J. and Stevens K.W.H., Zeitschrift für Physik B 20, 21 (1975).

7. Vitins J., Journal of Magnetism and Magnetic Materials, 5, 234 (1977).

8. Lugscheider W., Pink H., Weber K. and Zinn W., Zeitschrift für Angewandte Physik, 30, 36 (1970).

9. Görlich E., Hrynkiewicz H.U., Kmiec R., Latka K and Tomala K., Physica Status Solidi (b) 64, K 147 (1974).

MAGNETIC PROPERTIES OF THALIUM CHROMIUM CHALCOGENIDES

M. Rosenberg[+], H. Sabrowsky[++], and A. Knülle[+]

Ruhr-Universität, Lehrstuhl für Experimentalphysik VI[+] und Arbeitsgruppe Festkörperphysik[++]

Postfach 102148, 4630 Bochum, BR Deutschland

The magnetic properties of $TlCrX_2$ (X = S, Se, Te), Tl_3CrS_3, $TlCr_3S_5$, $TlCr_3Se_5$ and $TlCr_5S_8$ have been investigated. $TlCrS_2$, $TlCrSe_2$ and Tl_3CrS_3 order ferromagnetically with Curie temperatures in the range 60–140 K. The other compounds are antiferromagnetics with negative paramagnetic Curie temperatures, whereas $TlCrTe_2$ has a positive paramagnetic Curie temperature.

$TlCrS_2$ and $TlCrSe_2$ have a hexagonal layer structure with Cr ions sandwiched between S and Se-layers which in turn are separated by the large radius Tl-layers. The shorter Cr-Cr distance in the haxagonal planes gives rise to a stronger intralayer exchange interaction and therefore to a behaviour analogous to that of a quasi-twodimensional magnet. $TlCrS_2$ and $TlCrSe_2$ seem to represent the extremal case in the range of $ACrX_2$ chalcogenides with A = Cu, Ag, Na, Li, K, where with the increase of the lattice constant a in the hexagonal plane a trend towards stronger positive exchange interactions has been observed.

INTRODUCTION

Most of the $Tl_xCr_yX_z$ compounds with X = S, Se, Tl have a layer structure, i.e. with the Cr ions sandwiches between X-layers which in turn are separated by Tl-layers[1]. $ACrX_2$ compounds (with A = Li, Na, K, Ag) with a related structure have been investigated in the past and they usually exhibited at low

temperatures antiferromagnetic types of ordering[2-5]. Because of
the layer structure with shorter metal-metal distance in the layer
as between the metal layers the intralayer exchange interactions
have to be the strongest ones.

Therefore, magnetic ordering could also be expected for
most of the investigated Tl-Cr-chalcogenides, eventually with some
additional peculiar aspects due to the larger ionic radius of Tl^+
and to the particular types of structures in which some of these
compounds crystallize.

EXPERIMENTAL AND RESULTS

The preparation and the structure of $TlCrS_2$, $TlCrSe_2$,
$TlCr_3S_3$, $TlCr_5S_8$, $TlCr_3Se_5$, $TlCrS_3$ and Tl_3CrSe_3 were described
recently in [1]. The present paper deals with the magnetic properties
of powdered samples of these compounds.

For all the investigated compounds the magnetic
susceptibility has been measured in the temperature range 3-800 K
with a Brucker-Sartorius type Faraday balance.

In the case of the ferromagnetic compositions below
Curie temperature the dependence of the magnetization on the
magnetic field was recorded with a Foner-vibrating sample magnetometer at 4.2 K.

The temperature dependence of the reciprocal susceptibility
of all the investigated compounds is shown in Fg. 1-3. Only $TlCrS_2$,
$TlCrSe_2$, Tl_3CrS_3 exhibit a behaviour of ferromagnetic type,
whereas the other compounds are antiferromagnetic.

Table 1

Compound	T_N, T_c (K)	T_p (K)	μ_{eff} (μ_B)	μ_{FU} (μ_B)
$TlCrS_2$	125 (F)	120	3,59	1,99
$TlCrSe_2$	140 (F)	130	3,71	1,09
$TlCrTe_2$	140 (A)	91	4,47	
$TlCr_3S_5$	45 (A)	-259	10,29	
$TlCr_3Se_5$	55 (A)	-156	8,37	
$TlCr_5S_8$	82 (A)	-208	8,89	
Tl_3CrS_3	60 (F)	55,3	3,80	2,29

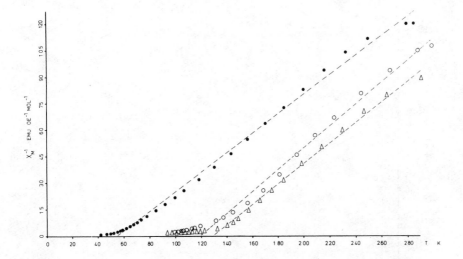

Fig.1 : Reciprocal molar susceptibility versus temperature:
ooo TlCrS$_2$, △△△ TlCrSe$_2$, ··· Tl$_3$CrS$_3$.

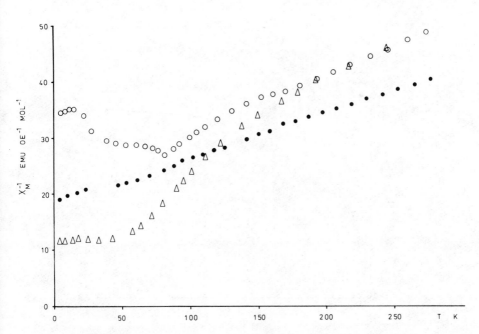

Fig.2 : Reciprocal molar susceptibility versus temperature:
ooo TlCr$_5$S$_8$, △△△ TlCr$_3$Se$_5$, ··· TlCr$_3$S$_5$.

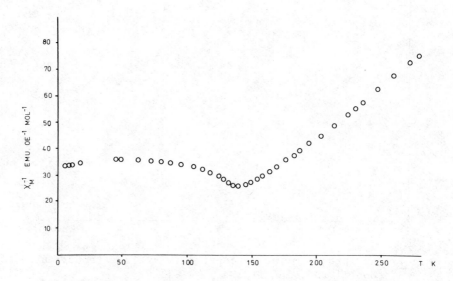

<u>Fig.3</u> : Reciprocal molar susceptibility versus temperature :
         ooo TlCrTe$_2$.

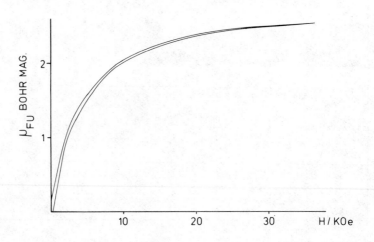

<u>Fig.4</u> : Magnetization curve of TlCrSe$_2$ at 4.2 K.

The experimental values of $\mu_{F,U}$ at 3 K for the ferromagnetic compositions, of the Curie and Néel temperatures T_c and T_N, of the asymptotic Curie temperatures T_p and of μ_{eff} in the paramagnetic range are given in Table 1. One can see that μ_{eff} in the Curie-Weiss region lies near to the spin-only value of 3,88 μ_B of the Cr^{3+} ion.

The magnetization curves M(H) of the ferromagnetic $TlCrSe_2$ and Tl_3CrS_3 are shown in Fig.4 and 5. In the case of $TlCrSe_2$ a not very abrupt change of 0,6 μ_B in the magnetization occurs between 30 and 40 kOe.

DISCUSSION

In the $TlCrX_2$ compounds the X-Cr-X sandwiches have the same f.c.c. structure as in the $ACrX_2$ chalcogenides (A = Cu, Ag, Na, Li, K) investigated by Bongers et al.[3], Van Laar and Engelsman[4], Engelsman et al.[5] They consist of sheets of $Cr-X_6$ octahedrons which share common edges. The distances between the octahedron-sandwiches are relatively large owing to the presence of the Tl^+ ions with a large radius. Such structures could be considered as giving rise to quasi-twodimensional magnetic systems because of the shorter Cr-Cr distance in the hexagonal planes of the X-Cr-X sandwiches as compared to the larger distances between these planes due to the large monovalent intercalated ions. Magnetic[3,9,10] Mössbauer[9] and neutron diffraction[5] studies of the $ACrX_2$ chalco-genides with A = Cu, Ag, Na, Li, K have shown that the tendency towards ferromagnetism starting from antiferromagnetic behaviour grows up as the lattice constant a increases. Actually with increasing radius of A the asymptotic Curie temperature changes from large negative to positive values.

In contrast to the other $MeCrX_2$ compounds with lower values of a and c and with antiferromagnetic types of magnetic order, the $TlCrX_2$ compounds which have also the greatest values of a and c exhibit a ferromagnetic type of order with a quite high value of the Curie temperature. The close correlation between the Cr-Cr distance within the hexagonal layers and the asymptotic Curie temperature can be understood according to Engelsman et al. in terms of Goodenough's model based on the competition between direct cation-cation and indirect cation-anion-exchange[11]. The direct Cr-Cr exchange across the common edges of adjacent CrX_6

(X = S, Se) octahedra involves the t_{2g} half filled orbitals of the Cr^{3+} ions, thus favouring an antiferromagnetic coupling of the spins. The indirect Cr-X-Cr exchange under an angle close to 90° involves a half filled t_{2g} orbital of one cation and an anionic p orbital, thus favouring the ferromagnetic coupling. Since the direct cation-cation exchange is much more sensitive to the distance between cations as the indirect one it seems plausible to expect that for large Cr-Cr-distance the indirect exchange predominates thus leading to a ferromagnetic arrangement of the Cr-moments within the Cr-layers. The magnetic moment per Cr-atom in the ferromagnetic compounds $TlCrS_2$, $TlCrSe_2$ and Tl_3CrS_3 in the paramagnetic region lies close to the spin-only value of 3,8 Bohr magnetons of the Cr^{3+} ion. But below the transition temperature the Cr-moment is smaller reaching in the highest used fields of 70 kOe values in the range 2.3-2.5 Bohr magnetons.

It is interesting to notice that similar values were obtained in the magnetically ordered phases of $NaCrX_2$, $AgCrX_2$, $Cr_{0.84}Te$ and the Cr-chalcogenides with spinel structure. The reduction of the Cr moment in the ordered phase seems therefore to generally occur in the Cr-chalcogenides with both cubic and hexagonal crystal structure independent on the nature of a second nonmagnetic metallic partner.

While $TlCr_3S_5$, $TlCr_3Se_5$ and $TlCr_5S_8$ are antiferromagnetic with still negative paramagnetic Curie temperatures, $TlCrTe_2$ is antiferromagnetic with a positive Curie temperature.

The antiferromagnetic properties of the Cr-chalcogenides with 3 and 5 Cr-atoms per formula unit could be interpreted in terms of a stronger influence of the direct exchange antiferro-magnetic interaction Cr-Cr due to the shorter distances Cr-Cr. Unfortunately the lack of structure data about $TlGrTe_2$ doesn't allow to speculate about the relative stronger antiferromagnetic character as compared to the ferromagnetic behaviour of $TlCrS_2$ and $TlCrSe_2$.

CONCLUSIONS

Owing to the large radius of Tl^+, many of the Tl-Cr-chalcogenides exhibit ferromagnetic ordering with Curie temperatures in the range 60-140 K. This result is in good agreement with earlier observations on g, Cu and alcaline-Cr-chalcogenides[5] where the increase of the ionic radius gives rise to a trend from anti-ferromagnetic ordering with negative paramagnetic Curie temperatures

towards antiferromagnetism with increasing positive paramagnetic
Curie temperatures. The critical Cr-Cr distance in the hexagonal
Cr-layers for occurrence of ferromagnetism seems to be 3.6 Å. The
occurrence of ferromagnetism could be at least qualitatively
explained as in[5] assuming that the increase of the Cr-Cr distance
will sharply decrease the direct negative interaction Cr-Cr in
favour of the positive indirect exchange Cr-X-Cr.

An increase of the number of Cr atoms per formula unit,
i.e. from $TlCrX_2$ towards $TlCr_3X_5$ and $TlCr_5X_8$ favours the
appearance of antiferromagnetism with negative paramagnetic
Curie temperatures.

In the ferromagnetic Tl-Cr-chalcogenides the Cr-moment
below the Curie temperature seems to be strongly field dependent
tending towards a value of about 2,5 μ_B, i.e. lower as the spin
only one of 3 μ_B of the parallely ordered Cr^{3+} spin-moments.

The reason for this decrease is not yet understood.
A possibility is that below the Curie temperature the type of
ordering is not collinear, i.e. there is a tendency to spin
canting in the hexagonal plane.

Further magnetic measurements in still higher fields
(H > 100 kOe) and NMR and neutron diffraction measurements are
highly desirable in order to explain this behaviour.

The growth of suitable single crystals and the
investigation of their magnetic properties could also contribute
to the understanding of the reduction in the value of the Cr-moment
eventually related to anisotropy effects.

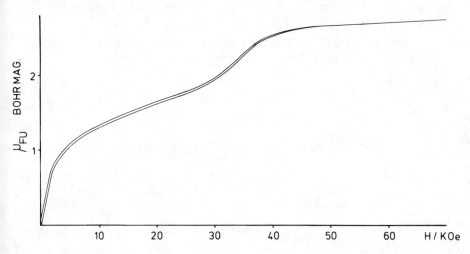

Fig.5 : Magnetization curve of Tl_3CrS_3 at 4.2 K.

Acknowledgement - The authors are indebted to the Deutsche
Forschungsgemeinschaft for supporting a part of present work.

REFERENCES

1. Platte, Chr. and Sabrowsky, H., Naturwissenschaften 62, 528
   (1975).

2. Bongers, P.F. and Enz, U., Commun. 4, 153 (1966).

3. Bongers, P.F., Van Bruggen, C.F., Koopstra, J., Omloo W.P.F.A.
   M., Wiegers, G.A. and Jellinek, F., J. Phys. Chem. Solids, 29,
   977 (1968).

4. Van Laar, B. and Engelsman, F.M.R., J. Sol. State Chem. 6,
   389 (1973).

5. Engelsman, F.M.R., Wiegers, G.A., Jellinek F. and Van Laar, B.,
   J. Sol. State Chem. 6, 574 (1973).

6. Gehle, E. and Sabrowsky, H., Z. Naturforsch. 30 b, 659 (1975).

7. Rüdorff, W. and Stegemann, K., Z. anorg. allg. Chem. 251, 376
   (1943).

8. Platte, Chr. and Sabrowsky, H., Naturwissenschaften 60, 474
   (1973).

9. Delmas, C., Menil, F., Le Flem, G., Fouassier, C. and
   Hagenmüller, P., J. Phys. Chem. Solids, 39, 51 (1978).

10. Delmas, C., Le Flem G., Fouassier, C. and Hagenmüller, P.,
    J. Phys. Chem. Solids, 39, 55 (1978).

11. Goodenough, J.B., In Magnetism and the Chemical Bond,
    Interscience, New York-London (1963).

n-TYPE CONDUCTION IN CHALCOGENIDE GLASSES OF THE Ge-Se-Bi SYSTEM

S. Vikhrov*, P. Nagels, and P.K. Bhat**

Materials Science Department, S.C.K./C.E.N.

B-2400 Mol, Belgium

Measurements of dc electrical conductivity, thermopower and Hall effect are carried out as a function of temperature on glasses of the $(GeSe_{3.5})_{100-x}Bi_x$ system. Within the composition limits studied (x = 8 to 14), the Bi doped $GeSe_{3.5}$ glasses show n-type conduction as evidenced by the thermopower measurements. The sign of the Hall coefficient is also negative. The Hall mobility is found to be very low ($\sim 10^{-2}$ $cm^2V^{-1}s^{-1}$ at 370 K) and slightly decreases with increasing temperature. The thermopower exhibits a higher activation energy than the conductivity. The results are discussed on the basis of the Mott and Davis model for the band structure of amorphous semiconductors.

It is well known that the electrical conductivity of chalcogenide glasses is only slightly affected by doping. Mott[1] suggested that this insensitivity should arise from the ease with which the amorphous structure can adapt itself to fulfil all the valency requirements of the impurities. Another essential feature of the transport properties of the chalcogenide glasses is the sign discrepancy between the thermopower and the Hall coefficient. Chalcogenide glasses, prepared in bulk form by quenching from the melt, generally show a positive sign of the thermopower. It is believed that the sign of the thermopower is a reliable indication of the conduction type in amorphous semiconductors. A positive sign indicates, therefore, that holes dominate the conduction process. In contrast to the thermopower, the sign of the Hall coefficient is usually negative. In a few exceptional cases an n-type thermopower has been observed in chalcogenide glasses.

Twadell et al.[2] found that Cl doping in amorphous Se gives rise
to this conduction type. In recent studies, Tohge and coworkers[3,4]
reported that $Ge_{20}Bi_xSe_{80-x}$ and $Ge_{20}Bi_xSe_{70-x}Te_{10}$ glasses show a
sign reversal from p to n at a bismuth concentration of 9 at. %.
The appearance of n-type conduction in these materials was
explained by these authors on the basis of the "charged dangling
bonds" model according to which the Fermi level is pinned near the
middle of the band gap by an equilibrium between positively and
negatively charged dangling bonds. The presence of negatively
charged Bi perturbs the equilibrium, and as a result unpins the
Fermi level. Because of their unique features, it seemed interesting
to us to study the Bi doped Ge-Se and Ge-Se-Te systems by performing
measurements of dc electrical conductivity, thermopower and Hall
effect as a function of composition and of temperature.

The glasses used in our study had a constant selenium
to germanium ratio equal to 3.5. Bismuth was incorporated in a
composition range from 8 to 14 at. %. The materials were prepared
in the conventional way by heating appropriate mixtures of the
pure elements in quartz tubes at 1050°C for 48 hours with continuous
rotation to ensure good homogeneization. The melt was then quenched
in ice-water. The chemical composition of the various glasses is
given in Table 1. X-ray diffraction and electron microscopic
examination indicated the absence of crystallinity in all the
materials. The electron diffraction patterns showed diffuse rings
characteristic for amorphous materials. The homogeneity of the as-
quenched glasses was examined by microprobe analysis. On a surface
of size comparable to that used in the electrical measurements
several areas were selected for a relative determination of the
three composing elements. This analysis did not show much variation
in composition. It is not excluded, however, that the material might
be composed of more than one amorphous phase on a scale smaller
than what can be revealed by the microprobe analysis. Some
indication in this direction was found in differential thermal
analysis experiments. The DTA curves of all the Bi doped samples
show a number of peaks which give evidence for the existence of
at least two glass transition, crystallinzation and melting
temperature in each composition. At the moment, it is not clear,
however, whether the starting material is multiphasic or whether
the material decomposes upon heating (around 300°C) into a number
of amorphous phase. This problem will be examined in further detail
by transmission electron microscopy and scanning electron
microscopy.

The experimental dc conductivity data of four samples of
the $(GeSe_{3.5})_{100-x}Bi_x$ system are shown in Fig.1. It can be seen
that the conductivity markedly increases when the Bi content
increases from 8 to 10 at. %. The addition of higher amounts of Bi

Table I. Physical Parameters of Electrical Conductivity and Thermopower
for Ge-Se based Glasses

Composition	E_σ (eV)	σ_0 ($\Omega^{-1} cm^{-1}$)	E_s (eV)	S_0 ($\mu V\ K^{-1}$)	$\dfrac{eS_0}{k}$
$(GeSe_{3.5})_{88}Sb_{12}$	1.60	1.3×10^3			
$(GeSe_{3.5})_{92}Bi_8$	0.92	4.4×10^3			
$(GeSe_{3.5})_{90}Bi_{10}$	0.645	1.5×10^2	0.76	+ 810	+ 9.4
$(GeSe_{3.5})_{88}Bi_{12}$	0.645	3.1×10^2	0.69	+ 400	+ 4.6
$(GeSe_{3.5})_{86}Bi_{14}$	0.655	1.7×10^3	0.705	− 280	− 3.2
$(GeSe_3Te_{0.5})_{90}Bi_{10}$	0.64	3.3×10^2	0.87	+ 530	+ 6.1

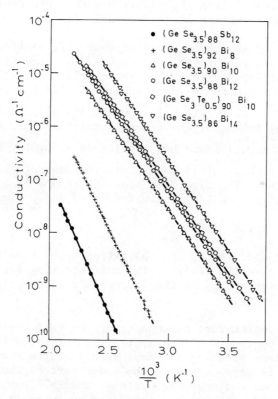

Fig.1 : Temperature dependence of the electrical conductivity of
four $(GeSe_{3.5})_{100-x}Bi_x$ glasses with x = 8, 10, 12 and 14,
a $(GeSe_3Te_{0.5})_{90}Bi_{10}$ glass and a $(GeSe_{3.5})_{88}Sb_{12}$ glass.

(10 to 14 at. %) has a less pronounced effect. It has generally
been observed that the incorporation of tellurium in chalcogenide
glasses leads to an increase in conductivity. For this purpose we
prepared a glass in which selenium was partly substituted by
tellurium. The conductivity data of this glass with composition
$(GeSe_3Te_{0.5})_{90}Bi_{10}$ are also represented in Fig.1. Surprisingly,
the substitution of tellurium for part of selenium has little
effect on the conductivity. In the same figure we also show a
chalcogenide glass in which bismuth was replaced by antimony.
This alloy, having the composition $(GeSe_{3.5})_{88}Sb_{12}$, exhibits a
much lower conductivity than the corresponding bismuth doped glass.
The conductivity of all compositions can be expressed by an
exponential relationship of the form :

$$\sigma = \sigma_o \exp (- E_\sigma/kT)$$

The values of the activation energy E_σ and of the pre-exponential
constant σ_o, deduced from the slopes and the intercepts at $1/T = 0$
of the straight lines are listed in Table I.

In Fig.2 the thermopower of four different compositions
is plotted versus reciprocal temperature. The sign of the thermopower of the Bi doped alloys is negative, in contrast with the
Sb doped glass for which p-type conduction was observed (S =
1100 μV K^{-1} at $10^3/T = 2.27$ K^{-1}). Due to its high resistance, only
one S value was taken for the alloy containing 8 at. % of Bi. The
thermopower is negative and has a value (S = -320 μV K^{-1} at $10^3/T$
= 2.65 K^{-1}), which is still lower than the ones of the alloys with
higher Bi content. The S vs. 1/T plots display a linear behaviour
and can be represented by the usual formula :

$$S = -k/e (E_s/kT + A)$$

The activation energies E_s and the extrapolated S values at $1/T = 0$
(S_o) are also given in Table I. The most striking feature is the
higher activation energy of the thermopower as compared to that
of the conductivity.

The temperature dependence of the Hall coefficient is
shown in the upper part of Fig.3. The sign of the Hall coefficient
is negative, thus yielding the same result as the thermopower. The
Hall mobility is represented in the lower part of the same figure.
Compared to other chalcogenide systems, the Hall mobility is
extremely low and exhibits a slight decrease with increasing
temperature.

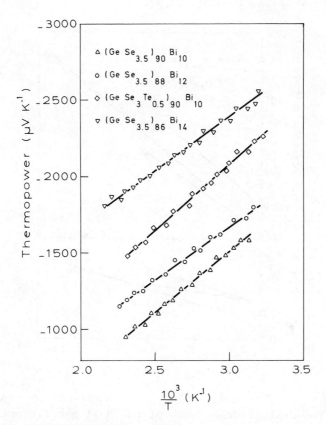

<u>Fig.2</u> : Thermopower versus reciprocal temperature of three
$(GeSe_{3.5})_{100-x}Bi_x$ glasses with x = 10, 12, 14 and a
$(GeSe_3Te_{0.5})_{90}Bi_{10}$ glass.

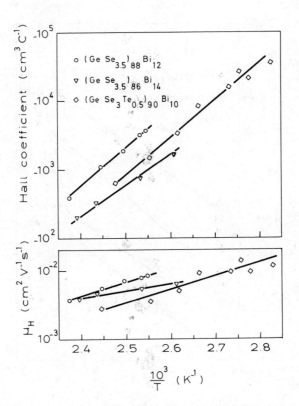

<u>Fig.3</u> : Temperature dependence of the Hall coefficient (upper part) and the Hall mobility (lower part) of two $(GeSe_{3.5})_{100-x}Bi_x$ glasses with x = 12, 14 and a $(GeSe_3Te_{0.5})_{90}Bi_{10}$ glass.

We shall interpret the electrical data on the basis of
the Mott and Davis model[5] for the energy dependence of the electronic
density of states, applicable to amorphous semiconductors. The
essential feature of this model is the existence of narrow tails
of localized states at the extremities of the valence and conduction
bands. The model furthermore assumes that missing chalcogenide atoms
leave dangling bonds on other atoms. These dangling bonds are
positively and negatively charged and form states of localized
nature near the middle of the gap. In a semiconductor the position
of the Fermi level is determined by the neutrality condition. Due
to the high and nearly equal concentration of the positively and
negatively charged defect states, the Fermi level will be pinned
near the center of the mobility gap. In most of the chalcogenide
glasses, the Fermi level lies somewhat closer to the mobility edge
of the valence band giving rise to p-type conduction. The appearance
of n-type conduction in the Bi doped samples must result from a
shift of the Fermi level towards the mobility edge of the conduction
band, due to a change in the relative concentration of the positively
and negatively charged dangling bonds.

We shall now examine the conductivity and thermopower
data on the basis of the Mott and Davis model. For unipolar
conduction by electrons in extended states the conductivity and
thermopower must show the same activation energy. For hopping
conduction in a band tail the activation energy derived from the
thermopower curve must be smaller than that of the conductivity.
In the Bi doped chalcogenide glasses we did not observe either one
of these features. On the contrary, the activation energy of the
thermopower is significantly higher than the one associated with
the conductivity. The apparently higher activation energy of the
thermopower can be explained if one takes into account a bipolar
transport mechanism. The dominating contribution in the transport
process arises from the conduction of electrons in the extended
states of the conduction band. The second contribution, which must
be a positive one, could originate from conduction of holes in the
extended states of the valence band. This two-channel model will
yield a more drastic drop of the thermopower towards higher
temperatures than that expected for unipolar conduction by
electrons. The very low value of the Hall mobility together with
its temperature dependence give further evidence for a bipolar
conduction process. For one type of charge carrier the random
phase model of Friedman[6] predicts a temperature-independent Hall
mobility. We observed, however, a decrease with increasing
temperature, which might be explained by an increasing contribution
of hole conduction in addition to the dominant electron conduction.

In the foregoing discussion it was tacitly assumed that
all the materials behave as homogeneous systems. Although it has
not been proved undoubtedly, this might not be the case. Further

investigation is needed in order to find out whether the Bi doped
Ge-Se glasses are mono- or multiphasic amorphous systems.

Acknowledgement - The authors wish to thank L. Van Gool for
his assistance in the sample preparations and the measurements.

REFERENCES

* Permanent address : Ryazan Radio Engineering Institute,
   Ryazan, USSR.

** Present address : Institut für Angewandte Physik, Technische
   Universität Wien, Wien, Austria.

1. N.F. Mott, Philosophical Magazine 19, 835 (1969).

2. V.A. Twaddell, W.C. Lacourse and J.D. Mackenzie, Journal
   non-crystalline Solids, 8-10, 831 (1972).

3. N. Tohge, T. Minami, Y. Yamamoto and M. Tanaka, Journal
   Applied Physics 51 (2), 1048 (1980).

4. N. Tohge, T. Minami and M. Tanaka, Journal non-Crystalline
   Solids 37, 23 (1980).

5. N.F. Mott and E.A. Davis, Electronic Processes in non-
   Crystalline Materials (Clarendon Press Oxford, 1979), p.209.

6. L. Friedman, Journal non-Crystalline Solids 6, 329 (1971).

DIRECT AND INDIRECT LCAO RECURSION METHODS FOR SURFACES

G. Biczó

Central Research Institute of Chemistry
Hungarian Academy of Sciences

H-1525 Budapest 114, P.O. Box 17, Hungary

In the LCAO representation our direct recursion (transfer matrix, T) method (DRM) applies strictly valid recurrence relations straight to the solution of the Hamiltonian matrix eigenvalue problem for the electronic system in bounded finite or infinite crystals and polymers with perturbed boundaries. We present the different reduced forms of the auxiliary equations of the DRM and introduce the notion of duality for these. Their relations to the equations of cyclic and half- infinite systems are given. The indirect recursion method (IRM) of Fromm and Koutecky (FK) utilize recurrence relations (strictly valid only for half-infinite perfect systems) to derive the bordering blocks of the resolvent (Green) matrix R. We completed their expressions with those of the inner blocks of R. Contrary to their statement, the inversion of matrix block B which describes the interactions between the consecutive sub-systems has similar role in both methods. We proved that the B^{-1}-free reduced forms of the auxiliary equations of the DRM for semi-simple T are identical to the FK equation of the IRM for the corner diagonal block of R multiplied by B'. Algebraic methods are proposed to eliminate the "B^{-1}-difficulty" of the transfer matrix method. The IRM was found generally much more complicated than the DRM is.

1. INTRODUCTION

We call an LCAO recursion method direct (DRM) if it is applied straight to the solution of the eigenvalue problem in equation and indirect (IRM) if first the resolvent of the eigen-equation is calculated. To investigate electronic problems of

341

polymers and crystals at their boundaries (ends and surfaces), we
have developed a DRM version at the simple LCAO level, outlining
its SCF generalization by the use of the already derived charge-
bond order matrices[1]. Very recently an IRM[2,3] was presented in
the Third International Congress of Quantum Chemistry (Kyoto, 1979)
by Koutecky. Since the two methods have very profound connections,
it may be of some interest to compare them. This is the object of
the present paper.

2. THE DIRECT RECURSION METHOD

The DRM has been presented hitherto in short
communications[1,4,5]. Only a special problem, the existence of
some "intermediate-" or "zigzag states" (ZZS) was published in
detail[6]. The complete formulation was circulated, however, in the
form of an unpublished preprint[7] which is even now available by
direct request from the author. We repeat here a few parts of the
above preprint.

We formulate the problem for a column of elementary units
(cells). The column is finite. It has a number of N units which,
naturally, may be composed of rows or layers of smaller primitive
elementary cells. Thus we treat and utilize the translational
symmetry of a 3-dimensional crystal only in the direction of one
of its crystallographic axes in the case of real crystals, for
example. In the simple LCAO (Hückel) description we apply first
neighbour approximation. So we can visualize our column by giving
the Hückel matrices of the units in one row and placing the beta-
interaction matrices above the blanks between the former :

$$F_{12} \quad F_{23} \quad \cdots \quad F_{r-1r} \quad B \quad B \quad \cdots \quad B \quad B \quad \overline{F}_{ss-1} \quad \cdots \quad \overline{F}_{21}$$

$$F_{11} \quad F_{22} \quad \cdots \quad \quad F_{rr} \quad A \quad \cdots \quad A \quad \overline{F}_{ss} \quad \cdots \quad \overline{F}_{11}$$

$$(1) \quad (2) \quad \cdots \quad (r)(r+1) \quad \cdots \quad (N-s)(N-s+1)\cdots \quad (N)$$

The Hückel matrix of the whole system is now a hypermatrix F with
the lower row of matrices in its diagonal blocks and with the
upper ones in the first superdiagonal. The subdiagonal contains
the transposes of the latter. Seemingly, our column contains a
number of r endperturbed units in its "penetration region" (PR)
at its left hand side. s is the number of the perturbed units at
its right side. In the translationally symmetrical "inner region"
(IR) we have a number of N-r-s repeated units. Let all these
matrix blocks be quadratic and or order n.

We decompose now the real symmetrical matrix eigenvalue
equation

$$Fc = \lambda c \ , \tag{1}$$

where c is a hyper-column-vector of order nN with the column-vector c_j in its j-th block position. In the IR, if B' is the transpose of B and det B $\neq$ 0,

$$B'c_{j-1} + Ac_j + Bc_{j+1} = \lambda c_j \qquad (j = r+1,\ldots,N-s)$$

and it is equivalent to the simple recurrence relation

$$x_{j+1} = Tx_j \ . \tag{2}$$

Here :

$$x_j = \begin{bmatrix} c_{j-1} \\ c_j \end{bmatrix}, \ T = \begin{bmatrix} 0 & I \\ D & G \end{bmatrix}, \ D = -B^{-1}B', \ G = B^{-1}(\lambda I - A)$$

0 and I being the zero and the unit matrix of order n. The trick giving equation (2) is due to Schmidt[8] and Hori and Asahi[9]. The latter authors called T transfer matrix and x_j state vector. (The most general version has been developed by Matsuda et al.[10] for vibrational problems). The definition of the recursion matrices for the PR is similar, only A and F must be replaced by, e.g. F_{jj} and F_{jj+1}, respectively, etc.

Contracting our recurrence equations of form (2) by successive substitutions we obtain the more condensed form

$$x_{j+1} = \begin{cases} {}^j R x_1 & \text{for } 1 \leq j \leq r \\ T^{j-r} R x_1 & \text{for } r < j \leq N-s \\ {}^j S T^{N-r-s} R x_1 & \text{for } N-s < j \leq N \end{cases} \tag{3}$$

Here ${}^j R = T_j \ldots T_2 T_1$, $R = {}^r R$, ${}^j S = \overline{T}_{N-j+1} \ldots \overline{T}_{s-1} \overline{T}_s$ and $S = {}^N S$. Completing (3) with the boundary conditions

$$x_1 = \begin{bmatrix} 0 \\ c_1 \end{bmatrix} \qquad \text{and} \qquad x_{N+1} = \begin{bmatrix} c_N \\ 0 \end{bmatrix} \tag{4}$$

it will be equivalent to the original eigenvalue problem, if there is no singular matrix among the first superdiagonal matrix blocks of F.

The lower block component of the last equation in (3) is the fundamental equation. Introducing the notation

$$M = \begin{bmatrix} M_{00} & M_{01} \\ M_{10} & M_{11} \end{bmatrix}$$

for the block decomposition of 2n by 2n matrices into four quadratic blocks, we obtain the first form

$$(ST^{N-r-s}R)_{11}c_1 = 0 \tag{5}$$

of the fundamental equation. It is a homogeneous linear equation. Its characteristic determinant is an algebraic polynomial of λ. The roots of this function are the permitted values of λ, i.e., the requested eigenvalues of (1). The appropriate non-zero vectors of (5) determine the other vector block of the requested hyper-eigenvectors of (1) by equation (3). The solution of (5), however, is not simple and its interpretation is also not too easy, since T has a complicated, not regularly antisymmetrical substructure.

A more convenient second form of the fundamental equation was published in the Vienna paper[1] as its equation (4). For its derivation we had to introduce the auxiliary equations

$$UT = \tau U \tag{6}$$

and

$$TQ = Q\tau \tag{7}$$

which are the left and the right eigenvalue problems of T written in compact matrix equation form when T is semisimple and may be regarded as normal form problems when T is defective. (See equations (2) and (3) in [1]). One row-component of, say (6) looks like

$$(u\ v) \begin{bmatrix} 0 & I \\ D & G \end{bmatrix} = t(u\ v) \tag{8}$$

Since det T = 1, T can have only nonzero eigenvalues. Therefore, we can decompose and reformulate (8) as

$$t^{-1}vD + vG = tv \quad \text{or} \quad \bar{v}(A + Bt + B't^{-1}) = \bar{v}\lambda, \tag{9}$$

if

$$\bar{v} = vB^{-1}. \text{ Similarly, } (A + Bt + B't^{-1})q = \lambda q \tag{10}$$

follows from (7). (Cf. equations (36) and (19) of [7]). Their common characteristic equation

$$\det(A + Bt + B't^{-1} - \lambda I) = 0 \tag{11}$$

is a reciprocal equation for t (see Lemma III in [7]), since the left hand side of (11) is seemingly a symmetrical function in t and t^{-1} when the determinant is evolved. Thus we can order the eigenvalues of T as it was done in [1] : $|\mu_k| \leqslant |\mu_{k+1}| \leqslant 1$ for

$k = 1,\ldots,n-1$, $\mu = \mathrm{diag}(\mu_k)$ and $\mu\nu = I$. Correspondingly, a special

decomposition was given for equations (6) and (7). Using definitions

$$U = \begin{bmatrix} X & Y \\ W & Z \end{bmatrix}, \quad Q = \begin{bmatrix} J & K \\ L & M \end{bmatrix}, \quad \text{and} \quad \tau = \begin{bmatrix} \mu & 0 \\ 0 & \nu \end{bmatrix} \quad \text{of } [1]$$

we arrive at the reduced form of the auxiliary equations (16c,a) and (17c,a) of [7] :

$$\nu YD + YG = \mu Y, \quad X = \nu YD \tag{12}$$

$$DJ\nu + GJ = J\mu, \quad L = J\mu \tag{13}$$

We introduce now $\overline{Y} = YB^{-1}$ into (12) and premultiply the first equation of (13) by B :

$$\nu\overline{Y}B' + \overline{Y}(A - \lambda I) + \mu\overline{Y}B = 0 \tag{14}$$

$$B'J\nu + (A - \lambda I)J + BJ\mu = 0 \tag{15}$$

These are the "B^{-1}-free" reduced auxiliary equations in condensed form for semisimple T. The appropriate equations for W, Z, K and M may be obtained from the above equations by their substitutions into the place of X, Y, Z and L and by exchanging μ and ν. These are, at the same time, condensed forms of the generalized eigen-value equations (9) and (10) of the present paper or those of equations (36) and (19) of [7].

For defective T, matrices μ and ν become matrices of Jordan normal form or at least one number 1 appears in the right upper block of τ or both phenomena occur. We do not discuss this otherwise very interesting case here[6], since the IRM to be compared has been developed only for the semisimple case in[3].

The "Characterization of the Solutions" is a section in[1]. Therefore, we give here only a few remarks on the interpretation of the auxiliary equations from [7]. REMARK I. Auxiliary equation (10) of the DRM is nothing but the well-known eigenvalue equation of of a perfect polymer or crystal (see, e.g. in [11] when the Born-von Kármán periodic boundary condition has been applied and if $|t| = 1$, say, $t = \exp(i\kappa)$. REMARK II. We may conclude from the above remark that there is at least one eigenvalue μ_k of T with unit absolute value if and only if λ lies in at least one of the "Born-von Kármán" type bulk bands. REMARK III. The notion of

duality can be defined for different forms of the auxiliary
equations which may be useful both for interpretational and
calculational reasons. Namely, let us call the pair of eigenvalue
equations (7) and (10) dual equations with respect to our original
"large" eigenvalue problem (1). Their relations are not symmetrical.
So we may call (7) the primal and (10) the dual pair of the primal.
Equivalently, the first equation of (12) or (13) may be regarded
also as primal (but now generalized!) eigenvalue equations in a
similar sense. The primal equation is an eigenvalue problem for
the eigenvalues μ_k when the value of λ is fixed. Its dual, (10),
may be regarded as an eigenvalue problem for λ when μ_k $(= \nu_k^{-1})$ is
fixed. The latter, in general, is not a Hermitian problem.

It is very important to investigate the intimate
connections of our dual equations since they have a key role and
characterize several essential properties of both the localized
and the bulk states. If, e.g., λ is a single eigenvalue of (10),
then generally, only one of its eigenvectors q may be used for
the construction of J as well as for Q. (We must keep in mind that
$\mu_k = \mu_k(\lambda)$. Accidentally, however, any of its other eigenvalues,
say λ', may be a common eigenvalue of $A + B\mu_k + B'\nu_k$ and $A + B\mu_m + B'\nu_m$
if $\mu_k(\lambda) = \mu_m(\lambda')$.) In the degenerate case, however, when λ has a
multiplicity of say v, all the v linearly independent eigenvectors
may be used, if they exist at all. The use of the dual equation
may be advantageous in actual calculations for bulk band regions,
since here, at least for one μ_k, it is a Hermitatian eigenvalue
problem (cf. REMARK III).

The primal equation may be regarded as a generalized
eigenvalue equation $(\lambda I - A)q = (B\mu_k + B'\mu_k^{-1})q$ of the real symmetrical
matrix $\lambda I - A$ with fixed λ, where μ_k is the generalized eigenvalue.
Apparently, the partly common eigenvector system of our dual
equations may characterize the degeneracy and other properties
of our original large eigenvalue problem (1). We must still mention
that they have common characteristic equation (11). For a deeper
understanding of these questions, however, we need further
investigations.

3. THE INDIRECT RECURSION METHOD

The IRM of[3] combines different methods to treat half-
infinite systems with perturbed PR and half-infinite IR. Inverting
by partitioning technique and using the Dyson equation, the total

Green matrix is expressed by the Green matrices of the non-
interacting PR and IR. By the same techniques, a special equation
is derived for the diagonal blocks of the latter Green matrix.

We give here an alternative, block-decompositional
derivation of the above equation. At the same time, we utilize
the main idea of their procedure[3]. Namely, that a half-infinite
matrix may be identical to its certain submatrices. Let

$$E = E(z) = zI - F_{IR} = \begin{bmatrix} E_{11} & E_{12}^{\infty} \\ E_{21}^{\infty} & E \end{bmatrix} ,$$

$$E_{11} = zI - A, \quad E_{12}^{\infty} = (-B\ 0\ 0\ 0\ ...), \quad E' = E$$

$$ER = I .\tag{16}$$

Here F_{IR} is half-infinite, otherwise its three main (block)
diagonals have the same matrix blocks as the IR of our original
matrix F. Seemingly, E is also block-tridiagonal and half-infinite.
R is the appropriate Green matrix (or resolvent) of the separated
perfectly translation symmetrical half-infinite IR. If

$$R = \begin{bmatrix} R_{11} & R_{12}^{\infty} \\ R_{21}^{\infty} & R_{22}^{\infty} \end{bmatrix}$$

is its appropriate decomposition, then we obtain from (16) :

$$E_{11}R_{11} + E_{12}^{\infty}R_{21}^{\infty} = I ,\tag{17}$$

$$E_{21}^{\infty}R_{11} + ER_{21}^{\infty} = 0\tag{18}$$

and two other equations of similar structures. Premultiplying (18)
by R we can eliminate

$$R_{21}^{\infty} = -RE_{21}^{\infty}R_{11}\tag{19}$$

from (17). Postmultiplying the resulted equation by B', taking into
account that $E_{12}^{\infty}RE_{21}^{\infty} = BR_{11}B'$ and introducing

$$H = R_{11}B' ,\tag{20}$$

we arrive at the equation of[3] in question :

$$BH^2 - E_{11}H + B' = 0\tag{21}$$

The sign difference in the middle term is due to the sign
difference in definition (20).

Equation (21) determines the corner block R_{11} of the
Green matrix. We can obtain the bordering blocks of it from the
blocks of equation (19) by their successive substitution and
taking into account definition (20) again :

$$R_{j+11} = H^j \, R_{11} \, . \tag{22}$$

We have to emphasize here that hypermatrix R has not the property
of E that it is identical to its certain own submatrices. Therefore, we have to complete the equations of[3] if we need the inner
blocks of R also. If we extend the block decomposition until the
first k-1 block rows and block columns, we can derive the following
relations :

$$R_{j+kk} = H^j \, R_{kk} \qquad (j,k = 1,2,3,\ldots) \tag{23}$$

and

$$R_{kk} = R_{11} + HR_{k-1k-1}H' = \ldots = \sum_{\bar{k}=1}^{k} H^{\bar{k}-1} R_{11} (H')^{\bar{k}-1} \tag{24}$$

Since R = R', these equations completely determine the Green
matrix R if equation (21) is solved for H and equation (20) is
applied to express R_{11}. It is important to keep in mind that just
for the last step the inversion of B is necessary.

4. COMPARISON

First we can observe that the validity of the recurrence
relations used in the[3] version of the IRM is restricted to half-
infinite systems having perfect translational symmetry. This is
not the case for the DRM which may be applied equally well both
for finite and infinite systems with, in addition, perturbed
boundaries.

Fromm and Koutecky[3] overcome. the difficulties arising
from the boundary perturbation by the use of another, quite
different approach (partitioning inversion) in the IRM in question.
If the PR is deep (r is great), then this method is dealing with
unnecessarily large matrices of order rn, instead of n, the
characteristic order of the DRM[5].

Matrix B which describes the interactions of the consecutive subsystems has a similar role in both methods. It is merely a formal difference of conventional origin that, on the one hand, T contains B^{-1} explicitly in the transfer matrix method and, on the other hand, that Fromm and Koutecky[3] leave the inversion for the last step, when equation (20) is required to calculate R_{11} from H and B. When det B = 0, both methods will have the "B^{-1}-difficulty". We can overcome it within the framework of both methods by standard algebraic methods (see in [7]). We proposed very recently a really "B^{-1}-free DRM" version[12].

The similarity between our auxiliary equation (15) and the Fromm-Koutecky-equation (21) is striking. We can give their relation exactly. For this reason, let us postmultiply equation (15) by μJ^{-1} and insert $J^{-1}J$ in between μ and μ in μ^2. We see that

$$H \leftrightarrow J\mu J^{-1} \tag{25}$$

The discussion of the uniqueness for relation (25) is a little bit cumbersome. For example, even if all eigenvalues μ_k are different, a given solution of (21) for H does not determine J uniquely, only up to a postmultiplying diagonal phase matrix. Nevertheless, the eigenvalues are uniquely determined.

We summarize these in two further remarks : REMARK IV. The reduced forms (14) and (15) of the auxiliary equations of the DRM are equivalent to the Fromm-Koutecky-equation (21) of the IRM that determines the corner diagonal block R_{11} of the resolvent matrix R corresponding to the (interacting!) terminal unit (elementary cell, layer, etc.) of a half-infinite linear system (polymer, crystal, etc.) having perfect (!) translational symmetry. REMARK V. The recursion (transfer) matrix T and the Fromm-Koutecky-matrix H have common eigenvalues which determine the way of propagation of the one-electron wave functions.

5. CONCLUSION

The characteristic hypermatrix E (similarly to the Hamiltonian F) of a half-infinite perfect polymer or crystal has perfect translation symmetrical block structure. Even in the nearest neighbour approximation, its inverse, the resolvent hypermatrix R is much more complicated. How many matrix blocks are multiplied within the expression of one single block? In R_{jk} just j+k-1 (see equations (23-24)), in E_{jk} only 1 or 0 :

1	2	3	...	2k-1	2k	...		1	1	0	0	0	0	...
2	3							1	1	1	0	0	
3		2k-1		2k				0	1	1	1		
:		2k						0	0	1			
2k-1								0	0				
2k								0					
:								:					

in R   and   in E

In addition, the matrix factors H and R_{11} in the former are in
themselves hardly derived structures. We can conclude, therefore,
that an IRM must be generally much more complicated than a DRM.

At the same time, the inclusion of perturbed boundaries
is natural within the DRM, but it requires the mixing of different
techniques for the IRM.

For finite, doubly bounded infinite, periodic and half-
infinite polymers and crystals we found some relations between
the auxiliary as well as fundamental equations of the methods
compared. We summarized these in five remarks.

Acknowledgements - The author is very grateful to Professor
J.T. Devreese Conference Chairman for giving the permission to
present this contribution in his absence. He grately thanks
Professor F. Beleznay for the presentation. He is very indebted
to Professor J. Koutecky for showing him the detailed text and
slides of the lecture in Kyoto and for sending its manuscript
before publication. Grateful acknowledgements are made to
Professor G. Schay for the very careful review of the manuscript.

REFERENCES

1. G. Biczó, Proc. 7th Int.Vac.Congr. and 3rd Int.Conf.Solid
   Surf. (Vienna : Berger and Söhne ed. Dobrozemsky et al.) (1977)
   Vol.I, pp.407-10.

2. O. Fromm, to be published (1980).

3. O. Fromm and J. Koutecky, Int. J. Quantum Chem., to be
   published (1980).

4. G. Biczó, Proc. ECOSS I Nederlands Tijdschrift voor Vacuum-
   techniek 16, 195 (1978).

5. G. Biczó, ibid., p.236.

6. G. Biczó, and I. Lukovitz, Int. J. Quantum Chem. 16, 21 (1979).

7. G. Biczó, "On the Calculation of Boundary (Surface) States and Bulk State Distortions at Different Boundaries", Preprint TCG-1-77 (1977, Budapest : CRIC HAS), pp.1-99+22.

8. H. Schmidt, Phys. Rev. 105, 425 (1957).

9. J. Hori and H. Asahi, Progr. Theoret. Phys. (Kyoto) 17, 523 (1957).

10. H. Matsuda, K. Okada, T. Takase and T. Yamamoto, J. Chem. Phys. 41, 1527 (1964).

11. G. Del Re, J. Ladik and G. Biczó, Phys. Rev. 155, 997 (1967).

12. G. Biczó, Abstracts of IVC 8, ICSS 4 and ECOSS 3, (1980), submitted.

CORE LEVEL SHAKE UP STRUCTURES OF N_2 ADSORBED ON NICKEL

SURFACES: CLUSTER MODELS

K. Hermann* and P.S. Bagus

IBM Research Laboratories

San Jose CA 95193, USA

Self-consistent Hartree-Fock-LCAO calculations on a linear NiN_2 cluster are used to explain the two-peak structure observed in core level photoemission from N_2 adsorbed on the Ni(100) surface. The two peaks are due to different N_2 core hole final states, a screened and an unscreened final state that can be clearly identified in the cluster model. The energy separation and intensity ratio (computed using the sudden approximation) of the two final states in NiN_2 are comparable with the experimental peak separation and intensity ratio for reasonable cluster geometries. A comparison with previous model studies on NiCO is used to explain the different shape of the core level spectra of N_2 and CO adsorbed on nickel surfaces.

Core level photoemission spectra from adsorbed molecules have proven to give useful information about the electronic structure of the adsorbate in the presence of the surface. The qualitative shape of adsorbate core level spectra has been investigated[1-3] by studying the time dependent relaxation process when an adsorbate core hole is created. The model based on an Anderson-type hamiltonian leads, at intermediate adsorbate-substrate interaction, to a two-peak structure. The peak at lower binding energy corresponds to a screened final hole state where screening of the adsorbate core hole occurs as a consequence of relaxation by partial filling of adsorbate levels that are empty in the initial state[4]. The peak at higher binding energy relates to a final core hole state where relaxation does not result in screening of the adsorbate core hole. The intensity of the unscreened peak with respect to that of the screened peak should increase with decreasing adsorbate-substrate coupling. The latter result is consistent with experimental photoemission (XPS) data of the two-peak structure observed for CO and N_2 adsorbed on transition metal surfaces[5].

In a previous paper[6] we have shown that for CO adsorbed on Ni surfaces the two-peak structure observed in XPS measurements of both C1s and O1s ionization can be understood by studying relaxed core hole states of a linear NiCO cluster model. This model already contains the essential features of the screening mechanism[4] described above. The two-peak structure arises due to relaxation to different final states, a screened final state at lower binding energy and an unscreened final state at higher binding energy. The energy differences between the two final states for both O1s and C1s ionization are comparable to the measured peak separations and the calculated relative intensities can reproduce the experimental peak ratios at reasonable cluster geometries.

In the present paper we extend our cluster study to molecular N_2 adsorption on Ni surfaces. It will be shown in the following that a linear NiN_2 cluster can provide a reasonable model for core ionization of adsorbed N_2. The two-peak structure observed in the adsorbate core level photoemission spectrum of the $N_2/Ni(100)$ system[5] is explained as due to different final hole states analogous to the CO/Ni(100) system. The experimental energy separation of the two peaks and their intensity ratio can be reproduced by the cluster model reasonably well. Further, the difference in the XPS data for core ionization from adsorbed CO and N_2 is explained in our model by the fact that, compared to adsorbed $N_2, 2\pi^*$ backbonding contributes more strongly to the adsorbate-substrate coupling of adsorbed CO in accordance with the Schönhammer-Gunnarsson (SG) model[1-3].

In the NiN_2 (Ni-N-N) cluster we denote the nitrogen atom nearest to the Ni by N_a and the one furthest away by N_b. For the N_a-N_b distance a constant value of 2.074 bohr from experiments on free N_2[7] is used whereas the $Ni-N_a$ distance is varied about the sum of the atomic radii, $d_o = 3.80$ bohr, in order to study its effect on the $Ni-N_2$ interaction. Similar cluster geometries have been used in studies on the binding energy shifts of chemisorbed N_2[8]. A initio Hartree-Fock-LCAO calculations are performed using extended contracted Gaussian basis sets[9].

From calculations on different electronic configurations of NiN_2 the energetically lowest ^{3} state with open shell structure $11\sigma^2 \, 5\pi^1 \, 1\delta^3$ is considered as initial state for ionization processes. Here the N_2-type orbitals retain much of their free molecule character suggesting a N_2 notation with a tilde added : $3\tilde{\sigma}_g, \tilde{1}\tilde{\pi}_u$ etc. We shall, however, denote the molecular N_2 orbitals with $C_{\infty v}$ rather than $D_{\infty h}$ (the full symmetry group of the molecule) designations in order to facilitate comparison with CO. (Thus $2\sigma_g \equiv 3\sigma$, $1\pi_u \equiv 1\pi$, etc.). A detailed analysis of the cluster electronic structure[8] shows that in NiN_2 both the N_2 $\widetilde{4\sigma}$ and $\widetilde{5\sigma}$ orbitals are involved in the bonding to Ni by admixing Ni 3d contributions and are bonding to about the same extent as can be seen from the Ni-N_a overlap population. The singly occupied 5π cluster orbital in $NiN2$ is characterized as a combination of Ni 4p- with N_2 $2\pi^*$-type contributions and thus models the well known $2\pi^*$ backbonding effect in the system.

The electronic structure of the NiN_2 $^3\Phi$ initial state is slightly different from that of the $^3\Delta$ cluster ground state. The 5π orbital of $^3\Phi$ being unoccupied in the $^3\Delta$ ground state explicitly accounts for Ni 4p - N_2 $2\pi^*$ coupling and therefore models the backbonding interaction of the N_2 adsorbate with the Ni 4sp band at the surface (in addition to quite small Ni 3d - N_2 $2\pi^*$ back-bonding allowed for in both $^3\Phi$ and $^3\Delta$). As a conseqence, the N_2 $2\pi^*$ backbonding effect (found to be essential for the intensities of core ionization processes[3,6]) is increased in $^3\Phi$ with respect to $^3\Delta$ resulting in a more plausible model for the surface situation. However, the Ni-N_2 bond characterized as a coupling of Ni 3d with N2 4σ and 5σ electrons is almost identical in both $^3\Phi$ and $^3\Delta$.

The Ni-N2 bonding found in our cluster model differs from the Ni-CO bonding in linear NiCO[10] in mainly two respects. First, in NiN_2 two adsorbate orbitals, $\widetilde{4\sigma}$ and 5σ, are responsible for the bonding compared to one orbital, 5σ, in NiCO. However, the bond strength in NiN_2 seems to be considerably smaller than in NiCO as indicated by a comparison of population analysis as well as from studying the binding energy shifts of the respective adsorbate valence orbitals discussed elsewhere[8]. Second, as a consequence of the weaker Ni-N2 bonding there is, compared to NiCO, a decreased $2\pi^*$ backbonding in NiN2. For example, the $2\pi^*$ occupancy of $^3\Phi$(NiCO) at Ni-C distances where the experimental photoemission intensity results are reproduced is about 0.6 electrons[6] whereas for $^3\Phi$(NiN_2) at respective Ni-N_a distances (-4.2 bohr) the $2\pi^*$ occupancy is only about 0.3 electrons. It is mainly this difference in backbonding that causes the different satellite peak intensities of the core level spectra for the two adsorbates in our cluster models. This effect should also be present in the real surface situation.

In order to determine relaxed N_2 core ionization potentials of the NiN_2 cluster we compute the self-consistent wavefunctions of the respective ionized states. The ionization potential (IP) is then given by the difference of the (self-consistent) initial- and final-state total energies. Thus, the IP results include all final state relaxation effects within the Hartree-Fock cluster approach. We restrict ourselves in the following to quartet, high spin, final states since, for the present purpose, multiplet splitting[8,10] seems to be neglegible. There are two kinds of nitrogen core hole states in NiN_2 : the N_a 1s hole states corresponding to the removal of a 1s electron from the nitrogen center nearest to the Ni and the N_b 1s hole states where the 1s electron is removed from the nitrogen center further away from the Ni. However, it will be shown that the differences between equivalent N_a 1s and N_b 1s hole states are quite small and can be neglected in a first approximation.

Starting from the $^3\Phi$ initial state we find an energetically lowest N_a 1s core hole final state $^4\Phi_1(N_a 1s)$. Here the occupied orbitals are only slightly modified by relaxation except for the 5π orbital. This backbonding orbital described as a mixture of Ni 4p and N_2 $2\pi^*$ contributions in the initial state becomes, in the $^4\Phi_1(N_a 1s)$ final state, a pure N_2 $2\pi^*$-type orbital. As a result, there is an effective charge transfer from the Ni towards the N_2 part of the cluster partially screening the core hole. So this $^4\Phi_1(Na 1s)$ final state is a cluster equivalent of the screened final state in the SG-model[1-3]. The occupied orbitals of the energetically second lowest relaxed final state $^4\Phi_2(N_a 1s)$ are again very similar to those of the initial state except for 5π. Now the backbonding orbital of the initial state becomes nearly pure Ni 4p-type and a Mulliken population analysis of $^4\Phi_2(N_a 1s)$ gives a charge distribution $Ni^{+0.1}N_2^{+0.9}$ indicating the absence of screening charge transfer towards the N_2 part of the cluster. Therefore, the $^4\Phi_2(N_a 1s)$ final state is equivalent to the unscreened final state in the SG model.

The valence orbital structure of the relaxed core hole final states in NiN_2 is almost independent of whether the N 1s electron is removed from the N_a or from the N_b center. In particular, the screening mechanism described above is equally present in the N_b 1s final core hole states. Thus, we find a screened $^4\Phi_1(N_b 1s)$ and an unscreened $^4\Phi_2(N_b 1s)$ final hole state in analogy to the $^4\Phi_1(N_a 1s)$ and $^4\Phi_2(N_a 1s)$ states.

Figure 1 shows the computed N_2 core IP's for the $^4\Phi_1$ and $^4\Phi_2$ final states as well as the IP difference ΔIP defined by

$$\Delta IP = IP(^4\Phi_2) - IP(^4\Phi_1)$$

as a function of the Ni-N_a distance in the cluster. The IP results for respective N_a and N_b core holes differ by less than 0.1 eV over the whole distance range considered. Therefore, fig. 1 gives in each case only one IP curve for both N_a 1s and N_b 1s ionization. The IP's of the $^4\Phi_1$ and $^4\Phi_2$ final states decrease with increasing Ni-N_a distance while the IP difference ΔIP remains almost constant at 8 ± 0.2 eV. This ΔIP value is comparable with the energy separation of the two-peak structure observed for the N_2/Ni(100) system where ∿5.5 eV is obtained[5].

The relative intensity contributions for N_2 core ionization resulting in a $^4\Phi_1$ or $^4\Phi_2$ final hole state of NiN_2 are computed using the sudden approximation[6,11]. Our results are given in figure 2. While figure 2a shows the relative intensities

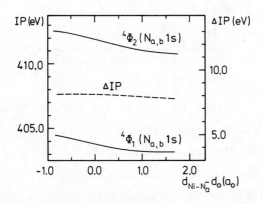

Fig. 1: Ionization potentials, IP (solid curves, left energy scale) of the screened $^4\Phi_1(N_{a,b}1s)$ and unscreened $^4\Phi_2(N_{a,b}1s)$ core hole states in NiN_2 as a function of the Ni-N_a distance. The IP difference IP of the two final states is also shown (dashed curve, right energy scale).

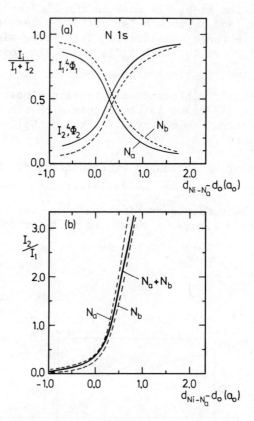

Fig. 2: Relative intensities I_1, I_2 (a) and intensity ratio I_2/I_1
(b) of the screened $^4\Phi_1$ and unscreened $^4\Phi_2$ final hole
states in NiN_2 as a function of the $Ni-N_a$ distance. The
curves for N_a1s and N_b1s ionization are denoted respectively
For definition of the intensity ratio $(I_2/I_1)_c$ denoted as
$N_a + N_b$ see text.

$I_1/(I_1 + I_2)$ of the screened $^4\Phi_1$ final state and $I_2/(I_1 + I_2)$ of the unscreened $^4\Phi_2$ final state figure 2b gives the intensity ratio I_2/I_1 as a function of the Ni-N_a distance. The intensity results for N_a and N_b core ionization (denoted by N_a and N_b respectively in fig. 2) differ very slightly from each other due to small variations in the relaxation process which can be disregarded for the present purpose. At a Ni-N_a distance d_o the photoemission intensity contribution I_1 due to relaxation to the screened $^4\Phi_1$ final state is larger than that, I_2, of the unscreened $^4\Phi_2$ final state. However, if the Ni-N_a distance is increased only slightly both intensities are of the same magnitude and at larger distances the intensity I_2 for the unscreened final state dominates (cp. fig. 2a). The latter result is due to the fact that screening charge transfer from Ni to N_2 by relaxation becomes more difficult if the N_2 is pulled away from the Ni.

Figure 2b also contains an intensity ratio $(I_2/I_1)_c$ for combined emission from N_a and N_b (denoted by $N_a + N_b$) defined as:

$$(I_2/I_1)_c = \frac{I_2(N_a) + I_2(N_b)}{I_1(N_a) + I_1(N_b)} .$$

This quantity seems to be plausible to be compared with experimental photoemission data since in the experiment separate emission from the adsorbate N_a or N_b part is quite difficult to distinguish[5]. Fig. 2b shows that the intensity ratio $(I_2/I_1)_c$ varies rather strongly with distance. A variation of 0.4 bohr about d_o (which seems to be a plausible distance range for the surface situation) leads to $(I_2/I_1)_c$ values between 0.2 and 1.3. Therefore, the present model provides a rather crude estimate of the peak intensity ratio to be expected in the experiment. The intensity ratio I_2/I_1 of the unscreened and screened N 1s peaks taken from the N_2/Ni(100) experiment[5] is ∿1.0 which corresponds in our NiN_2 model to a quite reasonable Ni-N_a distance being larger than d_o by only 0.3 bohr. This slightly increased distance with respect to d_o might (as in our NiCo model calculations[6]) indicate that at d_o the NiN_2 initial state contains a slightly overemphasized $2\pi*$ backbonding compared to the surface situation which is reduced by an increased distance.

It is interesting to compare the adsorbate core
ionization results of the present NiN_2 model with those of a

previous study on $NiCO$[6]. In both models we find that relaxation
to the respective energetically lowest adsorbate core hole final

state $(^4\Phi_1)$ results in a screening charge transfer from the Ni

towards the adsorbate part of the cluster. The screening is mainly
achieved by increasing the adsorbate $2\pi^*$ contribution to the
backbonding 5π cluster orbital in the final state. This screening
mechanism is essentially absent in the energetically second lowest

adsorbate core hole states $(^4\Phi_2)$. The difference in total energy

between respective screened and unscreened final states ($\Delta IP \simeq 9$ eV
for C1s, $\simeq 7$ eV for O1s ionization in $NiCO$[6], $\Delta IP \simeq 8$ eV for N1s
ionization in NiN_2) is comparable with the energy separation of
the two-peak structures observed in the XPS experiment of the
adsorbates on the Ni(100) surface[5]. The two models differ, however, in the actual size of adsorbate $2\pi^*$ contributions to the
backbonding in the initial state. The contributions are
considerably smaller in NiN_2 than in $NiCO$ at comparable geometries.
As a consequence, the relative intensities for adsorbate core
ionization behave somewhat different in the two models. It is
obvious from Fig. 2b and Fig. 2 of ref. 6 that the intensity ratio
I_2/I_1 of the screened and unscreened final core hole states

increases more rapidly with distance for N 1s ionization in NiN_2
than for C 1s or O 1s ionization in $NiCO$. Thus, at comparable
cluster geometries I_2/I_1 is larger for N_2 than for CO core

ionization. This is consistent with experimental photoemission
results of the adsorbate systems[5]. The measured intensity ratio
of the unscreened to the screened peak for N 1s ionization of
adsorbed N_2 is ~ 1.0, for C 1s and O 1s ionization of adsorbed
CO $\sim 0.3-0.4$.

In conclusion, our calculations on NiN_2 show that the
two-peak structure observed in the core level photoemission
spectrum of the N_2/Ni(100) adsorption system may be appropriately
described by a small cluster model. The two peaks are explained
as due to relaxation to different final core hole states, a
screened and an unscreened state where the screening mechanism
is analogous to the one proposed in the SG model[1-3]. Both the
energy separation and intensity ratio of the two peaks observed
in the experiment are obtained by our cluster model for reasonable
geometries. Further, a comparison with our previous cluster study
on $NiCO$[6] suggests that the differences in relative intensity of
the two peaks observed in core level ionization spectra of N_2 and
CO adsorbed on the Ni(100) surface can be explained by a smaller
$2\pi^*$ backbonding in adsorbed N_2 compared to CO.

REFERENCES

* Present address: Institut für theoretische Physik B TU
  Clausthal. 3392 Clausthal-Zfd., and Sonderforschungsbereich
  126, Göttingen/Clausthal, West Germany.

1. K. Schönhammer and O. Gunnarsson, Solid State Commun. 23, 691
   (1977).

2. O. Gunnarsson and K. Schönhammer, Phys. Rev. Letters 41, 1608
   (1978).

3. O. Gunnarsson and K. Schönhammer, Surface Science 80, 471
   (1979).

4. N.D. Lang and A.R. Williams, Phys. Rev. B16, 2408 (1977).

5. J.C. Fuggle, E. Umbach, D. Menzel, K. Wandelt and C.R. Brundle,
   Solid State Commun. 27, 65 (1978).

6. P.S. Bagus and K. Hermann, Surface Science 89, 588 (1979).

7. See e.g. Handbook of Chemistry and Physics, ed. P.C. Weast,
   56th Ed., CRC Press, Cleveland 1975.

8. P.S. Bagus, C.R. Brundle, K. Hermann and D. Menzel, J. Electr.
   Spectr. Rel. Phen., in press;
   K. Hermann, P.S. Bagus, C.R. Brundle and D. Menzel, Phys. Rev.
   B, to be published.

9. The basis sets used here are identical to those of ref.8.

10. K. Hermann and P.S. Bagus, Phys. Rev. B16, 4195 (1977).

11. T. Åberg, Phys. Rev. 156, 35 (1967).

LOW ENERGY ION SCATTERING STUDIES OF SODIUM SEGREGATION IN ZnO

C. Creemers[1], H. Van Hove, and A. Neyens

Fysico-chemisch laboratorium
Katholieke Universiteit Leuven

Celestijnenlaan 200G, B-3030 Leuven, Belgium

The segregation of Na and K, native sub ppm bulk impurities in
ZnO is studied on its polar surfaces using photoemission and low
energy ion scattering techniques. On the $(000\bar{1})$-face, the
resulting lowering of the electrostatic surface energy acts as
a strong additional segregation force. The angular dependence
of the ion intensities, backscattered from the segregated Na,
shows that the Na atoms are slightly protruding out of the $(000\bar{1})$-
surface.

1. INTRODUCTION

Clean polar surfaces of ZnO are known to strongly
interact with foreign atoms[2]. These phenomena, referred to as
stabilosorption[3], are explained by an electrostatic surface
theory[4]. According to this model the stabilizing structural surface charge may be accommodated in extrinsic surface states,
resulting in a lowering of the surface energy as compared to
intrinsically stabilized surfaces.

In a series of UPS-measurements on the ZnO polar
surfaces[5], low work function situations were systematically
observed whenever the previous cleaning cycle included an annealing
step, and yet not impurities could be detected by Auger spectroscopy. Low Energy Ion Scattering Studies were therefore undertaken
in order to identify the hidden surface impurities responsible
for the lowering of the work function and to study their possible
segregation.

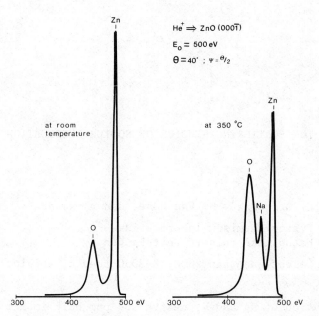

<u>**Fig. 1a:**</u> Low energy ion scattering spectrum on ZnO (000$\bar{1}$) after argon bombardment.

<u>**Fig. 1b:**</u> The same spectrum after annealing at 350°C.

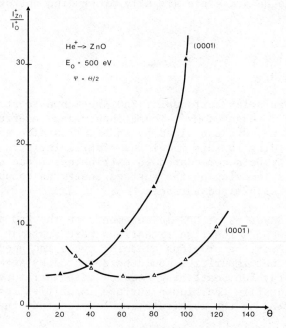

<u>**Fig. 2:**</u> Zn/O peak height ratio as a function of scattering angle θ.

2. EXPERIMENTAL

The ion scattering apparatus mainly consists of four parts : a differentially pumped ion gun, a straight through Wien filter for mass selection, a decelerator that allows for the final energy setting of the primary ions impinging on the surface and, finally, the rotatable 180° electrostatic energy analyzer for the scattered ions. Further technological details are published else-where[6].

In the scattering event the ions lose energy according to the simple binary collision formula

$$\frac{E}{E_o} = [\frac{\cos\theta + \sqrt{\gamma^2 - \sin^2\theta}}{1 + \gamma}]^2$$

where $\gamma = m(atom/m(ion)$ is the ratio of the atomic mass of the surface atoms to that of the impinging ions; θ is the scattering angle, E_o is the primary ion energy and E represents the final energy of the ions after scattering. An energy spectrum for a certain constant scattering angle θ is therefore equivalent to a mass spectrum of the very surface atoms.

Due to the high neutralization probability for the noble gas ions when interacting with the surface, only those ions colliding with the atoms of the outermost surface monolayer, have a measurable probability of retaining their charge after the collision event. This, however is the main reason why Low Energy Ion Scattering (LEIS) or Ion Scattering Spectroscopy (ISS) can be considered to be an absolutely surface sensitive technique.

3. RESULTS

The polar terminations of four undoped ZnO samples of different origin (Aachen-Erlangen) and of different shapes (thin disc, pencil) were analyzed. After the usual argon ion bombardment (2 KeV, 1 μAcm^{-2}, 1 hr) only the two peaks, corresponding exactly to the oxygen and zinc masses respectively, are observed. These results are represented in figure 1a. As might be expected for non-centro-symmetrical crystals of the Wurtzite type, the Zn/0 peak height ratio is different for the two polar surfaces of ZnO. The evolution of this peak height ratio as a function of the scattering angle θ is given in figure 2.

After a rather mild annealing in the temperature range from 250 to 400°C, a huge peak appears at the energy loss position corresponding to sodium. At the same time the zinc peak height

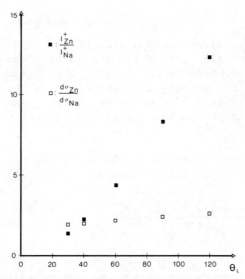

<u>**Fig. 3:**</u> Zn/Na peak height ratio as a function of scattering
          angle θ, and the calculated ratio of the scattering cross
          sections for the same scattering conditions: He$^+$,
          E_o = 500 eV, ψ = θ/2.

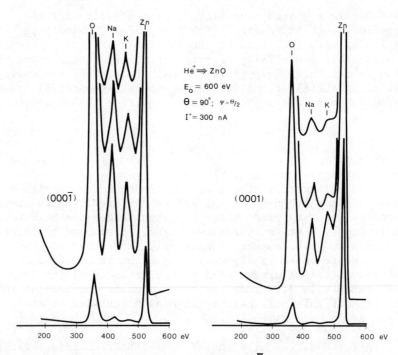

<u>**Fig. 4:**</u> Successive spectra of the (000$\bar{1}$) and (0001) Na and K
          enriched surfaces (the time interval is about 5 min).

decreases while that of the oxygen peak increases, as exemplified
in figure 1b. For increasing scattering angles θ the sodium peak
height strongly decreases : the variation of the Zn/Na peak
intensity ratio as a function of the scattering angle θ is shown
in figure 3. Figure 4 reveals that, for scattering angles beyond
a value of approximately 60°, also potassium can be detected as
a surface impurity.

On all investigated surfaces the observed phenomena are
of a similar nature, but as becomes apparent from figure 4, they
are always much more pronounced on the (0001̄)-termination. The
segregated sodium impurity on the (0001̄)-face withstands ion
bombardment for longer periods than on either the (0001̄)- or the
(101̄0)-face. Slightly Li doped ZnO however, did not show any
segregating bulk impurity.

The UPS-results confirm these experimental observations.
The presence of sodium is the main reason for the low value of the
work function after cleaning as well as for the impossibility to
adsorb cesium in a stable configuration, even on the (0001̄) face.
As is indicated in figure 5, cesium only shows up as a large
structureless bump on the low energy part of the energy distribution
curve.

4. DISCUSSION

Sodium is known to be a low level (ppb) bulk impurity
in ZnO single crystals. Our experiments clearly indicate that,
upon heating, sodium segregates to form a major constituent of
the outermost top layer of ZnO[7]. Auger spectroscopy will not reveal
this impurity because the host Zn-peaks at 906, 916, 939, 994 and
1017 eV strongly overlap with the Na-peaks at 923, 951, 963 and
990 eV. For the same reason the usually observed (1×1)-LEED
pattern for the polar surfaces of ZnO is not at all suspected to
be impurity generated.

Although all our experiments were performed in ultra
high vacuum conditions (10^{-10} Tør), we strongly believe that the
increase of the O-peak height in the ISS-spectrum must be
ascribed to a partial oxidation of the segregated sodium atoms
on the surface : the interaction of this impurity, which is known
to be one of the most reactive alkalimetals, with the oxygen from
the residual gas atmosphere, can be considered to be an essential
constituent of the driving force for the observed segregation
phenomenon.

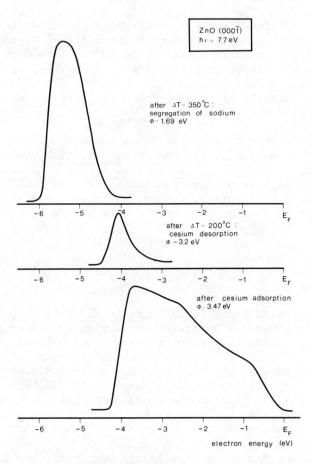

Energy distribution curves on ZnO (000$\bar{1}$): structureless cesium adsorption and sodium induced low work function situation after annealing at 350°C.

The preferential strong bonding of sodium on the $(000\bar{1})$-surface has to be correlated with the important role electropositive elements apparently play in stabilizing the out-of-balance charge distribution on $(000\bar{1})$ polar surfaces. This conclusion is further corroborated by the pronounced lowering of the surface electrostatic energy which, together with the oxidation phenomenon, is believed to be mainly responsible for both the more abundant segregation and the stronger bonding of sodium on the O-surface of ZnO.

The rapid decrease of the Na-peak height with increasing scattering angle θ, can be taken to imply that sodium protrudes out of the $(000\bar{1})$-face.

On labeling scattering from substrate atoms and from segregated species by the respective subscripts s and i, the peak height ratio I_s^+/I_i^+ of the scattered ion intensities is given by the formula

$$\frac{I_s^+}{I_i^+} = \frac{P_s^+}{P_i^+} \frac{d\sigma_s}{d\sigma_i} \frac{N_s}{N_i} (1 - \alpha \sigma_i N_i) \tag{1}$$

In this formula P_s^+, P_i^+ represent the probabilities that an impinging ion will not be neutralized on interacting with either a substrate or a segregated atom; $d\sigma_s$, $d\sigma_i$ are proportional to the differential scattering cross sections; N_s, N_i are the densities of scattering centres for one monolayer of atoms; σ_i is the total scattering cross section for the segregated atoms and α stands for a shadowing factor defined as the ratio of the area of the substrate that is shadowed by one atom i to the total scattering cross section of that atom (σ_i).

In order to evaluate the cross sections one needs a model for the interaction potential. The Molière approximation to the Thomas-Fermi potential (using Firsov's screening length) seems to yield the best results, especially in the case of ion-atom couples of low atomic number Z[8]. In all these calculations the cross section values for σ_i and $d\sigma_i$ were taken from Robinson's tables[9] and are reproduced in figure 3. If one assumes that the neutralization probability factor P_s^+/P_i^+ does not depend to strongly on θ, the experimentally observed strong decrease of $I^+(Zn)/I^+(Na)$ in the lower θ range (see figure 3) can only be accounted for by a drastic increase in the shadowing factor α, indicating that the segregated sodium atoms protrude out of the

surface. Such a conclusion is not hampered by the fact that the
substrate surface, instead of being a monoatomic monolayer, is in
fact a O-Zn double layer. Figure 2 provides convincing evidence
that on the ZnO (000$\bar{1}$)-face and at least in the range of the θ
values considered, there is no systematic shadowing of the Zn-
atoms by the O-atoms or vice versa.

5. REFERENCES

1. Aspirant N.F.W.O. Belgium.

2. B.J. Hopkins, R. Leysen and P.A. Taylor, Surface Science 48,
   486 (1975).

3. L. Monard, H. Van Hove and A. Neyens, Comptes rendus des
   Travaux du Troisième Colloque de Physique et Chimie des
   Surfaces, Grenoble 1977. Le Vide, supplément au N° 185 (1977).

4. R.W. Nosker and P. Mark, Surface Science 20, 421 (1970).

5. F. Humblet, H. Van Hove and A. Neyens, Surface Science 68,
   178 (1977).

6. To be published.

7. Also H.H. Brongersma, private communication.

8. E. Taglauer and W. Heiland, Surface Science 47, 234 (1975);
   H.H. Brongersma and T.M. Buck, Surface Science 53, 649 (1975);
   P. Bertrand and F. Delannay, Ph. D. thesis (1976), Université
   Catholique de Louvain, Louvain-la-Neuve.

9. M.T. Robinson, ORNL report n° 4556.

FORMATION OF Sb- AND Bi-CLUSTERS IN He-ATMOSPHERE

J. Mühlbach, E. Recknagel, and K. Sattler

Fakultät für Physik, Universität Konstanz

7750 Konstanz, Germany

Clusters in the size ranges Sb_1-Sb_{500} and Bi_1-Bi_{280} have been grown by inert gas condensation. The particle masses are analyzed by an electronic time of flight spectrometer. The Sb_n-spectrum is characterized by a sequence of tetramer-clusters Sb_{4i} (i = 1,2,...). The Bi_n-spectrum shows high abundances for Bi_5 and Bi_7, and low abundances for the sequence Bi_6, Bi_9, Bi_{12} and Bi_{15}. In the low size range (n $\lesssim$ 20) the individual electronic properties of the cluster elements and in the high size range (n $\gtrsim$ 20) the nucleation statistics is decisive for the size distributions.

Antimony and bismuth are semimetals[1] and crystallize in the so-called arsenic or A7 structure, which is slightly distorted from the simple cubic[2]. The free atom configurations are $s^2 p_{1/2}^2 p_{3/2}$, the valence s electrons being bound 8-10 eV more tightly than the valence p-electrons[3].

Theoretical band calculations show s-like bands several eV below the Fermi level E_F, and p-like bands near E_F[4,5], which has been confirmed by XPS[6] and UPS[7] experiments. XPS data of bismuth additionally show that the valence band structure of the liquid phase is very similar to that of the solid and very different from that of free electrons[8]. The tendency to tetrahedral coordinated covalent bonding causes the vapour of Sb and Bi to consist of clusters up to the tetramer[9-11].

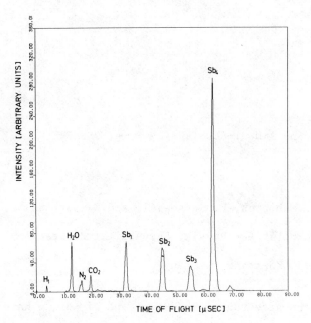

Fig.1 : Time of flight mass spectrum of Sb-clusters (Langmuir
evaporation).

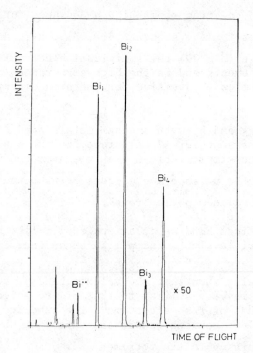

Fig.2 : Time of flight mass spectrum of Bi-clusters (Langmuir
evaporation).

The formation of these clusters originally does not occur in the vapour, despite the fact that thermodynamic equilibrium reactions between different cluster sizes can change the relative abundances. Langmuir evaporation shows, that Sb_1-Sb_4 and Bi_1-Bi_4 are directly emitted from the solid and liquid (Figs.1 and 2). The relative abundances of different cluster sizes depend on the evaporation temperature[12]. Neither Knudsen[9-11] nor Langmuir evaporation[12] yield clusters with more than four atoms.

We report on cluster formation in inert gas atmosphere whereby for the first time Sb- and Bi)clusters with sizes exceeding the tetramer have been grown. Particles have been detected in the size ranges Sb_1-Sb_{400} and Bi_1-Bi_{280}. The mass analysis is performed by time of flight (TOF) technique.

The metal vapour effuses from an oven into a condensation chamber which contains He-gas cooled to LN_2 temperature. Thereby the vapour is supersaturated and condensation to clusters occurs. From the condensation region the particles pass into the high vacuum chamber of the TOF-spectrometer, and furthermore don't interact with each other. Thus, the particle growth is stopped and clusters containing only a few atoms can be detected. The integral intensity of the neutral cluster beam is measured by a film thickness monitor. For the size determination the clusters are ionized by a pulsed electron gun (electron energy $\sim$100 eV) accelerated into the drift section and detected by single ion analysis. Further details of the apparatus are described elsewhere[13].

If no He-gas is let into the condensation cell (residual gas pressure 10^{-7} Torr), the spectrum of initially emitted particles is to be seen : from antimony predominantly Sb_4 (Fig.1), from bismuth the particles Bi_1 and Bi_2 (Fig.2). These are the basic specimen for a further clustering. At low He-pressure in the condensation cell, $p_{He} \lesssim 0.1$ Torr, the clustering may occur, however, the scattering of the metal particles in the He-gas prevents the effusion into the spectrometer. For leaving the condensation zone a suitable flow characteristics is necessary, which occurs at $p_{He} \gtrsim 10$ Torr. Clusters up to a few hundred atoms (Figs.3 and 4) are formed and are detected with the TOF spectrometer.

Particles up to Sb_{500} and Bi_{280} have been detected, while single clusters up to Sb_{100} and Bi_{21} have been resolved. Broadening of the peaks is due to the gradient of the extracting

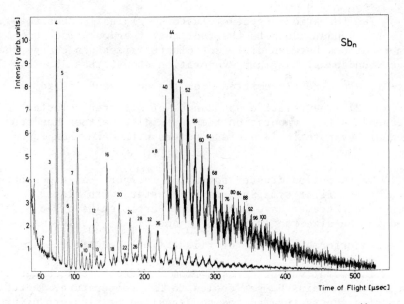

<u>Fig.3</u> : Time of flight mass spectrum of Sb-clusters (inert gas
          condensation).

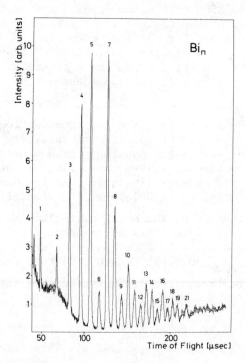

<u>Fig.4</u> : Time of flight mass spectrum of Bi-clusters (inert gas
          condensation).

electrostatic field is the formation region of the spectrometer.

For clusters M_n, n < 20, the mass distribution exhibits pronounced structures. In this size range the ionization cross section and the detector efficiency are expected to be independent from cluster mass and the TOF-spectra reflect superpositions of nucleation and stability distributions of the clusters. The spectra of Sb_n and Bi_n are completely different. In the case of antimony the initial cluster Sb_4 appears with the highest intensity. The Bi_n-spectrum however shows low rates of the initial particles Bi_1 and Bi_2, and high rates of the clusters Bi_5 and Bi_7, which originally have not been emitted from the solid and are not found in the vapour in thermodynamic equilibrium. Evidently, statistical nucleation theories[14-17], which predict the same distribution for any element, are not the proper description of cluster formation, at least for n < 20. The size distributions are mostly caused by the individual electronic properties of the original particles, mentioned above.

The clusters are bound systems between atom and solid. In both extremes, the symmetry of the valence electrons (for Sb and Bi) is p-like, even for different crystalline modifications (single crystals[6], amorphous[6] and liquid[8] structure). Therefore the same symmetry character is expected for the cluster wave function, and this seems to cause the great differences in the abundances of different cluster sizes.

Whereas the growth statistics exhibits long range influences on the size distribution, pronounced intensity differences of neighboured cluster peaks has to be attributed to different stability. Therefore, we conclude that e.g. the small intensity of Bi_6, Bi_9, Bi_{15} is due to low binding energies.

The Sb-spectrum is characterized by a sequence of clusters Sb_{4i} (i = 1,2,...), which must probably have been formed by nucleation of the initial tetramers. From Sb_{16} the intensity decreases nearly monotonously with increasing cluster size, however, this decrease is much weaker than predicted by nucleation theories, based on successive nucleation[18].

The high cluster abundances and the possibility of separating beams of particles with unique mass[13] henceforth enable the systematic size dependent investigation of Sb- and Bi-clusters.

Acknowledgement   -   This work was partly supported by the Deutsche Forschungsgemeinschaft.

REFERENCES

1. J.J. Hauser and L.R. Testardi, Physical Review Letters 20, 12 (1968).

2. M.H. Cohen, L.M. Falicov, and S. Golin, IBM Journal, July 1964, p.215.

3. C.C. Lu, T.A. Carlson, F.B. Malik, T.C. Tucker, and C.W. Nestor Jr., Atomic Data 3, 1 (1971).

4. D.W. Bullet, Solid State Communications 17, 965 (1975).

5. S. Golin, Physical Review 166, 643 (1968).

6. L. Ley, R.A. Pollak, S.P. Kowalczyk, R. McFeely, and D.A. Shirley, Physical Review B8, 641 (1973).

7. C. Norris and J.T.M. Wotherspoon, Journal of Physics F : Metal Physics 6, L 263 (1976).

8. Y. Baer and H.P. Myers, Solid State Communications 21, 833 (1977).

9. J. Kordis and K.A. Gingerich, Journal of Chemical Physics 58, 5141 (1973).

10. B. Caboud, A. Hoareau, P. Nounou, and P. Uzau, International Journal of Mass Spectrometry and Ion Physics 11, 157 (1973).

11. F.J. Kohl, Q.M. Uy, and K.D. Carlson, Journal of Chemical Physics 47, 2667 (1967).

12. J. Mühlbach, E. Recknagel, and K. Sattler, to be published.

13. K. Sattler, J. Mühlbach, A. Reyes Flotte, and E. Recknagel, Journal of Physics E : Scientific Instruments 13 (1980) in print.

14. R. Becker and W. Döring, Annalen der Physik 24, 719 (1935).

15. F. Kuhrt, Zeitschrift für Physik 131, 185 (1952).

16. D. Kashchiev, Surface Science 18, 389 (1968).

17. K. Binder and D. Stauffer, Advances in Physics 25, 343 (1976).

18. J. Feder, K.C. Russell, J. Lothe, and G.M. Pound, Advances in Physics 15, 111 (1966).

SERS FROM ADSORBED MOLECULES ON METAL SURFACES

Ralf Dornhaus[*] and Richard K. Chang

Department of Engineering and Applied Science

Yale University, New Haven, Connecticut 06520

Several key questions to distinguish between various microscopic models for surface enhanced Raman scattering (SERS) are discussed. Most experimental results stress the importance of bonding between adsorbates and metal and the influence of surface roughness, which can be well explained with mechanisms involving electron-hole pair excitations. The coupling mechanism to the adsorbed molecules could not be unambigeously identified up to now. It has to explain the strong intensity cariation of SERS from certain substrates and the change of Raman selection rules.

Although strongly enhanced Raman scattering (SERS) from adsorbates on certain metal surfaces has been the subject of intensive experimental and theoretical investigations over the last two years, no final understanding of the enhancement mechanism has been achieved up to now. Instead there are a large number of competing theoretical models and there is a somewhat confusing experimental situation. It is the intention of this letter to discuss certain key questions concerning the microscopic enhancement mechanism in the light of our own and other experimental results.

Perhaps the most important question in order to distinguish between several groups of theoretical models is whether proximity and relatively strong bonding forces between the adsorbate molecules and the metal are necessary to observe a strong enhancement. Despite the fact that some recent reports seemed to cast some doubt about the necessity of a proper chemical bond between adsorbates and metal we are still convinced that the vast majority of experiments performed up to now favour theories

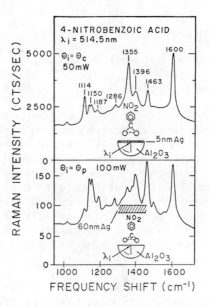

<u>Fig.1</u> : Raman spectrum of 4-nitrobenzoic acid adsorbed on an Ag
metal-island film and with 50 mW 514.5 nm excitation incident near the critical angle (upper half) and with a
60 nm thick Ag overlayer and 100 mW incident at the
surface-plasmon angle (lower half).

which rely on hybrid resonance rather than long range interactions
like coupling via surface plasmon excitations and similar effects.
Specifically it has been shown that
- submonolayer coverages of Ag on a relatively smooth Au-substrat
  induce enhanced Raman scattering without showing additional
  absorption due to bulk or surface plasmons or transverse
  collective excitation resonances /1/ ,
- adsorbed benzoic acid exhibits an enhancement only if it is
  chemisorbed to an Ag-layer but not with an Ag-overlayer /2/ ,
- it is possible to displace one adsorbate with another /3/ ,
- intensities and positions of vibrational modes in SERS depend
  strongly on the voltage drop across the thin ($\sim$ 10 Å) Helmholtz-
  layer in an electrochemical cell /4/, an effect which is not due
  to big changes in the number of adsorbed molecules /5/.
Recently /6/ we compared directly the Raman intensities from
monolayer adsorbates obtained via excitation of surface plasmons
by the method of attenuated total reflection (ATR) in different
configurations with intensities from adsorbates on thin metal-
island films (Fig.1). Using surface-plasmon excitation there was
no indication of a larger enhancement than expected from the well
known theory /7/. Adsorption on metal-island films (as evaluated
from SEM-pictures with a resolution of 30 Å), on the other hand,
gave intensities higher by more than an order of magnitude.

        In the latter case also the influence of polarization
of the incoming and scattered beam was completely different from
expectations for surface plasmon excitation. An interesting detail
is the difference in intensity ratios of certain modes for
adsorption on Ag-metal-island films and for adsorption with Ag-
overlayers (Fig.1). Mode positions generally are in reasonable
agreement with IR and inelastic electron tunneling data (where
available, /8,9/) with only small deviations by several wave-
numbers.

        Based on the arguments given above we think all models
relying on long range interactions as a major source of the large
enhancements observed can be ruled out. Left with models which
rely on a direct interaction between adsorbates and metal these
theories have to be tested by comparison with other experimental
results. While theories presented in /10,11/ seem to be in contradiction with specific experimental findings /12-14/ (and subject
of some conceptual criticism /1,15/) important aspects to criticize
others are the influence of surface roughness, the question of
selection rules and the dependence on the wavelength of the
incident radiation.

        To our knowledge nobody has ever observed SERS from a
smooth metal surface and it seems generally accepted by now that
sub-microscopic roughness is essential for SERS. It is one of the
shortcomings of models proposed in /16/ not to account for this

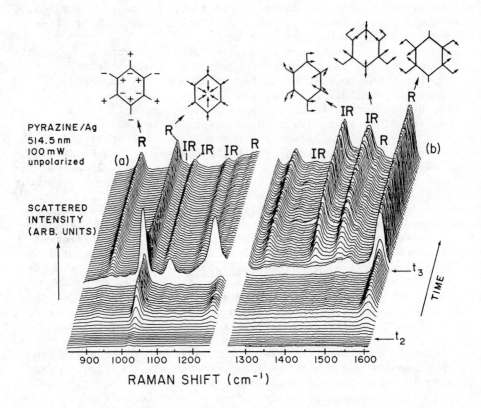

PYRAZINE/Ag
514.5 nm
100 mW
unpolarized

SCATTERED
INTENSITY
(ARB. UNITS)

RAMAN SHIFT (cm⁻¹)

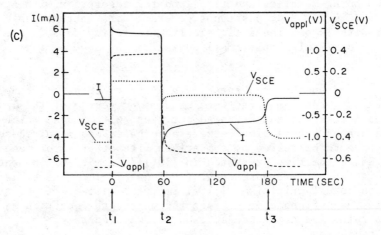

Fig.2 : Development of SERS from pyrazine adsorbed on an Ag
electrode during the first oxidation reduction cycle
using single pulse anodization as indicated in the lower
half of the picture.

roughness.

As has been pointed out /1,2,13-15,17,18/ the main effect of surface roughness is probably to lead to a breakdown in the conservation of momentum and thereby to an enhanced radiative-excitation and -recombination of electron-hole pairs in the surface region of the metal thus being responsible for the strong scattering continuum exhibited by rough surfaces. The breakdown in momentum conservation may well be induced by different types of surface structures (e.g. adatoms /13,14/, metal-islands /2/ or nodular formations /19/).

While qualitatively the part of electron-hole pair excitation in producing SERS seems clear, no quantitative calculations are available up to now. Similarly the important question after the microscopic coupling mechanism between excited eh-pairs and adsorbate molecules has only been discussed verbally /17,14/. Here selection rules and the dependence of the enhancement on the wavelength of the incident radiation may give valuable hints.

We have for the first time /4/ observed strong SERS from the nonpolar pyrazine molecule (D_{2h}-symmetry), which, as a free molecule, has its Raman and infrared active modes totally separated. A comparison of SERS spectra with earlier Raman and IR investigations /20/ reveals that the former exhibit almost all fundamentals, even those which should be only infrared and not Raman active (Fig.2). In a recent paper /21/ these results have been confirmed and interpreted by assuming a pyrazine-metal-complex with reduced symmetry (see also /22/). We have pointed out, however, that there is another possibility to account for the observation of Raman forbidden modes : electron-hole pair excitation models do not differentiate /17,13/ between Raman active and IR active modes, since the selection rules for electrons are different from those for photons. A definit conclusion would however require detailed knowledge about orientation, binding and identity of the molecules on the surface (molecule-metal-, molecule-halide-complex, temporary negative ion, influence of water, adsorption sites /23/), which is not yet available. Such knowledge would also facilitate an interpretation of the complicated mode development during the anodization procedure (Fig.2).

In the case of the much simpler CN-adsorbate some of these data are known and a partial understanding has been achieved /14,24-26/. To monitor the electrochemical kinetics (electron transfer followed by chemical reaction) voltammograms showing the electrochemical current-voltage relations were recorded. The reaction rates were affected by the dissolution of $Ag(CN)_n^{(n-1)-}$

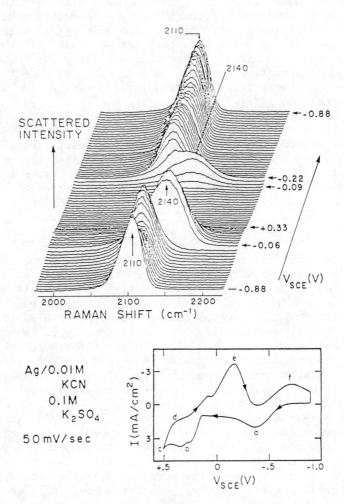

Fig.3 : Development of SERS from cyanide adsorbed on Ag during
        the second oxidation reduction cycle obtained with an
        optical multichannel analyser parallel to the voltammogram
        shown in the lower part. 100 mW of 514.5 nm excitation have
        been used. The SERS peaks correspond to different

$Ag(CN)_n^{(n-1)-}$ complexes adsorbed at the electrode /25/.

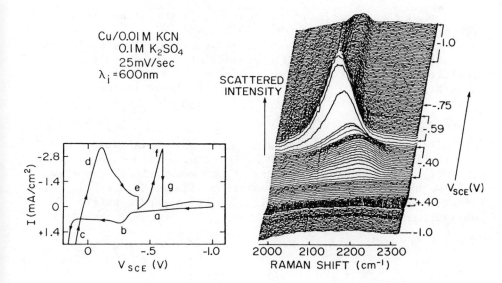

<u>Fig.4</u> : Development of SERS from cyanide adsorbed on Cu during
the first oxidation-reduction cycle consisting of a ramp-
hold activation procedure shown in the associated
voltammogram. The incident wavelength is 600 nm.

complexes (n = 1,2,3,4) at the electrode, the CN^- diffusion rate
in solution, and the diffusion of the various other salts in the
electrolyte. Since Raman data could be collected rapidly /25/
parallel to the voltammograms (Fig.3,4) it was possible to
identify and separate contributions of chemical reactions,
kinetics (adsorption and diffusion) and change in electron
density of the metal surface (see /24/). A more detailed analysis
of these results will be published elsewhere /25/. Recently we
have also been successfull to measure the growth and decay of
SERS from cyanide complexes adsorbed on a Cu-electrode /26/.
Discrete vibrational SERS peaks of the adsorbed Cu-cyanide
complexes /27/ were observed for wave-length λ_i longer than
575 nm (Fig.4), while in the region of d-band to Fermi level
transitions in Cu ($\lambda_i < 575$ nm) only a potential dependent
continuum inelastic scattering was detected. With $\lambda_i = 647.1$ nm
and a significant amount of electrochemical surface roughening
SERS intensities from cyanide on Cu are comparable to those
associated with Ag electrodes, peak intensities, however, decay
rapidly within a few minutes in the Cu case. This makes a
quantitative determination of the dependence of SERS intensities
on the incident photon energy very difficult.

        Our observations of a change in SERS intensities by
several orders of magnitude around the onset of d- to Fermi-
level transitions in Cu is consistent with the proposed electron-
hole pair excitation as the background intensity doesn't change
significantly. Every microscopic model for the coupling between
substrate and molecule has to explain this large intensity
variation.

Acknowledgements — We wish to acknowledge the valuable
contributions of R.E. Benner, M.B. Long, K.U.v.Raben, B.L. Laube,
F.A. Otter and I. Chabay to parts of the measurements reported
here. We have profited from discussions with A. Otto and T.
Furtak, they also provided us with copies of papers before
publication.

REFERENCES

* Present address : I. Phys. Institut der RWTH Aachen, 51 Aachen,
   Germany.

/ 1/ B.H. Loo and T.E. Furtak, to be published.

/ 2/ C.Y. Chen, E. Burstein, and S. Lundquist, Solid State
     Commun. 32, 63 (1979).

/ 3/ J. Billmann, J. Timper, A. Otto, Verh. DPG (VI) 15, 706 (1980).

/ 4/ R. Dornhaus, M.B. Long, R.E. Benner, and R.K. Chang, Surface
     Science 93, 240 (1980) and ref. therein.

/ 5/ W. Pütz and J. Heitbaum, Inst. Physikalische Chemie, Univ.
     Bonn, unpubl. res.

/ 6/ R. Dornhaus, R.E. Benner, R.K. Chang, and I. Chabay, Surface
     Science to be published.

/ 7/ A. Otto, Proc. Int. Conf. on Non-Traditional Approaches to
     the Study of the Solid-Electrolyte-Interface, Snowmass/
     Colorado 1979.

/ 8/ J. Kirtley and P.K. Hansma, Phys. Rev. B13, 2910 (1976).

/ 9/ M.G. Simonsen and R.V. Coleman, Phys. Rev. B8, 5875 (1973).

/10/ J.I. Gersten, R.L. Birke, and J.R. Lombardi, Phys.Rev.Lett.
     43, 147 (1979).

/11/ R.L. Birke, J.R. Lombardi, and J.I. Gersten, Phys.Rev.Lett.
     43, 71 (1979).

/12/ D.A. Weitz and A.Z. Genack, to be published.

/13/ A. Otto, J. Timper, J. Billmann, G. Kovacs, and I. Pockrand,
     Surface Science 92, L55 (1980).

/14/ J. Billmann, G. Kovacs, and A. Otto, Surface Science 92, 153
     (1980).

/15/ T.E. Furtak and J. Reyes, to be published.

/16/ S. Efrima and H. Metiu, Surface Science 92, 417 (1980) and
     ref. therein.

/17/ E. Burstein, Y.J. Chen, C.Y. Chen, S. Lundquist, and E. Tosatti,
     Solid State Commun. 29, 567 (1979).

/18/ E. Burstein, C.Y. Chen, and S. Lundquist, Light Scattering in
     Solids, J.L. Birman, H.Z. Cummins, and K.K. Rebane, ed.
     (Plenum Press, New York, 1979), p.479.

/19/ J.F. Evans, M.G. Albrecht, D.M. Ullevig, and R.M. Hexter,
     J. Electroanal. Chem. 106, 209 (1980).

/20/ K.K. Innes, J.P. Byrne, and I.G. Ross, J. Mol. Spectrosc. 22,
     125 (1967).

/21/ G.R. Erdheim, R.L. Birke, and J.R. Lombardi, Chem.Phys.Lett.
     69, 495 (1980).

/22/ H.A. Pearce and N. Sheppard, Surface Science 59, 205 (1976).

/23/ P.J. Hendra and M. Fleischmann, Topics in Surface Chemistry,
     E. Kay and P.S. Bagus, ed. (Plenum Press, New York, 1978),
     p.373.

/24/ J. Timper, J. Billman, A. Otto, and I. Pockrand, to be
     published.

/25/ R.E. Benner, R. Dornhaus, R.K. Chang, and B.L. Laube, Surface
     Science to be published.

/26/ R.E. Benner, K.U. von Raben, R. Dornhaus, R.K. Chang,
     B.L. Laube, and F.A. Otter, to be published.

/27/ M.J. Reisfeld and L.H. Jones, J. Molec. Spectr. 18, 222 (1965).

CONSEQUENCES OF DENSITY GRADIENTS ON SURFACE-FORCES

OF INTERFACES

A. Grauel

Department of Theoretical Physics
University of Paderborn

4790 Paderborn, Federal Republic of Germany

We investigate some features of the non-equilibrium behaviour of a fluid interface.

A non-equilibrium distribution of surface densities can change the mechanical properties of the interface and influences the interfacial stability[1]. We investigate the mechanical behaviour tangential to the interface of a mixture in the interfacial region. To that we consider the force density m_δ^A on the interface and relate m_δ^A to the gradients of the surface tension and the chemical surface potential. With this we show that these gradients drive the diffusive motion.

On the interface we have a mixture of λ fluids and we have to calculate 4λ fields as functions of the surface parameters u^1, u^2 and the time t, namely

the partial densities $\qquad\qquad \gamma_\delta(u^1,u^2,t)$

and the partial velocities $\qquad w_\delta^k(u^1,u^2,t)$. $\qquad$ (1)

For the determination of these fields we need field equations. On semipermeable fluid interface we have balance equations[2] of the partial mass densities and momenta

$$\partial_t \gamma_\delta - 2 \, w_\lambda^n K_M \gamma_\delta + (\gamma_\delta \, w_{\tau\delta}^A)_{;A}^i + [q_\sigma(v_\sigma^j - w_\lambda^j)e^j] = \pi_\delta \, , \quad (2)$$

$$\partial_t(\gamma_\delta \, w_\delta^k) - 2 \, w_{n\lambda} K_M \gamma_\delta w_\delta^k + (\gamma_\delta w_\delta^k w_\delta^A - T_\delta^{kA})_{,A} +$$

$$[q_\sigma v_\sigma^k(v_\sigma^j - w_\lambda^j)e^j - t_\sigma^{kj}e^j] = m_\delta^k + \gamma_\delta F_\delta^k, \qquad (3)$$

where $\delta = 1, \ldots, \lambda$ and we have assumed that the interface is material for particles of the surface mass density γ_λ. In the quantity $w_{n\lambda}e^i$ is the velocity normal to the interface and $K_M(\perp)$ is the mean curvature of the interface.

We denote the jump of the quantity ψ across the interface by $[\psi] = \psi^+ - \psi^-$. Both, ψ^+ and ψ^- are limit values of $\psi(x^i, t)$ on the surface as x^i from outside of the surface approaches a point $x^i(u^1, u^2, t)$ on the surface.

$\pi_\delta = \sum\limits_{r=1}^{n} \zeta_{\gamma\delta}^r m_\delta Z_r$ is the production of mass in the interface due to chemical reactions and consequently we have one part of the inter-action force m_δ^k du to chemical reactions and another part is due to the interaction of constituent δ with the other constituents. The numbers ζ_δ^+ are called stoichiometric coefficients and they specify how many molecules of the mass m_δ are created in the reaction r, and the number Z_r depends on the interfacial material.

The mass and the momentum of the mixture in the interface must be conserved, therefore, it holds

$$\sum_{\delta=1}^{\lambda} \pi_\delta = 0, \qquad \sum_{\delta=1}^{\lambda} m_\delta^k = 0 \qquad (4)$$

$\gamma_\delta F_\delta^k$ are the components of the force of an external field like gravitation which acts on the fluid particle of the constituent in the membrane. F_δ^k is a vector-valued quantity which we neglect in the further considerations. The quantity T_δ^{kA} is the stress on the interface.

In order to obtain field equations for γ_δ and w_δ^k from (3), we need constitutive functions which relate the quantities Z_r, T_δ^{kA} and m_δ^k to the surface fields and we assume that the constitutive functions are analytic functions of their variables γ_δ, w_δ^k, b_{AB} and g_{AB}.

For the time being, we are interested to discuss some consequences from the field equations of momenta tangential to the curved interface and, therefore, the following constitutive functions are of interest

$$T_\delta^{BA} = -\sigma_\gamma (\gamma_\alpha) \cdot g^{BA} \tag{5}$$

$$\underset{\tau\delta}{m}^A - \pi_\delta \underset{\tau\gamma}{w}^A = -\sum_{\zeta=1}^{\lambda-1} m_{\delta\zeta} \cdot (\underset{\tau\zeta}{w}^A - \underset{\tau\lambda}{w}^A) + \sum_{\zeta=1}^{\lambda} g^{AB} M_{\delta\zeta} \gamma_{\zeta,B} \tag{6}$$

where the coefficients $m_{\delta\zeta}$ and $M_{\delta\zeta}$ are functions of γ_α, K_M and $K_G = \det(b^A_B)$.

We conclude from an entropy principle on interfaces that the coefficient $M_{\delta\zeta}$ can be expressed as follows

$$M_{\delta\zeta} = \frac{\partial\sigma_\delta}{\partial\gamma_\zeta} - \gamma_\delta \frac{\partial\mu_S^\delta}{\partial\gamma_\zeta} , \tag{7}$$

where σ_δ is the surface tension and μ_S^δ is the chemical surface potential of the constituent δ in the mixture of the interface. The chemical surface potential μ_S^δ is continuous on a semipermeable curve on the surface. The quantity μ_S^δ is related with the specific free energy on the interface. We have :

$$\mu_S^\delta = \frac{\partial\gamma(\varepsilon_S - T_S\eta_S)}{\partial\gamma_\delta} . \tag{8}$$

If we know the specific free energy on the interface we can calculate the surface tension[3]

$$\sigma = \sum_{\delta=1}^{\lambda} \sigma_\delta = \gamma \sum_{\delta=1}^{\lambda} \frac{\partial(\varepsilon_S - T_S\eta_S)}{\partial\gamma_\delta} \tag{9}$$

With the constitutive function (6) we can rewrite the balance of momentum tangential to the surface. If we neglect the acceleration terms we shall obtain

$$g^{AB} \frac{\partial\sigma_\delta}{\partial u^B} = -[q_\sigma (\underset{\tau}{v}^A_\sigma - \underset{\tau}{w}^A_\delta)(v^j_\sigma - w^j_\lambda)e^j] -$$

$$\sum_{\zeta=1}^{\lambda-1} m_{\delta\zeta} \cdot (\underset{\tau}{w}^A_\zeta - \underset{\tau}{w}^A_\lambda) + g^{AB} \sum_{\zeta=1}^{\lambda} M_{\delta\zeta} \gamma_{\zeta,B} \tag{10}$$

If $v^A_{\tau\sigma} = w^A_{\tau\delta}$ and $M_{\delta\zeta}$ is equal to zero, we obtain

$$g^{AB} \frac{\partial\sigma_\gamma}{\partial u^B} = - \sum_{\zeta=1}^{\lambda-1} m_{\delta\zeta} \cdot (w^A_{\tau\zeta} - w^A_{\tau\lambda}). \tag{11}$$

We can see that the diffusive motion causes by the surface gradients of the surface tension.

Furthermore, if holds $v^A_{\tau\sigma} = w^A_{\tau\delta}$ and if we rewrite eq.(10) with equation (7) it will follow

$$g^{AB} \gamma_\delta \frac{\partial\mu^\delta_S}{\partial u^B} = - \sum_{\zeta=1}^{\lambda-1} m_{\delta\zeta} \cdot (w^A_{\tau\zeta} - w^A_{\tau\lambda}). \tag{12}$$

This shows that the diffusive motion causes by the chemical surface gradients. These results are similar to the considerations of mixtures of fluids in the bulk phases[4].
From this we conclude that the mechanical and chemical properties tangential to the interface may be determinant for the instability of interfacial layers.

REFERENCES

1. W. Dalle Vedove, P.M. Bisch and A. Sanfeld, J. Non-Equilib. Thermodyn. 5, 35 (1980).

2. A. Grauel, A two-dimensional Thermodynamic Field Theory, in Liquid Crystals of One- and To-Dimensional Order, W. Helfrich, G. Heppke (eds.), Springer Series in Chemical Physics, (Springer, Heidelberg 1980).

3. A. Grauel, Physica A (1980).

4. I. Müller, J. Non-Equilib. Thermodyn. 2, 133 (1977).

AN EXPERIMENTAL AND THEORETICAL STUDY OF THE ELECTRONIC

STRUCTURE OF THE W (111) SURFACE

F. Cerrina[*], G.P. Williams, J. Anderson, G.J. Lapeyre
Montana State Univeristy, Bozeman, Montana 59715 USA

O. Bisi, C. Calandra
Instituto di Fisica, Università di Modena
41100 Modena, Italy

We report results of synchrotron angular resolved photoemission
experiments from W (111) together with a theoretical analysis of
the data. For normal emission several different surface sensitive
features are found depending on the s or p polarization of the
incident radiation. The symmetry assignements derived from this
polarization dependence are compared with theoretical results
based on a tight binding calculation of the electronic structure
of a 16 layers thick slab. The same analysis is extended to off-
normal data collected in the mirror plane.

The W surfaces have received a large amount of
experimental and theoretical attention. Surface states have been
investigated on both the (001) and (110) surfaces[1-10]; no data
are however available for the (111) to our knowledge. We present
here some preliminary results of a joint theoretical and
experimental study of the (111) surface of tungsten.

Among the photoemission techniques, the Polarization-
dependent Angle-Resolved Ultra-Violet Spectroscopy (PARUPS) has
been successfully applied to investigate the existence and the
symmetry of the initial states[11-13]. The analysis is based on
the following result : the optical matrix element for the

transition $<f|\vec{A}.\vec{p}|i>$ is nonzero only if the integrand has a

component which is invariant with respect to the symmetry
operations of the crystal. The surface component of the wavevector

$\vec{k}\parallel$, conserved during the emission process is given by :

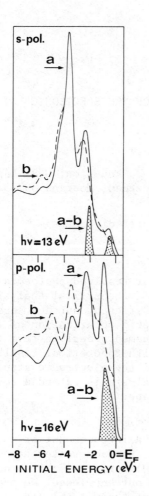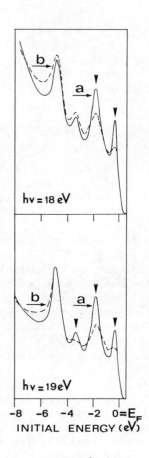

Fig.1   Fig.2

Fig.1 : Normal emission AREDC for W(111) for s (upper diagram)
        and p (lower diagram) polarized light.
        a : clean surface; b : hydrogen covered surface;
        a-b : difference curve indicating SSF.

Fig.2 : AREDC for θ = 42.3° and $\vec{k}_{/\!/}$ along the MP for two different
        photon energies. s-polarization radiation with $\vec{A}/\!/$ y.
        Vertical arrows indicate SSF. a : clean surface;
        b : hydrogen covered surface.

$$|k/\!/| = [E_f/(\hbar^2/2m)]^{1/2} \sin \theta \tag{1}$$

where θ is the polar angle of emission and E_f is the final state energy. Hermanson[14] showed that for emission normal to the surface, or confined to a mirror plane (MP) containing the surface normal, the final state is invariant under the crystal symmetry operations that leave $\vec{k}/\!/$ unchanged. The symmetry of the initial state is then the same as that of the dipole operator.

The theoretical approach consists in describing the crystal as a slab of 16 regularly stacked (111) planes. A para-metrized LCAO hamiltonian[4,15-17] obtained by a fit to non-relativistic APW bulk bands of Mattheiss[18], with s, p and d atomic orbitals in the basis set has been used. By including hopping integrals up to third neighbours, we were able to reproduce the bulk energies in 55 points of the bulk Brillouin zone within a r.m.s. of 0.16 eV.

The connection between the computed energies and wave-functions and the angle-resolved photocurrent emitted into the direction (θ,ϕ) by radiation of energy $\hbar\omega$ is given by[19] :

$$J(\theta,\phi,E_f(\vec{k}/\!/),\hbar\omega) = \frac{2\pi e}{\hbar} \left(\frac{e}{mc}\right)^2 \sum_i |<f|\vec{A}.\vec{p}|i>|^2$$

$$\times \delta(E_f(\vec{k}/\!/) - E_i(\vec{k}/\!/) - \hbar\omega) \tag{2}$$

where the sum covers the initial states. Provided that the energy dependence of the matrix elements is small the photocurrent at a given final energy turns out to be determined by the initial state density taken at a given $\vec{k}/\!/$ fixed by eq. (1).

The emission spectra were measured with a modified double-pass cylindrical mirror analyzer so that angle-resolved spectra with an angular acceptance of $\pm 2°$ have been obtained. Linearly polarized radiation emitted from the storage ring Tantalus I at the Wisconsin Synchrotron Radiation Center was monochromatized and focused on the sample. The experimental geometry allowed the use of s or p polarization for the normal emission case, while only s-polarization was used for off-normal experiments. Surface sensitive features (SSF) were detected by exposure to 0.1 L of hydrogen gas. Details of the experimental configuration are published elsewhere[9,20].

Figure 1 shows normal emission Angle-Resolved Energy-Distribution-Curves (AREDC) for s and p polarization. SSF are easily seen in both spectra as evidenced by the difference curves. For s-polarization two peaks are observed at the energies -0.5 and -2.0 eV. Two unresolved structures near the Fermi energy appear with p-polarized radiation at about -0.4 and -0.9 eV.

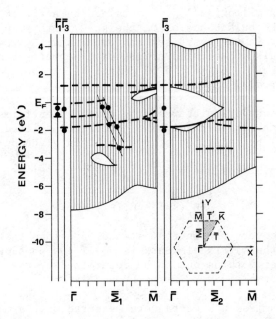

Fig.3 : PBBS of the (111) W surface, surface states (full lines)
and resonances (dashed lines) at $\bar{\Gamma}$ and along the MP
direction. Experimental data from Figs.1-2 are indicated
by full circles.

Moving the analyzer to "off-normal emission" these two peaks splits into two well resolved structures.

The dipole allowed initial state symmetries for normal and mirror plane emission from a BCC (111) surface are shown in Table 1. As in the case of W(001)[9], we neglect spin-orbit coupling and thus use single group labels.

Table 1 : Dipole-allowed initial state symmetry for a BCC (111) surface for different polarizations. The reference axes and the 2DBZ are shown in the inset of Fig.3.

Polarization Direction	Allowed Symmetry	Atomic Basis Functions
Normal Emission		
$\vec{A}/\!\!/ z$	$\overline{\Gamma}_1$	$s, z, 3z^2-r^2$
	$\overline{\Gamma}_2$	
$\vec{A}/\!\!/ x; \vec{A}/\!\!/ y$	$\overline{\Gamma}_3$	$\{ {x,y,yz,zx \atop xy, x^2-y^2} $
Mirror Plane Emission		
$\vec{A}/\!\!/ y; \vec{A}/\!\!/ z$	$\overline{\Sigma}_1$	$s, y, z, yz, x^2-y^2, 3z^2-r^2$
$\vec{A}/\!\!/ x$	$\overline{\Sigma}_2$	x, xy, zx

At this point the symmetry assignement of the SSF seen in the normal emission data is immediate. For s (p) polarized exciting radiation only emission from $\overline{\Gamma}_3$ ($\overline{\Gamma}_1$) is allowed.

We consider now the spectra collected at $\vec{k}/\!\!/ \neq 0$ belonging to the MP. From Table 1 we see that it is possible to discriminate between $\overline{\Sigma}_1$ and $\overline{\Sigma}_2$ symmetries by a suitable choice of the orientation of the $\vec{A}$ vector of the s-polarized radiation onto the crystal surface. For $\vec{A}$ parallel to the MP ($A/\!\!/$) only the $\overline{\Sigma}_1$ states can couple to the outgoing final state. Such a configuration was used to collect the AREDC spectra shown in Fig.2. In the upper panel ($\hbar\omega$ = 18 eV) two SSF are seen at -0.4 and -1.8 eV. For $\hbar\omega$ = 19 eV, lower panel, we see three SSF at -0.3, -1.6 and -3.4 eV.

In order to interpret the experimental data we show in Fig.3 the results of our calculation for $\vec{k}/\!\!/$ at the center of the two-dimensional Brillouin zone (2DBZ) and along the MP direction. The projected bulk band structure (PBBS) and the surface states together with the resonances are shown separately according to the symmetry. For clarity the surface features at $\overline{\Gamma}$ are plotted aside and indicated by a hyphen. We indicate also by full circles the experimental data taken from Figures 1-2. As far as normal emission is concerned, we identify the two peaks seen in the p-polarization spectra with the two surface resonances found at -0.08 and -0.98 eV. The true surface character of the observed SSF is also confirmed by the absence of dispersion of these peaks by changing the photon energy[20]. In the case of s-polarization spectra ($\overline{\Gamma}_3$ symmetry) the theory correctly predicts the higher binding energy state. This appears, in contrast to the $\overline{\Gamma}_1$ surface resonances, to be a true surface state as it lies in a symmetry gap at -1.68 eV. We do not find the $\overline{\Gamma}_3$ state near the Fermi energy which is instead observed in the spectra of Fig.1.

Several reasons may be suggested to account for this discrepancy between experiment and theory. From the experimental point of view one has to take into account the finite angular acceptance of the spectrometer; angular integrated spectra show a very high density of SSF just below the Fermi energy. From the theoretical point of view the model used could not be completely adequate to describe all the features of this surface.

Coming to the $\overline{\Sigma}_1$ direction a remarkable agreement between the computed surface resonance bands and the data taken from the AREDC of Fig.2 is found. This joint analysis of the $\overline{\Sigma}_1$ direction provides a picture of the electronic structure where four different flat surface resonance bands occur in the energy range from -3.5 eV to the Fermi level.

It is interesting to notice this new feature : the 6 surface bands departing from $\overline{\Gamma}$ (4 along $\overline{\Sigma}_1$ and 2 along $\overline{\Sigma}_2$) are characterized by having a single d-symmetry character. On going towards higher binding energies the symmetry sequence is yz, $3z^2-r^2$, $3z^2-r^2$, x^2-y^2 (for $\overline{\Sigma}_1$) and zx, xy (for $\overline{\Sigma}_2$). This may be understood in terms of the large surface atoms separations and the consequent atomic-like character of the W surface atoms.

These promising preliminary results confirm the power of PARUPS when substantiated by a theoretical analysis. Other measurements to extend the analysis to the $\overline{\Sigma}_2$ symmetry and to the remaining symmetry directions of the 2DBZ will be the subject of a further work[20].

Acknowledgement - This research was partly supported by AFOSR under Grant No 75-2872. The Wisconsin Radiation Cnter was supported by NSF under Grant No 144-F805. The cooperation of E. Rowe and the staff of the UWSRC is deeply appreciated. All the calculations have been performed at the Centro di Calcolo Elettronico, Università di Modena.

REFERENCES

* Permanent address : PULS, Laboratori Nazionali di Frascati, 00044 Frascati, Italy.

1. B.J. Waclawski and E.W. Plummer, Phys.Rev.Lett. 29, 783 (1972); B. Feuerbacher and B. Fitton, Phys.Rev.Lett. 29, 786 (1972).

2. W.F. Egelhoff, J.W. Linnett and D.L. Perry, Phys.Rev.Lett. 36, 98 (1976).

3. B. Feuerbacher and R.F. Willis, Phys.Rev.Lett. 37, 446 (1976).

4. N.V. Smith and L.F. Mattheiss, Phys.Rev.Lett. 37, 1494 (1976).

5. S.L. Weng, Phys.Rev.Lett. 38, 434 (1977); S.L. Weng, T. Gustaffson and E.W. Plummer, Phys.Rev. B 18, 1718 (1978).

6. C. Noguera, D. Sparnjaard, D. Jepsen, Y. Ballu, C. Guillot, J. Lecante, J. Paigne, Y. Petroff, R. Pinchaux and R. Cinit, Phys.Rev.Lett. 38, 1171 (1977).

7. E.G. McRae, Phys.Rev. B 17, 907 (1978); R.F. Willis, Phys. Rev. B 17, 909 (1978).

8. M. Kawajiri, J. Hermanson and W. Schwalm, Solid State Commun. 25, 303 (1978).

9. J. Anderson, G.J. Lapeyre and R.J. Smith, Phys.Rev. B 17, 2436 (1978).

10. M.W. Holmes, D.A. King and J.E. Inglesfield, Phys.Rev.Lett. 42, 394 (1979).

11. E. Dietz, H. Becker and H.F. Roloff, Phys.Rev.Lett. 37, 775 (1975).

12. G.J. Lapeyre, R.J. Smith and J. Anderson, J. Vac. Sci. Technol. 14, 384 (1977).

13. F.J. Himpsel and D.E. Eastman, Phys.Rev.Lett. 41, 507 (1978).

14. J. Hermanson, Solid State Commun. 22, 9 (1977).

15. D.C. Dempsey, W.R. Grise and L. Kleinman, Phys.Rev. B 18, 1550 (1978).

16. O. Bisi and C. Calandra, Surf. Sci. 74, 541 (1978).

17. O. Bisi and C. Calandra, Surf. Sci. 83, 83 (1979).

18. L.F. Mattheiss, Phys.Rev. 139, A 1893 (1965).

19. B. Feuerbacher and R.F. Willis, J. Phys. C 9, 169 (1976).

20. F. Cerrina, G.P. Williams, J. Anderson, G.J. Lapeyre, O. Bisi and C. Calandra, to be published.

HIGH FREQUENCY RESPONSE AND CONDUCTION MECHANISM

IN THICK-FILM (CERMET) RESISTORS

M. Prudenziati and B. Morten
Instituto di Fisica della Università
via Campi 213/A, 41100 Modena, Italy

C. Martini and G. Bisio
Laboratorio per i Circuiti Elettronici del C.N.R.
via All'Opera Pia, 16145 Genova, Italy

The high frequency response of thick-film (cermet) resistors (TFRs) has been investigated in the range 10^8–$1.8 \cdot 10^{10}$ Hz. The measurements allow us to establish an electrical equivalent circuit for TFRs, whose parameters are directly related to the microstructure of the films. In particular a high effective dielectric constant is obtained which gives information on the mean grain distance in the glassy matrix; the contribution due to grains and barriers to the material resistance is directly evaluated. The barrier resistance R_b has an a.c. component which appears to drop rapidly with frequency, a crude approximation being $R_b(\omega) \sim 1/\omega$.

INTRODUCTION

The use of thick-film technology for the manufacture of advanced electronic circuits is now firmly established[1]. In this technology the role played by thick-film resistors (TFRs) is relevant, thanks to properties such as the wide range of achievable sheet-resistivities, the low values for temperature coefficient of resistance (TCR) and the high power capability.

TFRs are produced by screen-and-firing processes of suitable inks containing submicron conductive particles (grains) of metal oxides and glass particles suspended in a fluid organic vehicle. After the screen process on a refractory substrate, usually alumina (Al_2O_3 96 %) in the desired pattern, the ink is

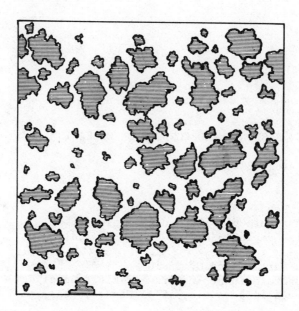

<u>Fig.1</u> : Schematic representation of the microstructure of thick-
film resistors. Shaded area represent metal-oxide particles
(grains) or cluster of particles, embedded in the dielectric
matrix.

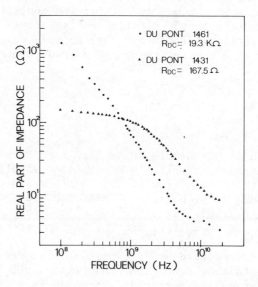

<u>Fig.2</u> : Frequency dependence of the real part of impedance of some
resistors from the Du Pont 1400 series.

dried in order to evaporate the most volatile components, fired at a temperature well above the glass softening point and then slowly, cooled down.

During the firing process the organic vehicle is supposed to burn off completely and the conductive grains remain embedded in the glassy matrix, surrounded by dielectric material. A schematic microstructure is shown in Fig.1[2]. The grains are randomly distributed in the matrix and have sizes and distances distributed on a wide range of values.

Structure and composition of these resistors appear quite complex and the mechanisms of electrical transport is not completely understood although an intensive work on this problem has been recently carried out[3-7].

In this study we report measurements of the a.c. response of TFRs which provide new relevant information on the electrical properties of these components and can be useful for a correct formulation of the transport mechanism in TFRs.

EXPERIMENTAL

We investigated samples obtained from several commercially-available ink series with sheet-resistivities ranging from 10^2 to 10^6 $\Omega/\square$. Large area patterns (2.5×2.5 cm) of resistor ink were screened and fired on 96 % alumina substrates and processed according to their respective manufacturer instructions. From that pattern a circular ring was scribed with a CO_2 laser, at dimensions fitting a 50 Ω transmission line (3 mm internal and 7 mm external diameter). A similar ring of alumina without resistor was prepared as a "blank sample".

The electrical contact between the sample and the transmission line was improved with a thin layer of silver-epoxy resin paste. Measurements in the frequency range 0.1 to 18 GHz were performed with an Automatic Network Analyzer (HP 8542A). The calibration of the system with the "blank sample" in the line, allows us to neglect effects of the substrate in the measured data on TFRs.

The measurements give the reflection coefficient Γ i.e. the ratio of reflected (E_r) to incident (E_i) signal, both in magnitude and in phase components. Γ is related to the impedance of the transmission line Z_o and of the sample Z through the relations :

$$Z/Z_o = (1+\Gamma)/(1-\Gamma) \qquad E_r/E_i = \Gamma/\theta_r \qquad (1)$$

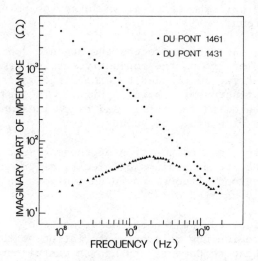

<u>Fig.3</u> : Modulus of the imaginary part of impedance of Du Pont 1400
resistors, as a function of frequency. The experimental
values of Im Z have negative sign.

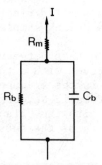

<u>Fig.4</u> : A lumped equivalent circuit for thick-film resistors. R_m
is the resistance of metal oxide grains. R_p and C_p are
the resistance and capacitance respectively, associated
to intergranular barriers.

The real Re Z and imaginary Im Z parts of the sample impedance were calculated by a computer programme for a large number of frequency points.

RESULTS AND DISCUSSION

Figures 2 and 3 show typical behaviours of the frequency dependence of the real and imaginary part of the impedance for some samples of the Du Pont 1400 series, a lower resistivity sample ($10^3 \Omega / \Box$) and a higher resistivity sample ($10^6 \Omega / \Box$).

Remarkable features of the experimental data are :
i) the drop in magnitude of the real part of impedance larger than one order of magnitude in the lower-resistivity resistors and about three orders of magnitude in higher-resistivity resistors;
ii) a trend to saturation of Re Z at very high frequencies ($>10^{10}$ Hz);
iii) negative values for the imaginary part of the impedance, with the modulus (Fig.3) increasing at lower frequencies and decreasing at higher frequencies.

We point out that these features are common to all TFRs examined, i.e. were observed in resistors with different structure and composition. In particular these findings are common to RuO_2-based resistors (e.g. 2700 ESL), pyroclore-ruthenate-based resistors (e.g. GERM 500 and Du Pont 1400 resistors) and iridium-dioxide-based resistors (e.g. 3800 ESL series)[8].

To discuss the experimental data is convenient to represent the TFRs by the equivalent circuit shown in Fig.4. Here R_b and C_b represent the resistance with the intergranular barriers (Fig.1) and R_m the resistance of the metal-oxide particles. The capacitance associated with the particles is small enough to be negligible.

The use of the equivalent circuit of Fig.4 does not a priori imply that R_b and C_b are frequency independent, as we will see later on.

It is easily seen that the real part of impedance of this circuit is given by :

$$Re\ Z = R_m + R_b \{1 + (C_b R_b \omega)^2\}^{-1} \tag{2}$$

while the imaginary part is :

$$Im\ Z = -\omega C_b R_b^2 / \{1 + (\omega C_b R_b)^2\} \tag{3}$$

hence for $\omega \to 0$, Re $Z = R_b + R_m = R_{dc}$,   Im $Z \to 0$
while for $\omega \to \infty$, Re $Z = R_m$,   Im $Z \to 0$
so that Im Z presents a minimum Im $Z = R_b/2$ at $\omega R_b C_b = 1$.

The experimental data (Fig.2 and 3) are well described by eqs. 2 and 3 as regards the negative sign of Im Z and its functional dependence on frequency, the decreasing behaviour of Re Z towards a saturation value.

However for a fitting of the data a frequency dependence of the barrier resistance has to be introduced. For this purpose R_b is splitted in two parallel resistances[9], of which one R_{bo} is independent of frequency and the other frequency-dependent according to $\{A\omega^n\}^{-1}$.

Table I summarizes the results of a computer fitting of experimental data for the resistors of the Du Pont 1400 series. The last column of the table reports the quantity $\alpha \equiv R_{bo}/(R_m + R_{bo})$ which expresses the fractional resistance of the barriers in d.c. operation, i.e. the contribution due to the barrier-resistivity to the total resistivity of the material[4].

It can be seen that this fractional resistance is very large, on the contrary to what supposed by some authors[3,4], hence the contribution due to metal grains is almost irrelevant. Consequently it can be easily seen that the metallic behaviour of the particles plays a negligible role in regard to the temperature-dependence of resistance of TFRs.

Let us briefly discuss also the results concerning capacitance values and frequency dependence at the barrier resistance.

The relative dielectric constant ε obtained from the capacitance values (Table I) and the sample geometries are in the range 250-600. These values reflect the two-phase structure of TFR materials[2] and give a mean value for the grain separation. In fact it has been shown[10] that the effective dielectric constant ε of a two-phase material (cermet) is approximately given by

$$\varepsilon = \varepsilon_1 (d+s)/s \qquad\qquad\qquad (4)$$

where ε_1 is the dielectric constant of the insulator-phase, d the diameter of the conductive grains and s the separation between these grains. Letting $\varepsilon = 7$[11] and $d = 1000 \div 2000$ Å in our resistors[1] (measured by X-ray diffraction line broadening[8]) we get $s = 10 \div 20$ Å for Du Pont 1431 resistors and $s = 30 \div 60$ Å for 1451 resistors.

The increasing values of C_b by decreasing sample resistivity (Table I) reflects the fact that lower resistivities are obtained with a greater volume fraction of conductive particles[4], which implies lower mean distances of particles themselves. The 1461 resistors are an exception in this trend; this is not surprising since the resistivity and TCR of inks at the extremes of one series are controlled not only by decreasing the volume fraction of conductive components but also by changing composition of glass and conductive component[8].

It can be observed that mean separations between grains so evaluated are consistent with electron tunnelling between grains themselves, supposed to be the basic transport mechanism for conduction in TFRs[4,5].

Tab.I : Best-fit values of the circuit parameters C_b, $R_b(\omega=0)$, R_m in fig.4. For the definition of A and α see text.

Sample	R_b (Ω)	R_m (Ω)	C_b (pF)	A $10^{-13}(\Omega^{-1})$	α
1431 $(10^3\ \Omega/\square)$	164.5	2.95	.60	1.2	0.9824
1441 $(10^4\ \Omega/\square)$	1625	1.08	.35	0.36	0.9993
1451 $(10^5\ \Omega/\square)$	–	–	.25	0.1	–
1461 $(10^6\ \Omega/\square)$	19320	1.0	.37	0.30	>0.9999

A frequency dependence of the a.c conductivity has been observed in an extremely wide variety of disordered materials[12,13]. In most cases the conductivity can be written as

$$\sigma(\omega) = \sigma_{dc} + \sigma_{ac}(\omega) \tag{5}$$

with the true a.c part of the conductivity of the type

$$\sigma_{ac}(\omega) \sim \omega^n \tag{6}$$

where n is close to unity.

A theoretical interpretation of this power law has been obtained by considering electric charges which move in the material from site to site by discontinuous "hopping" jumps.

This type of behaviour is not indeed unexpected in TFRs in view
of their disordered structure. On the contrary it seems to support
the hopping and percolative behaviour of electrons in TFRs,
suggested for an explanation of the temperature dependence of
resistance in these materials[5,6]. However we need further work
to clarify the role of conductive grains and/or localized states
in the glassy matrix, which can both play the role of sites for
the electron jumps. We observe here that the orders of magnitude
for the A values (Table I) are very similar to that observed in
discontinuous metal-film resistors[9].

CONCLUSION

        The study of the high frequency response of thick-film
resistors, extended to the GHz range, allows to identify a lumped
equivalent circuit for these components, whose parameters are
directly correlated to the microstructure of the films, consisting
of metal grains separated by thin dielectric layers. Particularly

interesting is the possibility to obtained direct information of
the contribution given by the grains to the total resistance of
the material, and to get an evaluation for the mean grain distance
by the capacitance of the samples.

        On the basis of the obtained results our immediate
physical conclusions with regards to the nature of the transport
mechanism in thick film resistors may be stated as follows.

        The results substantiate conduction models which
envisage current channels consisting of metal particles separated
by thin dielectric barriers; the barrier conductance is controlled
by tunnelling processes, and dominate the conductivity of the
material. On the other hand the contribution due to metallic
grains to the material resistivity amounts to a very small fraction,
generally lower than 10 %. This finding shows that a model for the
interpretation of the minimum in resistance as a function of
temperature of TFRs is untainable, at least in its present
formulation[4].

Acknowledgements - The authors are indebted to the Laboratory
of Magneti-Marelli, Pavia, for the preparation of some samples.
Measurements in coaxial line would be impossible without the
scribing systems and collaboration of Valfivre Laboratory in
Florence. This work was partially supported by Consiglio
Nazionale delle Ricerche of Italy.

REFERENCES

1. C.A. Harper Ed. "Handbook of thick-film hybrid micro-electronics" McGraw Hill, N.Y. (1974).

2. T.V. Nordstrom, C.R. Hills, Proceed. 1979 Intern.Microelec. Symposium, Los Angeles, 1979, pp.40-45.

3. R.W. West "Conduction mechanism in thick-film microcircuits" Final Technical Report, Purdue University, 1975.

4. G.E. Pike, C.H. Seager, J. Appl. Phys. 48, 5152 (1977).

5. F. Forlani, M. Prudenziati, Electro Comp. Sci. Technol. 3, 77 (1976).

6. M. Prudenziati, Alta Frequenza 46, 287 (1977).

7. R.M. Hill, Proceed. 2nd European Hybrid Microel. Conf. Ghent, Belgium (1979), p.95.

8. C. Canali, B. Morten, D. Malavasi, M. Prudenziati, A. Taroni, J. Appl. Phys. June 1980, in press.

9. B.W. Lieznerski, Thin Solid Films 55, 361 (1978).

10. J. Volger, in Progress in Semiconductors, A.F. Gibson, Wiley, N.Y. 1960, Vol.4, p.207.

11. B.M. Cohen, D.R. Ahlman, R.R. Shaw, J. Non Cryst. Solids, 12, 177 (1973).

12. F. Argall, A.K. Jonscher, Thin Solid Films, 2, 185 (1968).

13. A.K. Jonscher, J. Non Cryst. Solids, 8-10, 293, 1972.

GOLD-SILICON INTERFACE : ELECTRON ENERGY LOSS AND AUGER SPECTROSCOPIES VERSUS GOLD COVERAGE

S. Nannarone, F. Patella, C. Quaresima, A. Savoia,
F. Cerrina, M. Capozi, and P. Perfetti

P.U.L.S.-Laboratori Nazionali di Frascati

00044 Frascati, Italy

1 monolayer, 20 monolayers and 100 monolayers of gold have been deposited on Si(111)2×1. Electronic properties of such a system have been studied by Energy Loss Spectroscopy and Auger Spectroscopy versus annealing temperature. At moderate annealing temperature Si migrates through the film. Higher annealing temperature results in gold islands formation : the electronic structure of discovered surface layer seems to be quite independent from gold film thickness.

Energy Loss Spectroscopy (ELS) is a powerful tool to investigate electronic and vibrational properties of surfaces and interfaces. In particular, its sensitivity to thickness of the order of one monolayer (mL) makes it a good candidate to study the layer by layer formation of semiconductor-semiconductor and metal-semiconductor junctions. Some results obtained by our group using a second derivative ELS technique on the Au/Si(1) and the Ge/Si(2) interfaces have already been published; we present here an extension of the study of the Au/Si system carried out mainly by ELS and Auger Electron Spectroscopy (AES). We refer to reference (1) for the experimental details and for some review of works done on this field. In the Au/Si system, where the existence of a reacted layer and of a diffuse junction seems to be proven, the nature itself of the surface or interface electronic structure is almost completely unknown. Some structural properties of the Au/Si system are known. Gold at least at low coverages grows us as an amorphous film. Upon annealing at temperatures of the order of 400°C gold films undergo a formation process of islands (2).

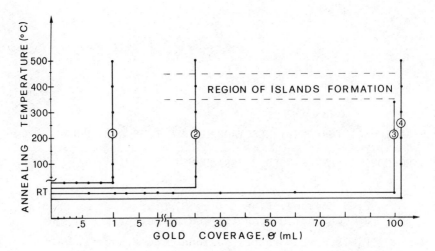

<u>Fig.1</u> : Summary of the work in the temperature of annealing-
coverage plane. The full lines represent paths followed.
The dots along them are θ-T points where EL and AES
measurements have been performed. θ = 1 mL coverage means
a number of atoms per square centimeter equal to the
substrate atomic surface concentration.

        The aim of this work is to follow the evolution with
temperature of the Au/Si system for different thicknesses of the
deposited gold film. The work is summarized in Fig.1. In the θ-T
plane, where T is the temperature of annealing for 30 minutes and
θ the coverage in mL. The paths we followed are reported (full
lines). Dots along them indicate where ELS, AES and Low Energy
Electron Diffraction (LEED) where performed. Paths 1, 2, 4 refer to the present while path 3 to the work of ref. (1). In that
work the evolution of the Au/Si system versus Gold thickness was
studied. It is worthwhile for the discussion of the present
results, to recall that such a study leads to the conclusion that
Au and Si react. The most involved gold electrons in the bond
formation are the d ones and their most characteristic electron
energy loss (what we will call the "d-loss") evolves from 4.4 eV
to 2.7 eV with increasing gold coverage. A plot of the "d-loss"
energy versus gold coverage indicate a gold-silicon reacted layer
of the order of 15 mLs. Coming back to Fig. 1 some information
about the surface morphology along paths 1, 2, 3, 4 can be desumed
by LEED, AES and Scanning Electron Microscopy (SEM). In the
temperature region between 350-450°C (dashed lines of the Fig.1)
Si (1619 eV)/Au (2024 eV) Auger peaks ratio increases very steeply.
This can be assumed as the proof of the formation of gold islands.
SEM analysis made on samples relative to paths 3 and 4 confirm
this hypothesis. The presence of LEED patterns (though on a strong
background) after annealing above 400°C also indicates that some
long range order has been restored. The EL studies we carried on

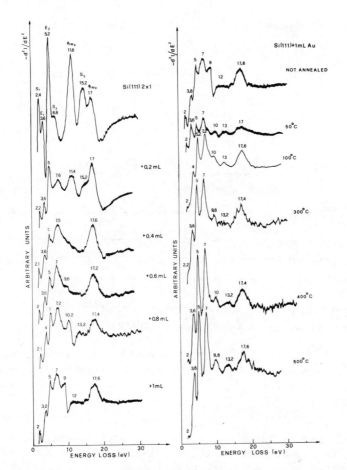

Fig.2 : Typical EL spectrum : left panel EL spectra versus Au
coverage (0 ÷ 1 mL region), right panel evolution of EL
spectrum with annealing of Si + 1 mL of Au.

show that (for paths 1, 2, 3 and 4) the electronic properties of
the surface region between the islands of Gold are similar to the
so called "Si rich phase", a surface phase typical of low θ.
These points will be clarified by the following discussion.

EL spectra of path 1 (as a function of θ up to 1 mL and
of temperature of annealing up to 500°C) are shown in figure 2.
For the structure of the clean Si surface we choose the usual
terminology (E_1, E_2 are due to electronic band to band transitions,
$\hbar\omega_s$ and $\hbar\omega_p$ are due to surface and bulk plasmon excitation, S_1,
S_2, S_3 are due to excitations involving surface electronic states).

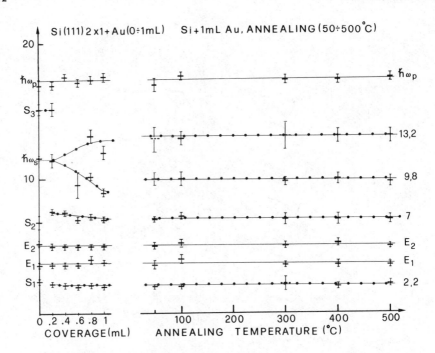

Fig.3 : Evolution of losses of Si + 1 mL of Au versus coverage
        and annealing temperature. Simple lines refers to bulk
        losses, dotted lines to losses attributed to surface
        layer.

In figure 3 the evolution of the losses with coverage and
annealing is summarized. Full lines indicate well defined loss
values, dotted lines indicate weak losses.

      Auger analysis performed on this system (and mL) shows
a constant ratio between the Si (92 eV) and the Au (69 eV) Auger
intensity peaks as a function of θ and T. Analysis of ELS spectra,
taken at different coverages, shows that : a) Si surface structures
disappear at 0.2 mL. S_1 seems to be replaced by a loss (at a

slightly lower energy) that could be attributed to an interface
states. b) Bulk features are evident up to 1 mL. c) Two new
"interface losses" appear at 7 eV and 2 eV. d) The disappearance
of ω_s, its replacement by the two losses at ~ 13 eV and ~ 9.5 eV
(for $\theta = 1$) and their behaviour in the range $\theta = 0 \div 1$ mL could
suggest that an interface loss, degenerate in energy with $\hbar\omega_s$,
grows up (4). We believe that this EL spectrum is
representative of the interface. Losses typical of this system
(apart Si bulk losses still detectable) will be classified as
interface losses. The annealing doesn't affect the main features
of the interface. The only effect, which occurs at very moderate

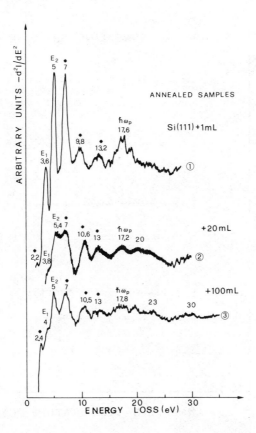

<u>Fig.4</u> : EL spectra of the systems Si + 1 mL, 20 mL, 100 mL of Au
           after an annealing of 500°C for 30 minutes. The stars
           indicate losses attributed to surface excitations.

annealing temperature, is the raising of about 1 eV of the losses
at 9 eV and 12 eV present at 1 mL coverage. From 100°C on the
lack of changes in the spectra indicates that the electronic
structure of the interface doesn't change within the ELS explored
depth. The spectrum obtained at the end of path 1 (1 mL annealed up
to 500°C) is reported in Fig.4 (curve 1) where it is compared with
the spectra obtained at the end of paths 2 and 3. The losses interpreted as interface losses are indicated by a star.

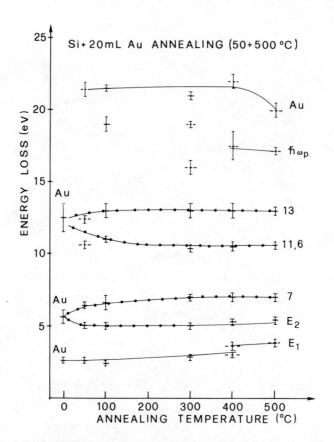

Fig. 5: Evolution of losses of Si + 1 mL of Au versus annealing
temperature. Simple lines refer to bulk losses, dotted
lines to losses attributed to surface layer. The losses
attributed to the gold film or to the gold islands are
indicated.

Results of path 2 (20 mL of Au over Si and subsequent
annealings) are summarized in figure 5. Such a system should be
characterized, before annealing, by a 15 mL thick reacted layer
plus 5 mL of gold. Auger analysis shows a nearly constant Si
(1619)/Au (2024) Auger peak ratio up to 300°C and its steeply
increase between 350 and 450°C. This last observation is
attributed to the phenomenon of gold islands formation, as
confirmed by SEM analysis. Making reference to Fig. 5 the following
points are of special interest:

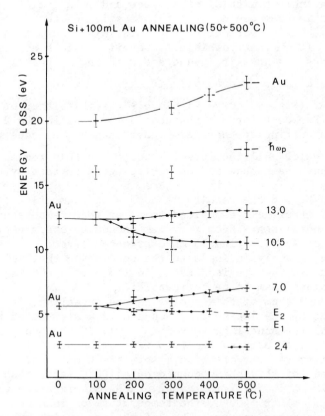

Fig. 6: Evolution of losses versus annealing temperature for
the system Si + 100 mL of Au. The full lines represent
bulk losses, the dotted lines represent the losses
attributed to surface layer. The losses, due to the
gold film or to the gold islands, are indicated.

a) the system, before annealing, is characterized by three losses
at 2.6, 5.6 and 12.5 eV. These losses are interpreted as due to
gold (5). This agrees with the hypothesis (and reinforces it) that
the reacted layer terminates at about 15 mL; after that a film
mainly gold-like begins to grow. b) Under moderate annealing
($\lesssim$ 100°C) the two losses at 5.6 and 12.5 eV both develop into new
componets. The energy losses observed in this case occur nearly
at the same energy of the annealed monolayer. (Fig. 3). The
migration of Si through the interface with Si enrichment of
surface layer (formation of Si rich phase) could take into account
this phenomenon. c) The 2.6 eV loss, the "d-loss," evolves

(disappearing at $\sim$400°C) towards higher energy loss for temperature $\lesssim$ 200°C. Enrichment with Si of surface and gold islands with some Si dispersed in them could explain this evidence too. d) Si bulk structure E_1, $\hbar\omega_p$ appear during the gold island formation process (T $\geq$ 350°). It is not possible to say if the loss at $\sim$5 eV is an interface loss (eventually degenerate with the E_2 bulk loss) or the E_2 Si bulk loss. Tentatively it is labelled E_2. e) The final loss spectrum (Fig. 4, curve 2) closely resembles that of the annealed 1 mL except for the absence of the loss in the 2 eV region. The losses occurring at energies higher than $\hbar\omega_p$ are due to the presence of gold. All the losses reported in this region in Figs. 4, 5 and 6 have been observed on various gold surface by ELS (5).

In Fig. 6 the results of path 4 are summarized; they refer to 100 mL of gold deposited on Si at room temperatures and successive annealings. Such a system should be characterized by about 85 mL of Au film plus 15 mL of reacted layer. As reported in Fig. 1 Auger analysis leads to the conclusion that islands of gold form, as in the previous case, in the 350°-450°C region. The following points seem to be noticeable:
a) The loss spectrum of this system before annealing, is characterized by the losses 2.6, 5.6 and 12.4 attributed to gold film (5). b) For annealing temperature $\sim$150°C the two gold losses at 5.6 and 12.5 eV both evolve into two components. The situation is very similar to that of point b) of the 20 mL case. The fact that the onset of this phenomenon occurs 100°C later than the case b) of 20 mL is still in favour of the Si migration through the Aug film. c) At $\sim$400°C, the temperature of islands formation, the Si bulk features appear. The 2.4 eV loss (which substitutes in this region the 2.6 eV loss of Au) is attributed to the region between islands. In fact a loss of $\sim$2 eV is typical of a "Si rich phase" (Fig. 4, curve 1). d) The 23 eV and 30 eV losses are due to gold (5). e) The spectrum after 500°C of annealing is reported in Fig. 4 (curve 3). The assignment of the losses is made on the same basis of the 20 mL case. In particular the same ambiguity is present for the E_2 loss. For the 2.4 eV loss see point c). EL spectra (apart losses of gold in the 20-35 eV region) of 20 mL and 100 mL after a 500°C of annealing show a close resemblance of that of 1 mL (see Fig. 4). This suggests that the electronic structure of the region discovered by gold agglomeration is similar to that of the "Si rich phase." Slight differences in electronic properties could be suggested by the absence of the loss in the 2-3 eV region in curve 2 (20 mL case).

The results of this work indicate that during the annealing process the migration of Si through the surface layers is the relevant process in the region of low annealing temperatures (T $\sim$ 100°C). This results in an enrichment with Si of the surface layers. At higher annealing temperature ($\sim$400°C) island formation dominates. The EL spectra suggest that the agglomeration of Au in islands discover a surface layer whose electronic properties closely resemble that of the "Si rich" (or interface) layer.

Acknowledgements – The authors are indebted to Prof. C. Calandra for helpful discussions. The technical assistance of R. Bolli, L. Siracusano, F. Campolungo, L. Moretto and M. Macri is highly appreciated.

REFERENCES

1. P. Perfetti et al., to be published in Solid State Commun.

2. C.M. Bertoni et al., to be published in Solid State Commun.

3. G. Le Lay et al., Thin Solid Films 35 (1976) 273.
    K. Oura and T. Hanawa, Surface Sci. 82 (1979), 202.

4. See for instance H. Ibach and J. E. Rowe, Phys. Rev. B 9 (1974) 1951.

5. J.L. Robins, Proc. Phys. Soc. (London) 78 (1961) 1177.
    J.F. Wendelken and D. M. Zehner, Surface Sci. 71 (1978) 178.
    G. Mc Elhiney and J. Pritchard, Surface Sci. 60 (1976) 397.

SURFACE AND INTERFACE ACOUSTIC PHONONS IN

ALUMINIUM-COATED SILICON

J.R. Sandercock[‡], F. Nizzoli[‡ §], V. Bortolani[§],
G. Santoro[§] and A.M. Marvin[¤]

[‡] Laboratories RCA Ltd., 8048 Zürich, Switzerland
[§] Istituto di Fisica and GNSM del CNR,
   Università di Modena, 41100 Modena, Italy
[¤] Istituto di Fisica Teorica, Miramare, Grignano,
   34014 Trieste, Italy

Measurements are presented of the Brillouin cross-section for a
thin aluminium film on a silicon substrate measured using a tandem
multipass interferometer. The measured spectra are compared with
the cross-section computed by assuming that the light is
inelastically scattered from the surface corrugation due to the
thermally excited acoustic phonons. By varying the thickness of
the aluminium film we observe different spectral features such
as the modified Rayleigh mode, the Sezawa modes and the continuum
of states of the substrate. The limiting case of a thick coating
is also discussed.

1. INTRODUCTION

The Brillouin scattering technique has recently proved
to be a successful method for studying thermally excited surface
phonons in the long wavelength limit[1]. The scattering from clean
surfaces has already been investigated in some detail both
experimentally[1-3] and theoretically[4-7]. It has been found that
two mechanism contribute to the measured spectra. In fact the
light is inelastically scattered both from the dynamical
corrugation of the surface due to the phonon field (ripple effect)[4]
and from the dielectric inhomogeneity of the medium (elasto-optic
effect)[5-7]. It has also been found that the spectral function
shows structures related to the surface modes (Rayleigh and
pseudo Rayleigh waves), to the bulk modes and to combination of

bulk waves with modes localized near the surface (the so-called
mixed modes)[4-7].

        This field of interest has been recently extended to the
case of coated surfaces by Rowell and Stegeman[8] for transparent
materials and by us for opaque media[9]. In particular our preliminary results concerning aluminium-coated semiconductors
revealed the effects of the coating on the Brillouin cross-section[9].

        The aim of this paper is to analyze the shape of the
spectrum as the thickness h of the coating is changing. The system
considered is crystalline silicon whose (001) face is covered with
evaporated polycrystalline aluminium.

2. THEORY

        We consider a laser beam impinging upon the film surface
with an angle θ to the surface normal. The cross-section is
computed in backscattering geometry, that is the scattered light
is assumed to be collected in a certain solid angle around the
incidence direction. Under these conditions the experiment probes
the Fourier components of the phonon field having a fixed momentum
parallel to the surface $(Q_{\parallel})$ equal to minus twice the parallel
component of the wavevector of the incoming light. Conversely,
due to the lack of translational invariance along the surface
normal, all the normal (complex) wavevectors $q_{\perp}$ of the phonon
field are mixed in the experiment and contribute to the spectra.

        We assume that the scattering cross-section is
dominated by the ripple-mechanism on the free surface of the film.
This is justified by the large value of the dielectric function
of aluminium[4-6]. Within this assumption the cross-section is
simply proportional to the spectral function of the normal
component of the phonon field, computed at the surface[4]. This
quantity can be numerically obtained for an isotropic film deposited over a semiinfinite cubic substrate by imposing mechanical
boundary conditions at the surface (stress free surface) and at
the interface (continuity of the stress and of the displacement
fields). For more details the reader is addressed to a forthcoming
paper dealing with Brillouin scattering from a coated surface in
a general situation[10].

        The spectral function shows two kinds of contributions.
For a frequency Ω less than the transverse threshold of the
substrate $\Omega_t(Si)$ only true surface and interface modes are allowed,
giving rise to a discrete spectrum. In our case - $Q_{\parallel}$ is directed
along the crystallotraphic direction [100] of the substrate - the

surface mode is a generalized Rayleigh wave whose phase velocity
ranges between the appropriate Rayleigh wave velocities of aluminium
and silicon. The modes which owe their existence to the presence
of the interface are called Sezawa waves[11] and their phase
velocity is in-between $v_t(Al)$ and $v_t(Si)$ - velocities of propagation
of the bulk transverse phonons with total momentum equal to $Q_\parallel$.
Both Rayleigh and Sezawa waves are polarized in the incidence
plane of the light (sagittal plane) and can contribute to the
ripple cross-section through their normal displacement components.
The second contribution to the surface spectral function of the
phonon field extends to frequencies $\Omega > \Omega_t(Si)$. It takes the form
of a continuous spectrum and is due to the mixed modes mentioned
in the introduction. Although this scattering channel is usually
weaker than the former one we know that for clean surfaces it is
not negligible[1] and in particular situations it is even very
important[5,6]. We will show that this also happens in the case of
coated surfaces.

3. EXPERIMENTS

        The spectra reported here have been measured with an
incident wavelength of 5145 Å and an angle $\theta = 70°$. A tandem
multipass interferometer (5 pass plus 2 pass) has been used,
as in the previous measurements for clean surfaces[1]. Further
details are given elsewhere[12].

        Figures 1-3 give the experimental spectra for three
increasing thicknesses of 390, 700 and 2000 Å. These values have
been derived through a comparison with the computed velocities
of the localized modes, and agree to within ± 100 Å with the
thicknesses measured independently using a profilometer. The
full lines represent the computed ripple cross-section obtained
by allowing a finite lifetime for the phonon modes. The relative
intensity of the computed cross-sections has been adjusted to fit
the strength of the lowest frequency experimental peaks. These
are clearly due in all cases to the generalized Rayleigh wave (RW)
present at any thickness. The phase velocity of RW is modified by
the presence of the Al-film and has almost approached that of Al
for 2000 Å. The other broad structure present at 390 Å (Fig.1)
is entirely due to the mixed modes previously discussed and
exemplifies how these modes can be important. The double-peak
shape of this structure is due to the presence, at nearly 21 GHz,
of a well defined resonance which, at a slightly larger thickness
($\simeq 400$ Å) transforms itself in a true Sezawa wave (SW). This
situation was found in the previous measurement[9] where, however,
the experimental resolution was lower. In that case the SW was not
resolved from the mixed modes.

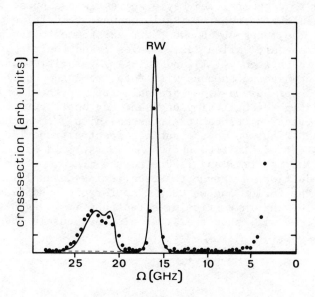

Fig. 1: Experimental (dots) and computed (full line) Brillouin cross-section for a 390 Å film of aluminum deposited on silicon. The broken line shows the spectrum without the contribution of the continuous part.

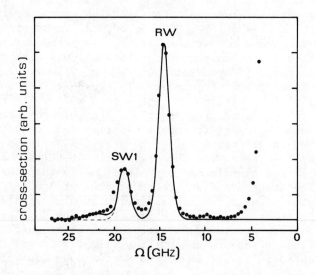

Fig. 2: As in Fig. 1. Film thickness 700 Å.

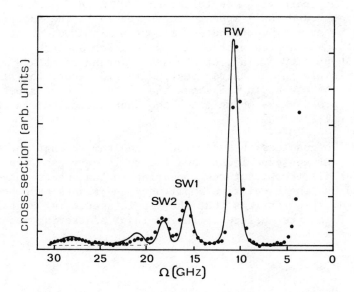

Fig.3 : As in Fig.1. Film thickness 2000 Å.

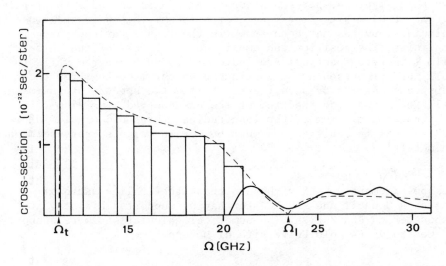

Fig.4 : Mixed mode spectrum of unsupported aluminium (broken line)
        compared with the spectrum of 1 μm of aluminium deposited
        on silicon (histogram plus full line). The Rayleigh wave
        is not shown.

By increasing the thickness (700 Å in Fig.2) a first order Sezawa wave (SW1) appears very clearly in the spectrum, followed by a small contribution from the mixed modes. Again this contribution is a precursor of a new SW which appears at a thickness of about 1200 Å. In fact the second order interface mode (SW2) is well visible in the third spectrum (Fig.3) where the film thickness is 2000 Å. In this case the mixed mode contribution shows two broad peaks which match two weak structures in the experimental spectrum.

To conclude this section we can remark that the overall agreement between experiments and theory is very good. In particular all the structures appearing in the spectra have been explained and their relative amplitudes and shapes have been reproduced. This confirms that the ripple scattering contribution from the film free surface is the dominant scattering mechanism for this system.

4. THICK COATING

As a final point we want to discuss the limiting case of a thick coating when $Q_{\parallel} h \gg 1$. In this case one expects that the spectral function of the film surface approaches that of a semi-infinite medium of the same material. This comparison is given in Figure 4, between unsupported aluminium and a thick film (1 µm) of aluminium deposited over silicon. The RW in the two cases coincides and has not been shown in the figure for the sake of simplicity. The position and amplitude of the SWs of the thick film (discrete spectrum) are indicated by an histogram while the full curve represents the continuous spectrum coming from the substrate. As can be seen the sum of the two spectra approximates well the mixed mode spectrum of the unsupported aluminium[4]. In particular a minimum at the longitudinal threshold of aluminium (23.45 GHz) is achieved, as must be for the ripple cross-section in metals[1,4].

Acknowledgement   -   FN wishes to thank the Italian National Research Council and the RCA for having supported his stay at Laboratories RCA Zürich. The numerical calculations were supported by Centro di Calcolo Università di Modena.

REFERENCES

1. J.R. Sandercock, Solid State Commun. 26, 547 (1978).

2. R. Loudon and J.R. Sandercock, J. Phys. C : Solid State Physics (1980) in press.

3. R.T. Harley and P.A. Fleury, J. Phys. C : Solid State Physics 12, L863 (1979).

4. R. Loudon, Phys. Rev. Lett. 40, 581 (1978).

5. V. Bortolani, F. Nizzoli and G. Santoro, Phys. Rev. Lett. 41 39 (1978).

6. A.M. Marvin, V. Bortolani, F. Nizzoli and G. Santoro, J. Phys. C : Solid State Physics 13, (1980) in press.

7. A.M. Marvin, V. Bortolani and F. Nizzoli, J. Phys. C : Solid State Physics 13, 299 (1980) and other references therein.

8. N.L. Rowell and G. I. Stegeman, Phys. Rev. Lett. 41, 970 (1978).

9. V. Bortolani, F. Nizzoli, G. Santoro, A. Marvin and J.R. Sandercock, Phys. Rev. Lett. 43, 224 (1979).

10. V. Bortolani, F. Nizzoli, G. Santoro, J.R. Sandercock and A.M. Marvin, to be published.

11. G.W. Farnell and E. L. Adler, Physical Acoustic vol. IX, pp. 35-127, Academic Press, New York (1972).

12. J.R. Sandercock, paper to be presented at the 7th International Conference on Raman Spectroscopy, Ottawa, Canada, August 1980.

OPTICAL EXCITATION OF QUADRUPOLE PLASMA OSCILLATIONS

IN ELLIPSOIDAL PARTICLES

A. Meessen and F. Gérils

Institut de Physique
Université Catholique de Louvain

B-1348 Louvain-la-Neuve, Belgium

The optical properties of thin granular films and colloids are usually described by means of the "dipole approximation", assuming that the light acts on ellipsoidal particles of extremely small dimensions compared to the wavelength of the incident light. A first order correction of this theory is obtained by considering somewhat larger ellipsoidal particles, so that the electric field of the incident light wave varies linearly instead of being constant in the vicinity of the particle. This leads actually to very concise analytical expressions, with well defined form factors for the corresponding quadrupole effects, when the directions of polarization and propagation of the light wave are specified with respect to the equatorial plane of arbitrarily shaped spheroidal particles. A more general expression for the effective dielectric constant of an ensemble of equally oriented spheroids can thus also be established.

The action of electromagnetic plane waves on axially symmetric ellipsoidal particles can only be described by means of the "dipole approximation", introduced by Rayleigh[1], when the particles are extremely small compared to the wavelength of the incident wave. While Mie[2] succeeded in solving the general problem of multipole scattering for spheres of arbitrary size, it seems that the corresponding problem for ellipsoidal particles is too complex to be solved in such general terms[3]. Stevenson[4] used therefore a perturbation method, where the wave number $k = \omega/c = 2\pi/\lambda$ is the expansion parameter, but his derivations were not very transparent and his results were questioned[5]. Oguchi[6] studied the scattering of micro-waves by spheroidal raindrops by means of numerical calculations with various approximations. Although these methods have been refined and extended by other authors[7-10], they remain cumberson and too particular to be easily applied to other situations. We tried therefore to get simple analytical expressions by going just one step further than the usual "dipole approximation".

To get more physical insight, we consider first the case of a sphere of radius a and dielectric constant $\varepsilon(\omega)$. The sphere is situated in vacuum and centered at the origin, the plane wave being polarized along the z-axis and propagating along the x-axis. The particle is assumed to be relatively small, so that the electric field of the incident wave can be considered as varying linearily in the vicinity of the particle : $E_o \exp(ikx) = E_o(1+ikx)$. This will lead to the "quadrupole approximation", while the "dipole approximation" is obtained by considering only E_o. The spatial variation of the scalar potential outsides the sphere is then given by

$$v^+(\underline{r}) = -E_o(1+ikx)z + p(z/r^3) + q(xz/r^5)$$

where p and q are the induced dipole and quadrupole moments. It is necessary, however, to consider also the spatial variations $A(\underline{r})$ of a vector potential with $\text{div } \underline{A} - ikV = 0$ and $\underline{E}(r) = -\text{grad } V(r) + ik\underline{A}(\underline{r})$, but it is sufficient to set $A_x^+ = -E_o z$, $A_y^+ = A_z^+ = 0$ outside the sphere to get $\underline{E}(\underline{r})$ up to first order in k. Inside the sphere we adopt potentials varying like the applied field, since they have to satisfy the same equations :

$$V^-(\underline{r}) = -E_o'z - E_o''ikxz; \quad A_x^- = -E_o'''z; \quad A_y^- = A_z^- = 0 .$$

Expressing these potentials in spherical coordinates and calculating the radial and tangential components of the electric field, we have to impose the boundary conditions $E_r^+ = \varepsilon E_r^-$, $E_\theta^+ = E_\theta^-$ and $E_\phi^+ = E_\phi^-$ for r = a. We get then

$$p = E_o a^3 \frac{(\varepsilon-1)}{(\varepsilon+2)} ; \qquad E_o' = \frac{3E_o}{(\varepsilon+2)} ;$$

$$q = ikE_o a^5 \frac{(\varepsilon-1)}{(2\varepsilon+3)} ; \qquad E_o'' = ikE_o \frac{(\varepsilon+4)}{(2\varepsilon+3)} ; \quad E_o''' = E_o .$$

The results of the "dipole approximation" are thus preserved, while the first order corrections in k correspond to Mie's theory.

Replacing the sphere by an ellipsoid of revolution, we have to distinguish three fundamentally different cases :

(t,t), (t,n) and (n,t). The first index indicates the direction
of the electric field E and the second index the direction of
propagation of the incident plane wave, using the notation t and n
respectively for the directions which are "tangential" or "normal"
to the equatorial plane of the particle. The case (t,t) is thus
obtained with the previous assumptions for the wave, when the
symmetry axis of the particle is taken along the y-axis. Considering
the case of an oblate spheroid, we define the excentricity by

$e = c/a$, with $c^2 = a^2 - b^2$, where a and b are respectively the semi-
axes in the equatorial plane and along the symmetry axis.

To get the allowed spatial variations of the scalar and
vector potentials inside and outside the particle it is sufficient
to solve Laplace equations, up to first order in k. Using
ellipsoidal coordinates (ξ,η,ϕ) defined by

$$x = c\xi\eta\sin\phi \; ; \; y = c\sqrt{(\xi^2-1)(1-\eta^2)}; \quad z = c\xi\eta\cos\phi \, ,$$

we can transpose the previous expressions of the potentials :

$$V^+ = -E_o(\xi+\alpha G)c\eta\cos\phi - ikE_o(\xi^2+\beta H)c^2\eta^2\sin\phi\cos\phi$$

$$V^- = -E_o'c\xi\eta\cos\phi - ikE_o''c^2\eta^2\sin\phi\cos\phi$$

$$A_x^+ = -E_o c\xi\eta\cos\phi \; ; \; E_x = -E_o'''c\xi\eta\cos\phi \; ; \; A_y^{\pm} = A_z^{\pm} = 0 \, .$$

α and β are constants, while G and H are functions of ξ which
vanish for $\xi \to \infty$ and which are such that $\Delta V = 0$ in ellipsoidal
coordinates. This leads to the equations $G'' + p_1 G' + q_1 G = 0$
and $H'' + p_1 H' + q_2 H = 0$, with

$$p_1 = \frac{2\xi^2-1}{\xi(\xi^2-1)} \; ; \; q_1 = \frac{1-2\xi^2}{\xi^2(\xi^2-1)} \; ; \; q_2 = \frac{2(2-3\xi^2)}{\xi^2(\xi^2-1)} \, .$$

Since these Sturm-Liouville equation are satisfied by the
particular solutions $G = \xi$ and $H = \xi^2$, we can derive associated
solutions :

$$G = \xi \int_{\xi}^{\infty} \frac{d\xi}{\xi^3\sqrt{\xi^2-1}} = \frac{\xi}{2}\left(\tan^{-1}\frac{1}{\sqrt{\xi^2-1}} - \frac{\sqrt{\xi^2-1}}{\xi^2}\right)$$

$$H = \xi^2 \int_{\xi}^{\infty} \frac{d\xi}{\xi^5\sqrt{\xi^2-1}} = \frac{\xi^2}{2}\left(3\tan^{-1}\frac{1}{\sqrt{\xi^2-1}} - \frac{(3\xi^4-\xi^2-2)}{\xi^4\sqrt{\xi^2-1}}\right)$$

The asymptotic behaviour of these functions $(G \to 1/3\xi^2 = c^2/3r^2$
and $H \to 1/5\xi^3 = c^3/5r^3)$ allows us to define the values of the

induced dipole and quadrupole moments : $p = -E_o\alpha c^2/3$ and
$q = -ikE_o\beta c^3/5$. The constants α and β are determined by imposing
the adequate boundary conditions for the components of the
electric field at the surface of the ellipsoidal particle :
$E_\xi^+ = \varepsilon E_\xi^-$, $E_\eta^+ = E_\eta^-$ for $\xi = 1/e$. The result can be stated more
generally, by considering also the (n,t) case, where
$E = E_o(1+ikx)$; $E_x = E_z = 0$, and the (t,n) case, where
$E_x = E_o(1+iky)$; $E_y = E_z = 0$, when the symmetry axis of the
spheroid is always the y-axis. The potentials and the corresponding
function G and H are modified accordingly, with the same boundary
conditions. We can assert then that an electric field component
$E_j(1+ik_s x_s)$, where $j,s = n,t$ and $x_s = x,y,z$, leads to non vanishing
components of the dipole and quadrupole moments given by

$$p_j = \frac{(V/4\pi)(\varepsilon-1)E_j}{1+(\varepsilon-1)d_j}$$

$$q_{js} = \frac{(3a^2/5)(V/4\pi)(\varepsilon-1)ik_s E_j \mu_{js}}{1+(\varepsilon-1)g_{js}}$$

V is the volume of the spheroid and f_j corresponds to the form
factors obtained by Rayleigh[1], while the new form factors for
the "quadrupole approximation" are given by

$$g_{tt} = \frac{1}{4e^4}\left(\frac{3\sqrt{1-e^2}}{e}\tan^{-1}\frac{e}{\sqrt{1-e^2}} + 2e^4 + e^2 - 3\right)$$

$$g_{nt} = g_{tn} = \frac{(2-e^2)}{2e^4}\left(-\frac{3\sqrt{1-e^2}}{e}\tan^{-1}\frac{e}{\sqrt{1-e^2}} - e^2 + 3\right)$$

with $\mu_{tn} = (1-e^2)$ and $\mu_{tt} = \mu_{nt} = 1$. These results can be extended
to prolate spheroids, with $e^2 = 1 - (b/a)^2 < 0$ when $b > a$. Since
e is then imaginary, we redefine the excentricity e so that
$e^2 = 1 - (a/b)^2$, which implies the substitution $e(1-e^2)^{-1/2} \to ie$.
We get then

$$g_{tt} = \frac{1}{4e^4}\left(\frac{3(1-e^2)}{2e}\log\frac{1+e}{1-e} + 5e^2 - 3\right)$$

$$g_{nt} = g_{tn} = \frac{(2-e^2)}{2e^4}\left(3 - 2e^2 - \frac{3(1-e^2)}{2e}\log\frac{1+e}{1-e}\right) .$$

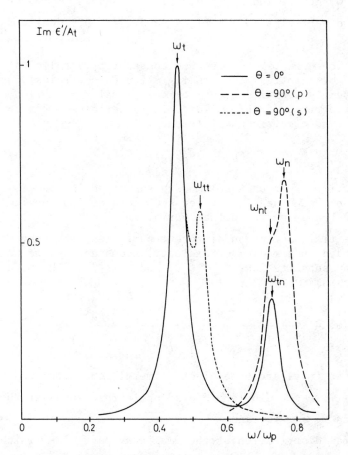

Absorption spectrum of a colloid of identical and equally oriented spheroids of free electron metals suspended in vacuum. A_t is the amplitude of the absorption peak at ω_t. The angles of incidence and the polarizations s and p are chosen to achieve the best separation between dipole resonances (ω_t, ω_n) and quadrupole resonances $(\omega_{tt}, \omega_{nt} = \omega_{tn})$. The parameters are specified in the text.

Instead of the sum rule $2f_t + f_n = 1$, we get $2g_{tt} + g_{nt}(2-e^2)^{-1} = 1$.
For a sphere $f_n = f_t = 1/3$ and $g_{tt} = g_{nt} = g_{tn} = 2/5$. The quadrupole
form factors can be easily calculated. For disk-like particles
$(b/a \to 0)$, we get f_t and $g_{tt} \to 0$, while f_n and $g_{nt} \to 1$. For
needle-like particles $(b/a \to \infty)$, f_t and $g_{tt} \to 0,5$, while f_n and
$g_{nt} \to 0$.

The effective dielectric constant of a colloidal
suspension of identical and equally oriented spheroidal particles
is an extremely useful concept for the global description of the
optical properties of the colloid. It is defined by the components
of the average electric displacement :

$$D_j = E_j + 4\pi N\ (p_j - \sum_s \partial_s q_{js}) = \varepsilon'_j E_j$$

when N particles per unit volume are suspended in vacuum, mutual
interactions being negligible for a colloid of low density. The
derivation bears on the spatial variation $\exp\ (i \sum_s k_s x_s)$ of the
polarizing field acting on different particles. When the particles
are suspended in a medium of dielectric constant ε_a, we have to
perform the substitutions $E_j \to \varepsilon_a E_j$ and $\varepsilon \to \varepsilon/\varepsilon_a$. The resulting
expression is

$$\varepsilon'_j = \varepsilon_a + \frac{NV(\varepsilon - \varepsilon_a)}{1 + (\frac{\varepsilon}{\varepsilon_a} - 1)f_j} + \frac{NV(3/5)a^2 k_s^2 (\varepsilon - \varepsilon_a)\mu_{js}}{1 + (\frac{\varepsilon}{\varepsilon_a} - 1)g_{js}}$$

When the particles are free electron metals characterized by ω_p
and $\gamma = 1/\tau$, we get dipole and quadrupole resonances at the
frequencies $\omega_j = \omega_p \sqrt{f_j}$ and $\omega_{js} = \omega_p \sqrt{g_{js}}$ for $\varepsilon_a = 1$.
The spectral distribution of the imaginary part of E'_j which
defines the absorption spectrum is represented in figure 1 for
oblate spheroids $(b/a = 0,4)$ that are relatively large $(a/\lambda = 0,15)$,
with $\gamma/\omega_p = 0,05$. At normal incidence $(\theta = 0)$ there exists only
a component E_t leading to a dipole resonance ω_t and a quadrupole
resonance ω_{tt}. At grazing incidence $(\theta = 90°)$ there is only a
component E_n with ω_n and ω_{nt} resonances for p polarization with
respect to the equatorial plane of the particles. For a
polarization and grazing incidence, there exists only a component
E_t with ω_t and ω_{tt} resonances. At oblique incidence, one would
get composite spectra, according to the relative importance of
the E_t and E_n components.

REFERENCES

1. Lord Rayleigh, Phil. Mag. 12, 81 (1881); 44, 28 (1897).

2. G. Mie, Ann. der Physik, 25, 377 (1908).

3. F. Möglich, Ann. der Physik, 83, 609 (1927).

4. A.F. Stevenson, J. Appl. Phys. 24, 1143 (1953).

5. J.M. Greenberg et al., Electromagnetic Theory of Antennas I,
   Ed. Jordan, Mc Millan, p.81 (1963).

6. Y. Oguchi, J. Radio Research Lab. 7, 467 (1960); 20, 79 (1973).

7. P.C. Waterman, Proc. IEEE 53, 805 (1965).

8. J.A. Morrison et al., Bell. Syst. Techn. J. 53, 955 (1974).

9. S. Asano and C. Yamamoto, Appl. Optics, 14, 29 (1975).

10. C. Warner and A. Hizal, Radio Sci. II, 921 (1976).

11. P. Rouard and A. Meessen, Progress in Optics, XV, 78 (1977).

FACTOR GROUP SPLITTING AND MULTIPOLE INTERACTIONS IN MOLECULAR CRYSTALS

Pradip N. Ghosh[*]

Physical Chemistry Laboratory
Swiss Federal Institute of Technology

CH-8092 Zurich, Switzerland

An electrostatic interaction potential has been used to explain the factor group splitting in several molecular crystals composed of small inorganic molecules. The potential function is expanded in terms of molecular multipole moments. It is shown that the dipolar coupling mechanism is not sufficient to reproduce the observed splitting in these crystals and that the quadrupolar interactions play a dominant role. In case of hydrogen halide and alkaline earth hydroxide crystals dipole-dipole, dipole-quadrupole and quadrupole-quadrupole interactions can explain the splitting. The values of the quadrupole moment derivatives obtained from such studies are compared with those obtained from ab initio calculations. The possibility of obtaining quadrupole moment derivative from the observed factor group splitting will be discussed. The role of quadrupolar interaction in the case of cyanogen bromide crystal will also be examined.

Molecular crystals are composed of molecules or molecular groups such that the energy of intermolecular interaction is very small compared to the binding energy of individual atoms in the molecule. The presence of small intermolecular interaction can lead to noticeable modification in the internal vibration of molecules and may be a useful source of information of different molecular parameters. These properties show up considerably when excited states of the crystal are studied[1].

The intermolecular forces can cause shift and splitting of internal modes. In solids, in general, there are two effects which produce the splitting - (1) the distortion of the molecule by the local crystalline field and (2) coupling between similar

435

vibrations in different molecules. The first effect will remove
the degeneracy that might be already present in the free molecule
and is due to the lowering of symmetry of the crystalline
environment. This is known as site group splitting. The second
effect will cause a splitting of each internal mode into p modes,
where p is the total number of molecules or molecular groups in
a primitive unit cell. The splitting is the result of interaction
between molecular groups which occupy different but equivalent
lattice sites in a primitive unit cell. It should be emphasized
that although an analysis of the oscillators forming the unit
cell will lead to the correct splitting and symmetry of the
split-up modes, the intermolecular interaction should be extended
to all the molecules which form the crystal. This splitting is
known as factor group splitting.

       Davydov[1] proposed the theory of molecular excitons to
interpret the splitting in the electronic excitation region. The
theory was later extended to the vibrational exciton region by
Hexter[2]. In the vibrational exciton model the factor group splitting
is due to the interaction of vibrational modes which are coupled to
each other through interaction of molecular moments. It is
generally assumed that the splitting arises from the interaction
of transition moments[3]. If the transition is dipolar in origin
the dipole-dipole interaction is responsible for the splitting.
The transition moment interaction model has been used successfully
for a number of molecular crystals. But it has been found recently
that in case of a number of simple molecular crystals the dipolar
interaction mechanism cannot explain the splitting although the
transition is dipolar in origin. In the case of alkaline earth
hydroxides[4] $Ca(OH)_2$ and $Mg(OH)_2$ and lithium hydroxide[5] crystals

it was found that dipole-dipole interaction cannot account for the
splitting. In the case of hydrogen halide crystals we[6] found that
dipolar interaction cannot explain the large magnitude of the
observed factor group splitting. We had shown that the quadrupole-
quadrupole interaction has an appreciably large contribution to
the splitting in the HX crystals. Similar dominant contribution
of the quadrupole-quadrupole interaction was also found by us in
the case of alkaline earth hydroxide crystals[7]. In this report
we discuss the possible role of quadrupole interaction in the
factor group splitting in molecular crystals, in general. In the
earlier work[6] on hydrogen halide crystals spherical lattice sum
was used for dipolar interactions. In the present work we have
used planewise summation method[8] for these interactions. We
present a comparison of the quadrupole moment derivatives
obtained by us from the experimental value of the factor group
splitting with those obtained from ab initio calculations. It is
proposed that in many such crystals the factor group splitting
may be used as an experimental source of information of $(\delta Q/\delta r)$
in absence of any other method to determine these quantities

experimentally. If H_m is the hamiltonian of a free molecule then the hamiltonian of the same group in a crystal will be

$$H = H_m + V_1 + V_2 \tag{1}$$

V_1 causes a shift of the free molecule frequency and produces site group splitting, this is the so-called static-field effect. V_2 produces factor group splitting and is known as the dynamic field effect.

In order to evaluate the effect of V_2 we assume the unperturbed vibrational hamiltonian as

$$H_m + V_1 = H_o = \sum_{\ell\alpha} H_{\ell\alpha} \tag{2}$$

having the unperturbed eigenfunction

$$\psi^o = \sum_{\ell\alpha} \phi^o_{\ell\alpha} \tag{3}$$

The exciton wave function is

$$\psi^e = N \sum_{m\beta} A_\beta \phi^e_{m\beta\ell} \prod_{\alpha \atop m\beta} \phi^o_{\ell\alpha} \exp(iqm) \tag{4}$$

$\phi^e_{\ell\alpha}$ and $\phi^o_{\ell\alpha}$ are excited and ground state wave functions respectively of the α-th oscillator in the ℓ-th unit cell, $\tilde{q}$ is the reciprocal lattice wave vector.

At $q = 0$ the perturbation method will lead to

$$\left| F_{\alpha\beta} + (\nu_e - \nu)\delta_{\alpha\beta} \right| = 0 \tag{5}$$

where

$$F_{\alpha\beta} = \sum_m \int \phi^{e*}_{\ell\alpha} \phi^o_{\ell\alpha} V_{\ell\alpha,m\beta} \phi^{o*}_{m\beta} \phi^e_{m\beta} \, d\tau \tag{6}$$

ν_e is the frequency of the unperturbed oscillator. Thus for a p-molecular unit cell the splitting is into p modes and the magnitude of the splitting is determined by $F_{\alpha\beta}$ which is a lattice sum of the interaction of one molecule with all others. For a correct evaluation of $F_{\alpha\beta}$ (equation 6) one should consider those parts of the intermolecular potential which are simultaneously functions of the vibrational coordinates of the interacting molecules.

We assume the intermolecular potential of coulombic origin and start with Buckingham-type multipole expansion

$$V = V^{\mu\mu} + V^{\mu Q} + V^{QQ} + \ldots V^{\mu\alpha\mu} + V^{\mu\alpha Q}$$

$$+ V^{Q\alpha Q} \ldots + V^{\alpha\alpha} + V^{\alpha\alpha}{}_a + \ldots \tag{7}$$

μ, Q, α and α_a are the multipole moments characterizing individual molecules. These parameters are functions of vibrational coordinates. Hence, in principle, all these terms could contribute to the splitting. They depend on inverse power of R, the intermolecular distance and relative orientations of the molecules. Usually the lattice sum for the dipolar interaction (R^{-3}) has the largest value and quadrupolar $(\sim R^{-5})$ lattice sum is less by one order of magnitude. But the contribution to $F_{\alpha\beta}$ depends also on the square of the multipole moment derivatives.

We have evaluated lattice sums upto terms of R^{-6} order of magnitude for the HCl and HBr crystals. It is found that the lattice sums for dipolar and quadrupolar interactions are almost of the same order of magnitude. An ab initio calculation[9] shows that for the HCl molecule the value of $(\delta Q/\delta r)$ is more than twice that of $(\delta\mu/\delta r)$. Hence although the fundamental transition is electric dipolar in origin and higher multipoles have a negligible effect in the band intensity, they may have a large contribution to the splitting if the molecules possess large values of these moments and their vibrational coordinate derivatives.

The lattice sums are calculated on the basis of crystal structure data which are known accurately. The values of $(\delta\mu/\delta r)$ are also known accurately in the crystal phase from the infrared intensity measurements. Hence we attempted to fit the experimental values of factor group splittings in the HCl and HBr crystals and their deuterated counterparts by adjusting the values of $(\delta Q/\delta r)$. Similar fits were also made by us[7] for the observed factor group splittings in $Ca(OH)_2$, $Ca(OD)_2$ and $Mg(OH)_2$ by adjusting the value of $(\delta Q/\delta r)$. The values of $(\delta Q/\delta r)$ obtained from such fits are listed in Table 1 together with the values reported from ab initio calculations. It is found that the use of planewise summation method instead of spherical lattice sums[6] for dipolar interactions in the HCl and HBr crystals do not affect the values of $(\delta Q/\delta r)$ significantly. This may be because of large contributions from the quadrupolar interactions. It is seen from Table 1 that the agreement of the values of $(\delta Q/\delta r)$ derived from factor group splitting with those derived from ab initio calculations is good.

In the case of BrCN crystal having two molecules in a primitive unit cell the IR and Raman measurements show that there is practically no splitting[11,12], although factor group analysis leads to a splitting. This prompted us to evaluate the factor

Table 1 : Values of $(\delta Q/\delta r)$ in debye

Molecular Group	Factor Group splitting	Ab initio calculations
HCl	4.25 $\pm$ 0.05	4.54[a]
HBr	4.80 $\pm$ 0.00	--
OH	4.00 $\pm$ 0.030[b]	2.53[c]

a) Ref.9   b) Ref.7   c) Ref.10

group splitting from the multipole moments. We evaluated the lattice sums for both the dipole-dipole (μ-μ) and quadrupole-quadrupole (Q-Q) interactions. These lattice sums have similar orders of magnitudes but opposite signs. Since the value of $(\delta\mu/\delta q)$ is rather accurately known from experiment[11] we can calculate the magnitudes of the splittings from the dipolar interaction. It is found that this leads to a very large value of the splitting. Hence we tried to treat the problem in the same way as in the HCl crystal and by adjusting the value of $(\delta Q/\delta q)$ we reproduced the splitting. The values of $(\delta Q/\delta q)$ are rather large, there is practically no way of comparing these values. But these molecules have very large values of quadrupole moments. The relevant results are summarized in Table 2.

Table 2 : Calculations in BrCN crystal

	Factor group splitting (cm^{-1})		
	observed (a)	calculated	
		μ-μ	μ-μ+Q-Q
$\Delta\nu_1$	1.0	30.0	1.0
$\Delta\nu_3$	-3.0	40.5	-3.0
Multipole moment derivatives (e.s.u. $gm^{-1/2}$)			
$(\delta\mu/\delta q_1)$ = 31.1[b]		$(\delta\mu/\delta q_3)$ = 70.8[b]	
$(\delta Q/\delta q_1)$ = 67.4[c]		$(\delta Q/\delta q_3)$ = 153.0[c]	

(a) Ref.12   (b) Ref.11   (c) adjusted values

To summarize we conclude that in a calculation of the factor group splitting one should consider the interaction of multipole moments that exist even before the vibrational excitation, in addition to the interaction of transition moments. It is further

proposed that if the value of $(\delta\mu/\delta q)$ is known accurately from
IR intensities one can evaluate $(\delta Q/\delta q)$ from the observed value
of the factor group splitting. In absence of any other method to
determine these quantities experimentally the factor group
splitting may be a useful source of information of these
quantities.

REFERENCES

* On study leave from the Department of Physics, University of
  Calcutta, Calcutta 700009, India

1. A.S. Davydov, Theory of Molecular Excitons (Plenum, New York)
   1971.

2. R.M. Hexter, Journal of Chemical Physics 33, 1833 (1960).

3. J.C. Decius, Journal of Chemical Physics 49, 1387 (1968).

4. P. Dawson, Journal of Physics and Chemistry of Solids 36,
   1401 (1975).

5. R.E. Frech and J.C. Decius, Journal of Chemical Physics 54,
   2374 (1971).

6. P.N. Ghosh, Journal of Physics C : Solid State Physics 9,
   2673 (1976).

7. P.N. Ghosh, Solid State Communications 19, 639 (1976).

8. F.J. De Wette and G.E. Schacher, Physical Review 137, A78
   (1965).

9. R.K. Nesbet, Journal of Chemical Physics 41, 100 (1964).

10. W.J. Stevens, G. Das, A.C. Wahl, M. Krauss and D. Neumann,
    Journal of Chemical Physics 61, 3686 (1974).

11. A.R. Bandy, H.B. Friedrich and W.B. Person, Journal Chemical
    Physics 53, 674 (1970).

12. M. Pézelet and R. Savoie, Journal of Chemical Physics 54,
    5266 (1971).

CRITICAL SLOWING DOWN OF ORIENTATIONAL FLUCTUATIONS AND SINGLE
MOLECULE REORIENTATION RATE AT AN ORDER-DISORDER TRANSITION
OF A MOLECULAR CRYSTAL

R.E. Lechner
Hahn-Meitner-Institut, Bereich C-1
Glienicker Strasse 100, 1000 Berlin 39

and

Université de Rennes
Groupe de Physique Cristalline, ERA au CNRS n° 070015
Campus de Beaulieu, 35042 Rennes Cédex, France

The relation between single molecule reorientations and critical
orientation fluctuations, occurring near structural order-
disorder phase transitions in certain molecular crystals, is
studied by considering the neutron incoherent scattering function
$S_S(\vec{Q},\omega)$ for a two-site rotational jump model. An expression for
$S_S(\vec{Q},\omega)$ is derived in terms of the single molecule rotational
correlation time τ_R, the mean residence time τ_o of a molecule
within an orientational cluster and the fraction C of molecules
participating in cluster formation at any given instant. It is
concluded that the critical slowing down of the relaxation of
orientational clusters near the phase transition should be
observable by incoherent neutron scattering via the development
of a narrow quasielastic component within $S_S(\vec{Q},\omega)$.

1. INTRODUCTION

Structural order-disorder phase transitions in certain
molecular crystals are related to the existence of multiple
potential wells for the orientation of molecules or molecular
groups. Whereas in the low-temperature phases each molecule or
group is fixed in one of the wells, at high temperature a
dynamical equilibrium is established between the different wells
corresponding to different orientations or conformations of the

molecules. The exchange between these different configurations
takes place by random reorientations of the molecules with a
rotational correlation (or "residence") time τ_R, which can be
measured for instance by quasielastic incoherent neutron
scattering[1].

Near the phase transition, clusters of the orientationally
ordered structure may develop within the disordered phase. These
clusters are of a dynamic nature and may be characterized by a
relaxation time τ_C, which is proportional to the inverse energy
width of quasielastic coherent neutron scattering measured in such
systems. When approaching the phase transition critical slowing
down of this relaxation is observed : the coherent quasielectric
intensity and τ_C diverge at the transition.

This phenomenon has been studied in methane $(CD_4)^2$ and
in para-terphenyl $(C_{18}D_{14})^3$.

In the present context the term "cluster" is used for a
region of arbitrary (a priori unknown) shape, where temporarily
the order of the low-temperature phase is established. We may
consider a molecule as a member of such a region as long as its
orientation (which corresponds to the local order) remains
unchanged for a time much longer than τ_R. Let us call this a
"cluster state" of the molecule. The fluctuations of orientational
correlation characterized by the cluster relaxation time τ_C may be
described by the motion of cluster walls. This motion is responsible
for the variation with time of cluster size, shape and location
and causes a finite mean residence time τ_o of a molecule within an
orientational cluster; τ_o may be considered as the cluster state
life time of the molecule. A molecule which is encountered by a
moving wall, by definition leaves the cluster. The molecule may
then either remain in the cluster wall for some time, return into
the original cluster or become a member of a new cluster. While it
stays in the wall the mean residence time of the molecule in one
of the allowed orientations is τ_R. This means that it is performing
reorientations with a correlation time τ_R. We will call this
situation a "wall state" of the molecule and define the wall state
life time of a molecule, τ_W, as the mean residence time of a
molecule in such a state. The cluster wall might have some finite
thickness, the minimum being two monomolecular layers corresponding
to the "surfaces" of two adjacent clusters. The mean volume of the
clusters and the mean thickness of the cluster walls defined in
this way determine the fraction C of molecules which at any given
instant are in a cluster state. We may call this fraction the
dynamic concentration of cluster molecules.

The critical phenomena observed in quasielastic coherent
neutron scattering experiments[2,3] are related to the divergence

of oriented cluster dimensions which ultimately leads to orientational long range order. We may expect that with growing cluster sizes the mean residence time τ_o of a molecule within a cluster will increase and that it will also diverge at the phase transition similarly to the cluster relaxation time τ_C. Since τ_o is a single molecule property, incoherent neutron scattering is an appropriate method to study the critical phenomena mentioned above by measurements of this parameter.

In the present paper we derive an expression for the neutron incoherent scattering function $S_S(\vec{Q},\omega)$ in terms of τ_R, τ_o and C for the simplest possible case, namely the case of a double well potential. We show that the critical phenomenon should be observable by incoherent neutron scattering via the development of a narrow quasielastic component within $S_S(\vec{Q},\omega)$ near the transition, with a width essentially governed by the cluster state life time τ_o.

2. RATE EQUATIONS

In order to simplify the problem, we will consider a molecular crystal where each molecule at any given instant may have one of two well-defined, equivalent, but distinguishable orientations. Transitions between these occur by rotation about one single axis, when the molecule is in a wall state. Let us assume that the time needed for such a transition is small as compared to the residence time τ_R. This means that we are using the rotational jump model[1]. Since incoherent neutron scattering observes the motion of single particles, this is equivalent to random hopping of each atom between two different sites with a jump rate of $1/\tau_R$. Let us assign the subscripts 1 and 2 to the two possible sites of an atom which are at a distance 2a from each-other. In each site, at a given instant, the atom - as a part of the molecule - is in one of two possible states, a cluster state or a wall state, for which we will use the subscripts C and W, respectively. Thus, for our present purpose, we will distinguish between four different states of each atom, which may be represented by the probabilities for the atom of being in the respective states : W_{1C}, W_{2C}, W_{1W} and W_{2W}.

It is evident that transitions between the four states will not occur in a completely arbitrary way. We can say for instance that direct transitions between two different cluster states are forbidden, since a molecule has to pass by a wall region in order to reach another cluster. This fact is expressed in the following reaction equation for the dynamical equilibrium between the four states :

$$\underline{1} \qquad W_{1C} \begin{array}{c} 1/\tau_o \\ \rightleftharpoons \\ 1/\tau_W \end{array} W_{1W} \begin{array}{c} 1/\tau_R \\ \rightleftharpoons \\ 1/\tau_R \end{array} W_{2W} \begin{array}{c} 1/\tau_W \\ \rightleftharpoons \\ 1/\tau_o \end{array} W_{2C}$$

where $1/\tau_o$, $1/\tau_R$ and $1/\tau_W$ are the relevant transition rates. More explicitly this equilibrium is described by the appropriate rate equations which correctly take into account the coupling between the different states :

$$\underline{2} \qquad \partial W_{1C}/\partial t = - XW_{1C} + Z.W_{1W}$$

$$\partial W_{1W}/\partial t = XW_{1C} - (Y + Z)\ W_{1W} + Y.W_{2W}$$

$$\partial W_{2W}/\partial t = Y.W_{1W} - (Y + Z)\ W_{2W} + XW_{2C}$$

$$\partial W_{2C}/\partial t = Z.W_{2W} - XW_{2C}$$

where $X = 1/\tau_o$, $Y = 1/\tau_R$ and $Z = 1/\tau_W$.

Solving the system of equations $\underline{2}$ yields the following eigenvalues :

$$\underline{3} \qquad \lambda_1 = 0$$

$$\lambda_2 = -(X+Z)$$

$$\lambda_3 = -\left(\frac{X}{2}+Y+\frac{Z}{2}\right) + \left[\left(\frac{X}{2}+Y+\frac{Z}{2}\right)^2 - 2XY\right]^{1/2}$$

$$\lambda_4 = -\left(\frac{X}{2}+Y+\frac{Z}{2}\right) - \left[\left(\frac{X}{2}+Y+\frac{Z}{2}\right)^2 - 2XY\right]^{1/2}$$

The solution of $\underline{2}$ is given by the conditional probabilities :

$$\underline{4} \qquad W_j^i(t) = \sum_{k=1}^{4} a_k E_k$$

for an atom (or a molecule) to be in state j at time t, if it was in the initial state i at t = 0; i and j run over the four possible states 1C, 1W, 2W and 2C.

The coefficients a_k which are functions of the eigenvalues, are determined by the initial state; the corresponding expressions are lengthy and will not be given here. The time dependence of the conditional probabilities is contained in the functions E_k, which are :

$$\underline{5} \qquad E_1 = 1$$

$$E_2 = \exp(t.\lambda_2)$$

$$E_3 = \exp(t.\lambda_3)$$

$$E_4 = \exp(t.\lambda_4)$$

The time Fourier transforms $F(E_k)$ of these exponential functions, which are required for calculating the scattering function $S_S(\vec{Q},\omega)$ are the Lorentzians $L_k(\omega)$:

$\underline{6}$

$$F(E_k) = L_k(\omega) = \frac{1}{\pi} (-\lambda_k) (\lambda_k^2 + \omega^2)^{-1}$$

3. THE INCOHERENT SCATTERING FUNCTION

Since for our present purpose we are only interested in the small energy transfer region we will now calculate the incoherent neutron scattering function for the above described model without taking account of other types of atomic motion such as lattice vibrations, which generally exist in crystals. So far as the quasielastic region is concerned, such effects can easily be introduced at a later stage via Debye-Waller-type structure factors[1].

The incoherent scattering function $S_S(\vec{Q},\omega)$ is obtained as the space and time Fourier transform

$\underline{7}$

$$S_S(\vec{Q},\omega) = \frac{1}{2\pi} \int_{-\infty}^{\infty} e^{-i\omega t} e^{i\vec{Q}\vec{R}} G_S(\vec{R},t) \, d\vec{R} dt$$

of the Van Hove self correlation function $G_S(\vec{R},t)$ which itself for our model is given by

$\underline{8}$

$$G_S(\vec{R},t) = \sum_{i=1}^{4} W_i(\infty) \sum_{j=1}^{4} W_j^i(t) \cdot \delta(\vec{R}-\vec{R}_{ji})$$

where $\vec{R}_{ji}$ are the vectors joining initial with final sites.

This expression is simplified by the fact that the two states belonging to a particular site (for instance site 1) obviously have the same vectors. Furthermore the symmetry of the problem (the two different possible sites are assumed to be equivalent) requires the following conditions on the infinite time probabilities, which are the weighting factors for the initial states :

$\underline{9}$

$$W_1(\infty) = W_4(\infty) = \frac{Z}{2(X+Z)} \quad (=W_{1C}(\infty)=W_{2C}(\infty))$$

$\underline{10}$

$$W_2(\infty) = W_3(\infty) = \frac{X}{2(X+Z)} \quad (=W_{1W}(\infty)=W_{2W}(\infty))$$

From the equations $\underline{3}$ to $\underline{10}$ the following expression is derived for $S_S(\vec{Q},\omega)$, the quasielastic incoherent scattering function for a single crystal representing our model :

$\underline{11}$ $\qquad$ $S_S(\vec{Q},\omega) = \frac{1}{2}[(1 + \cos 2\vec{Q}\vec{a})\delta(\omega) + (1 - \cos 2\vec{Q}\vec{a})S_L(\omega)]$

where $2\vec{a}$ is the vector joining the two sites ($|\vec{a}| = a$), and

$\underline{12}$ $\qquad$ $S_L(\omega) = F_3 \cdot L_3(\omega) + F_4 \cdot L_4(\omega)$

with

$\underline{13}$ $\qquad$ $F_3 = (\lambda_3 - \lambda_2) \cdot \lambda_4/\lambda_2(\lambda_3 - \lambda_4)$

$\qquad\qquad$ $F_4 = (\lambda_4 - \lambda_2) \cdot \lambda_3/\lambda_2(\lambda_4 - \lambda_3)$

It is seen that $S_S(\vec{Q},\omega)$ contains a purely elastic term with an elastic incoherent structure factor (EISF) governed by the distance between the two sites. The quasielastic term consists of two Lorentzians with weighting factors determined by the eigenvalues λ_2, λ_3 and λ_4. It is interesting to introduce the concentration C of cluster molecules into these expressions. Since the life times are proportional to the corresponding concentrations, we have

$\underline{14}$ $\qquad$ $\tau_0/\tau_W = C/(1-C)$

This can be used to obtain the coefficients F_3 and F_4 in the following form :

$\underline{15}$ $\qquad$ $F_3 = (1 + \lambda_3\tau_0(1-C))/(1 - \lambda_3/\lambda_4)$

$\qquad\qquad$ $F_4 = (1 + \lambda_4\tau_0(1-C))/(1 - \lambda_4/\lambda_3)$

Let us now consider two interesting limiting cases :

i) $\qquad$ $\tau_0 \rightarrow 0$ and $\tau_W \rightarrow \infty$

$\qquad$ In this limit the expression $\underline{11}$ for the scattering function becomes

$\underline{16}$ $\qquad$ $S_S(\vec{Q},\omega) = \frac{1}{2}[(1 + \cos 2\vec{Q}\vec{a})\delta(\omega) + (1 - \cos 2\vec{Q}\vec{a}) \cdot L(\frac{2}{\tau_R})]$

where $L(2/\tau_R)$ is a Lorentzian of half-width-half-maximum equal to $2/\tau_R$. This is the scattering function for the simple two-site jump model[1], as it should be, when the cluster state life time tends to zero and the wall state life time to infinity.

ii) $\qquad$ $\tau_0 \rightarrow \infty$ and $\tau_W \rightarrow \infty$

$\qquad$ In this case one obtains

$\underline{17}$ $\qquad$ $S_S(\vec{Q},\omega) = \frac{1}{2}[(1 + \cos 2\vec{Q}\vec{a})\delta(\omega) + (1 - \cos 2\vec{Q}\vec{a}) \cdot S_E(\omega)]$

Table 1 : Ratio of widths of the 2 Lorentzians : λ_4/λ_3

τ_o/τ_R C	1	2	4	10
0.1	2.43	4.32	8.25	20.11
0.2	2.96	4.71	8.92	20.92
0.5	5.82	6.85	10.41	22.15
0.8	33.24	18.20	19.07	31.30
0.9	70.00	46.98	38.47	42.99

where

18          $S_E(\omega) = C.\delta(\omega) + (1-C).L(2/\tau_R)$

This corresponds to an experimental situation where the effective behaviour of the crystal can be described by that of a mixture of two kinds of molecules : one kind having fixed orientations and therefore contributing a $\delta(\omega)$-function to $S_E(\omega)$; these are the cluster state molecules (concentration C). The other kind of molecules are "permanently" performing a two-site jump motion with a jump rate of $1/\tau_R$ and therefore contribute a Lorentzian term to $S_E(\omega)$; these molecules are in a wall state (concentration (1-C)). In fact precisely this situation will arise, when the experimental energy resolution is good enough to resolve the rotational jump rate $1/\tau_R$ but not sufficient to observe the cluster state decay rate $1/\tau_o$. Obviously this also requires that $\tau_R \ll \tau_o$, a condition which may well be fulfilled near the phase transition.

In the general case the interesting physical information is contained in the two Lorentzian terms of equation 12. The first term, F_3L_3, reflects to a large extent the cluster state life time τ_o, whereas the second term, F_4L_4, is mainly governed by the residence time τ_R (equation 18 shows the extreme case). A few numerical examples given in Table 1 for the ratio λ_4/λ_3 of the two Lorentzian widths may illustrate this. It is seen that critical slowing down will be observed via the width of the Lorentzian L_3. From the experimental point of view it is important that the

Table 2 : Fractional Weights, F_3, of the narrow Lorentzian L_3
(fraction of total Lorentzian intensity):

τ_o/τ_R c	1	2	4	10
0.1	0.31	0.176	0.13	0.10
0.2	0.52	0.34	0.29	0.23
0.5	0.85	0.72	0.62	0.55
0.8	0.99	0.96	0.92	0.87
0.9	0.997	0.99	0.98	0.95

Lorentzian to be studied has non-negligible intensity. We therefore give in Table 2 the fractional weights F_3 of the narrow
Lorentzian L_3 for the same examples.

Acknowledgements — The author would like to thank J.L. Baudour,
H. Cailleau, H. Dachs and J. Meinnel for a number of useful
discussions.

REFERENCES

1. See for instance : A.J. Leadbetter and R.E. Lechner in "The
   Plastically Crystalline State", Ed. J.N. Sherwood, J. Wiley
   and Sons (1979), p.285-320.

2. W. Press, A. Hüller, H. Stiller, W. Stirling and R. Currat,
   Phys. Rev. Letters, 32, 1354 (1974).

3. H. Cailleau, A. Heidemann and C.M.E. Zeyen, J. Phys. C :
   Solid State Phys. 12, L411 (1979).

POSITRON ANNIHILATION STUDIES OF VACANCIES AND

POLYVACANCIES IN MOLECULAR CRYSTALS

O.E. Mogensen*, M. Eldrup*, D. Lightbody**,
and J.N. Sherwood**

* Chemistry Department, Risø National Laboratory
  DK-4000 Roskilde, Denmark

** Department of Pure and Applied Chemistry
   University of Strathclyde
   Glasgow, G11XL, Scotland

The use of the positron annihilation technique (PAT) in the
investigation of vacancies and phase transformations in
molecular crystals is discussed.

In recent years, the positron annihilation technique
(PAT) has proved to be an efficient tool in the study of vacancies
and vacancy clusters in metals. Positrons injected into a metal
tend to become trapped at these electron deficient regions thus
increasing the lifetime of the positron with respect to its bulk
lattice value. The sensitivity of the positrons to the presence
of such defects has allowed the determination of vacancy
formation and migration energies.

In many molecular solids a certain fraction of the
injected positrons form a positron-electron bound state,
positronium (Ps), before they annihilate. This species can exist
in two states, the shortlived para-Ps (spins antiparallel) and
the long-lived ortho-Ps (spins parallel).

Extensive PAT studies on pure, doped and irradiated
ice have shown that o-Ps undergoes vacancy trapping in a similar
way to the positron/metal case. This has rendered the assessment
of vacancy formation and migration energies possible and the
phenomenon of vacancy clustering has been observed[2].

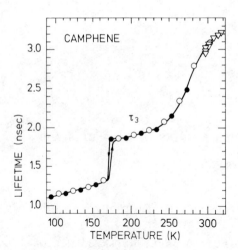

The temperature dependence of the average ortho-Ps lifetime
for a dl-camphene sample.

These studies open the way for a detailed investigation
into the vacancy properties of molecular solids to supplement the
rather sparse existing information.

We have recently started PAT investigations into a
series of molecular solids which display a solid-solid transition
to an orientationally-disordered (plastic) phase which persists
to the melting point[3]. In this phase Ps trapping has been observed
and vacancy formation energies for succinonitrile[4,5], dl-camphene[4]
and adamantane[6] have been obtained.

The solid-solid transition has also received attention,
with hysteresis and impurity effects being clearly noted[7].

By way of illustration, the variation in ortho-Ps lifetime
with temperature for a dl-camphene sample is shown in Fig.1. The
large charge in lifetime at approximately 175 K represents the
solid-solid transition while at high temperatures the sigmoidal
variation in lifetime is indicative of ortho-Ps trapping in
vacancies. At the lower temperature range of this plastic phase
the ortho-Ps exists, prior to trapping, in a bulk state. As the
vacancy concentration increases with increasing temperature, so

too does the rate of trapping of the ortho-Ps into the vacancies, where it displays a longer lifetime. The lifetime variation described by the sigmoidal curve is therefore an average of the bulk lifetime and the longer vacancy lifetime. A saturation effect occurs close to the melting point, with nearly all the ortho-Ps annihilating in the trapped, vacancy state. From the shape of the sigmoidal curve and by the application of a suitable model to describe the trapping process, the vacancy formation energy may be determined[4,5,6].

The financial support of S.R.C. and NATO are most gratefully acknowledged.

REFERENCES

1. Positrons in solids, edited by P. Hautojaervi (Springer, Berlin, 1979).

2. O.E. Mogensen and M. Eldrup, J. Glaciology 21 (1978) 85. M. Eldrup, O.E. Mogensen and J.H. Bilgram, J. Glaciology 21 (1978) 101.

3. The Plastically Crystalline State, edited by J.N. Sherwood (Wiley, 1979).

4. M. Eldrup, O.E. Mogensen and J.N. Sherwood, Proc. Fifth Conf. Positron Annihilation, Lake Yamanaka, Japan, 1979, R.R. Hasiguti and K. Fujiwara, Editors (Japan Inst. of Metals, Sendai, 1979).

5. M. Eldrup, N.J. Pedersen and J.N. Sherwood, Phys. Rev. Letters 43 (1979) 1407.

6. D. Lightbody, M. Eldrup and J.N. Sherwood, Chem. Phys. Letters 70 (1980) 487.

7. M. Eldrup, D. Lightbody and J.N. Sherwood, Faraday Disc. 69 (1980) 000

BRILLOUIN SCATTERING STUDY OF THE PHASE TRANSITION OF BENZIL

M. Boissier, R. Vacher
Université des Sciences et Techniques du Languedoc
34060 Montpellier (France)

J. Sapriel
Centre National d'Etudes des Télécommunications
92220 Bagneux (France)

The Brillouin scattering technique is used for the first time to study the phase transition at 84 K of Benzil crystals. All the elastic constants have been determined vs temperature in the paraelastic phase. The velocities of the longitudinal and transverse acoustic waves have been deduced as a function of the wave vector in the principal planes. The transverse mode of lowest energy which propagates along the x axis is the soft acoustic mode related to the same order parameter as the Brillouin-zone-center E soft optical mode.

Benzil ($C_6H_5COCOC_6H_5$) is a molecular crystal which belongs to the class 32 at room temperature. Below 84 K it undergoes a phase transition and becomes monoclinic (point groupe 2). Symmetry considerations show that in the low temperature phase the crystal is ferroelastic as well as ferroelectric. Actually the ferroelastic behaviour is evidenced by a domain structure described in Ref.1. A detailed temperature dependent Raman study[1] in the range (5 K - 300 K) led to the discovery of a soft E optical mode in the high temperature phase which splits into two components A and B in the monoclinic phase. One can also expect the occurrence of a soft acoustic mode in this ferroelastic crystal; yet no study of the acoustic modes versus temperature has been published till now in benzil. Brillouin scattering is probably the most suitable method to perform such investigations[2].

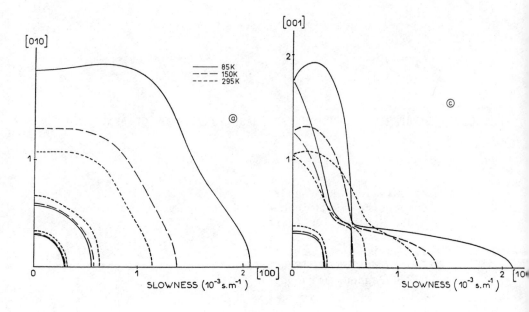

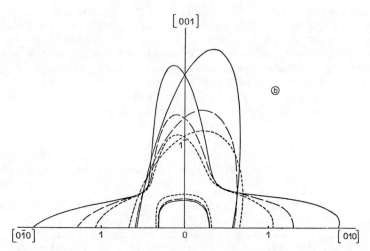

Fig. 1: Slowness curves in benzil in the trigonal phase at 3
         temperatures :   295 K, 150 K and 85 K.
         $1/v_L < 1/v_{T1} < 1/v_{T2}$

         where v_L, v_{T1} and v_{T2} are the acoustic velocities
         (L = longitudinal, T1 and T2 : transverse).

         1-a: Slowness curves in the x-y plane
         1-b: Slowness curves in the y-z plane
         1-c: Slowness curves in the x-z plane

An interferential spectrometer of high resolving power $R > 10^7$ was constituted by a double pass plane Fabry-Pérot used as a monochromator, followed by a spherical Fabry-Pérot interferometer with a free spectral range of 1.48 GHz and a finesse of 50. The maximum resolving power of this tandem was equal to 2×10^7 and the total contrast was about 10^7; 200 mW of a single frequency argon-ion laser was directed upon the sample.

Two samples were used in our experiment. One was oriented along the principal axes of the trigonal phase and the other has its faces perpendicular to the z axis and to the bissecting directions of the x and y axes of this phase. Back-scattering as well as 90° scattering measurements were performed on these samples. The crystal was cooled between 295 K and 85 K by a continuous flow of helium gas which allowed a homogeneous temperature regulated to better than 0.1 K during all the measurement time. The systematic procedure for the determination of the complete set of elastic constants from the Brillouin line shifts is described elsewhere[2]. The refractive indices have been measured for the 5145 Å and 4880 Å laser lines and the density ρ calculated from crystallographic measurements and expansion coefficients data of Ref.3. In the determination of elastic constants the piezoelectric effect, which is rather weak in benzil[4] has been completely neglected. Once obtained the complete set of elastic constants, the resolution of the secular equation :

$$\det\left|c_{ijk\ell}s_{j}s_{\ell} - \rho v^2 \delta_{ik}\right| = 0 , \qquad (1)$$

$$i,j,k,\ell = 1,2,3 ,$$

gives the velocity v of the 3 acoustic waves[5,6] propagating along the direction parallel to the unit vector $\vec{s}$; $c_{ijk\ell}$ are the rigidity tensor components. The velocities of the quasi-longitudinal and of the two quasi-transverse waves are v_L, v_{T1} and v_{T2}, respectively ($v_L > v_{T1} > v_{T2}$). The slowness (1/v) for the three modes is plotted in polar coordinates for 3 values of the temperature (295 K, 150 K, 85 K) in the principal planes x-y, y-z, x-z (see fig. 1-a, 1-b, 1-c). The lowest transverse velocity v_{T2} softens markedly with temperature. The softening is particularly important in the x direction for the acoustic mode labelled γ_3 which corresponds[2] to the combination of the elastic constants :

$$\gamma_3 = \frac{1}{2}\left\{c_{44} + c_{66} - \left[(c_{44} - c_{66})^2 + 4\,c_{14}^2\right]^{1/2}\right\} .$$

The variation of γ_3 with temperature T is principally due to the behaviour of c_{44} which is the elastic constant associated to the

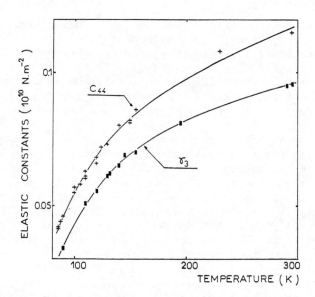

Fig.2 : c_{44} and γ_3 variations as a function of temperature.

The relative accuracy was 0.1 % for velocities measured
in backscattering experiments.

unpolarized transverse waves propagating along the z axis. γ_3 and
c_{44} are plotted vs T in the trigonal phase (Fig.2). Extrapolation
of the γ_3 curve below 85 K gives a cancellation of the velocity for
$T_o \simeq 45$ K. Following Aubry and Pick[7] the γ_3 acoustic mode nas been
predicted[1,8] to soften more than the E optical mode which
corresponds to the same order parameter located at the Γ point
of the Brillouin zone. This is in good agreement with the Raman
experiments[1] where the cancellation of the E soft mode is
obtained at $T'_o \simeq 0$ K. The difference between T_o and T'_o gives an
information on the magnitude of the coupling between the acoustic
mode and the optic mode. The non-vanishing of γ_3 at the transition
temperature is probably a consequence of the first order character
of the transition which has been conjectured[9] on the basis of the
occurrence of discontinuities in the spontaneous quantities,[3-10]
the coexistence of the two phases at the transition[1] and the form
of the specific-heat anomaly[11]. In addition to the primary order
parameter located at the Γ point, a secondary one is expected at
the Brillouin zone edge[9] corresponding to $\vec{k}_m = (0,1/2,0)$. The
effect of the occurrence of this later parameter which actually
produces a multiplication of the unit cell, is not clear on our

Brillouin measurements in the paraelastic phase. Complementary experiments on the acoustic modes in the low-temperature phase are now under investigations on a higher contrast Brillouin set up which could reduce the inelastic spurious light issued from scattering on ferroelastic domain walls[12].

REFERENCES

1. J. Sapriel, A. Boudou and A. Périgaud, Physical Review B, 19, 1484 (1979).

2. R. Vacher and L. Boyer, Physical Review, B6, 639 (1972).

3. G. Odou, M. More and V. Warin, Acta Crystallographica, A34, 459 (1978).

4. S. Haussühl, Acta Crystallographica, 23, 666 (1967).

5. B.A. Auld, "Acoustic fields and waves in solids" (Wiley, New-York, 1973) Vol.I.

6. J. Sapriel, "Acoustooptics" (Wiley, 1979) p.18.

7. S. Aubry and R. Pick, Journal de Physique (Paris) 32, 657 (1971).

8. P.B. Miller and J.D. Axe, Physical Review, 163, 924 (1967).

9. J.C. Tolédano, Physical Review B, 20, 1147 (1979).

10. P. Esherick and B.E. Kohler, Journal of Chemical Physics, 59, 6681 (1973).

11. A. Dvorkin and A. Fuchs, Journal of Chemical Physics, 67, 789 (1977).

12. J. Sapriel, Physical Review B, 12, 5128 (1975).

EXCITATION AND RADIATIVE PROPERTIES OF SURFACE EXCITON STATES IN ORGANIC MOLECULAR CRYSTAL: EXPERIMENTAL EVIDENCES AND QUANTUM MECHANICAL INTERPRETATION

M. Orrit, J. Bernard, J. Gernet, J.M. Turlet and Ph. Kottis

Centre de Physique Moléculaire Optique et Hertzienne (LA 283 du C.N.R.S.) - Université de Bordeaux I 33405 Talence (France)

In the low temperatures reflection, fluorescence and excitation spectra of the $|\vec{a},\vec{b}|$ face of anthracene monocrystal, we report the presence of structures shifting to lower energies under crystal surface coating. We interpret these structures as transitions involving the two Davydov components of surface and sub-surface exciton states.
In a second part, a pure quantum mechanical approach of the radiative properties of a finite system is presented. This investigation supports our experimental assignments and accounts for the existence of both radiatively stable and unstable surface states. Our calculations predict the interference quenching of the surface emission by the bulk reflectivity as the lack of resonance of the surface state diminishes.

INTRODUCTION

The surface and the first few layers of an optically excited molecular crystal, may have drastic influence on the spectral and temporal behavior of this crystal. Consequently, the concepts of collective excitation (exciton) and propagation (polariton), built up for infinite or cyclic crystals, have to be reconsidered in order to account for the finite size properties of the system : such as finite life-time of the polariton states and high radiative unstability of exciton states localized on the crystal surfaces. Indeed, for a particular crystal face, the interplane electron exchange terms are

sufficiently small to be neglected, the fact that surface
molecules undergo different interactions than the bulk molecule
may lead to bidimensional surface and sub-surface localized,
exciton states ("Site Shift Surface Exciton") with resonant
energies different (generally higher in organic molecular crystal)
from their bulk counterparts. We report in this paper the results
of an experimental and theoretical investigation that shows strong
evidences of such surface states and of their radiative, quasi-
superradiant (lifetime $\tau \sim 10^{-12}$ sec), desactivation in organic
crystals of the anthracene type.

EXPERIMENTAL PART

        We recorded, at low temperature (1.7 K) and high
resolution (0.4 cm^{-1}), the reflectivity, the fluorescence and
the excitation polarized spectra of sublimation grown mono-
cristalline anthracene excited, quasi-normally to its $\vec{a},\vec{b}$
surface, in the region of the first singlet-singlet transition
($\sim$ 25 100 cm^{-1}). The results of these three types of experiments,
done quasi-simultaneously on the same sample in the two cases of
an uncoated crystal surface (liquid helium-anthracene interface)
and of a crystal surface coated with a condensed gas (N_2) film,
may be summarized as follows :

1 - The "quasi-metallic" $\vec{E}||\vec{b}$ reflection spectrum (due, according
to the polariton theory, to the strong coupling of the $\vec{b}$ polarized
excitons with photons) presents two intense structures I
(25 301 cm^{-1}) and II (25 103 cm^{-1}) that have fluorescence
counterparts (weak emissions completely $\vec{b}$ polarized) at the same
energies on the "high frequencies" side of the bulk fluorescence
spectrum (see figure 1). In both experiments, these structures I
and II shift to lower energies (78 cm^{-1} for I and 3 cm^{-1} for II)
when the crystal surface is coated with a layer of condensed gas
(which partially restores the lacking interplane interactions of
surface molecules), while all the other features of the reflection
and fluorescence spectra remain unaffected (see figure 2). This
surface coating experiment allows us to interpret very consistently
these transitions (not observed for light incident on the other
crystal faces) by the presence of surface and sub-surface exciton
states with energies resonances respectively 207 cm^{-1} and 9 cm^{-1}
above the bulk resonant state (25 094 cm^{-1}). Moreover, from these
spectra, we can estimate for these involved excited states,
radiative bandwidths (12 cm^{-1} for I and 0,8 cm^{-1} for II) that
agree with the very short lifetimes expected for surface states
($\tau \sim 1$ p sec), (see theoretical part), and explain nicely why
such energy levels, above those of the bulk, have not their
radiative channel quenched by non-radiative transitions to the
bulk.

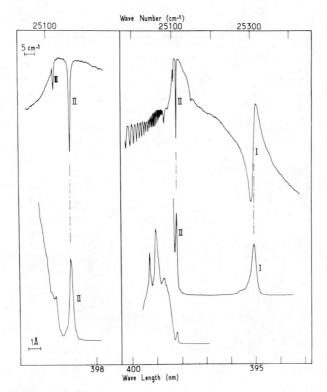

Fig.1 : $\vec{E}||\vec{B}$ reflectivity (top) and fluorescence (for two slit
widths) of the same anthracene monocrystal at 1.7 K. On
left a scanning, at higher resolution, of the region of
structure II.

2 - In the $\vec{E}||\vec{a}$ reflection spectrum (of much less intensity
owing to the weaker oscillator strength of this transition), we
observed a small and broad maximum ($\sim$ 25 523 cm^{-1}) which also
shifts to lower energy (roughly by the same amount than structure
I) when the crystal surface is coated. This is the evidence that
this maximum is associated with the $\vec{a}$ polarized Davydov component
of the surface state exciton. In order to confirm the presence
of this $\vec{a}$ polarized surface exciton state, we recorded the
excitation spectrum of the emission I ($\vec{b}$ polarized) for $\vec{a}$ and $\vec{b}$
polarized exciting light. We observed indeed, an important maximum
in the intensity of the emission I when the excitation is $\vec{a}$
polarized and centered on the energy (25 523 cm^{-1}) of the observed
$\vec{E}||\vec{a}$ reflectivity maximum. This means that in this case, we excite
preferentially the surface through its upper Davydov component and,
after relaxation between the two components, we observe finally
the emission of the lower surface state (see figure 3).

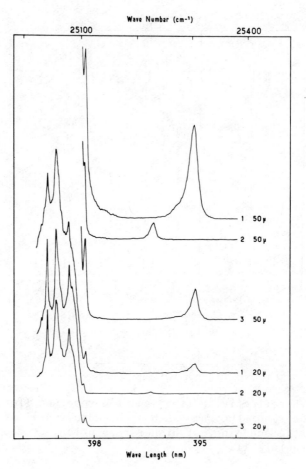

Fig.2 : $\vec{b}$ polarized fluorescence spectra of an anthracene
monocristal at 6 K for two monochromator slit widths
(50 µ and 20 µ).
1. before N_2 condensation on the $[\vec{a},\vec{b}]$ crystal surface
2. after N_2 condensation
3. after removal of the frozen N_2 layer by warming to
100 K in a stream of the gas

From these important results, we can deduce a surface Davydov
splitting of 221 cm^{-1} similar as theoretically expected (the inter-
planes exchange terms being negligible) to its bulk counterpart
(224 cm^{-1}) estimated from reflectivity (see figure 3).

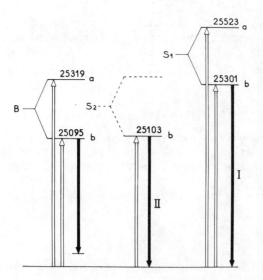

Fig.3 : Interpretation of the observed transitions in terms of
bulk (B), surface (S_1) and sub-surface exciton states
(vacuum values in cm^{-1}; a and b refer to the
polarization of the state; I and II to the structures
of figure 1).

THEORETICAL PART

        We established a quantum model of one and two-dimensional
rigid lattices of point dipoles (e.g. polymers, moleculars chains
and layers, crystal surfaces, ...) in order to derive their optical
properties. Using a two levels approximation for the molecules,
or the unit cells constituting the lattice, translational
invariance leads to define Frenkel excitons as the material
excited states involved in optical transitions. Both discrete
and finite aspects can be disregarded in this first step, because
the principal phenomena are due to dimensionality effects. As the
coupling of photons with excitons must also obey translational
symmetry in the lattice directions, the photon continuum to which
a given exciton is coupled has a dimension complementary to that
of the material lattice (instead of 3 for an isolated dipole) i.e.
1 for a surface exciton, 2 for a chain exciton. The interaction
hamiltonian has the standard form of a dipole field interaction
energy. $\vec{K}$ will denote the (1 or 2-dimensional) wave vector of the
exciton; the photons of the continuum are then defined by their
polarization and their wave vector $\vec{K} + \vec{q}$, with $\vec{q}.\vec{K} = 0$.
In a perturbation theory framework, we calculate the Green function
of excitons coupled to the field, which gives for instance the
eigenstates of the coupled system, the evolution of an initial
excited material state, etc... The shift operator (or effective-
mass operator) of the exciton has the following expression for

surface excitons (similar results hold for one-dimensional
excitons, with a convenient change of energy scale) :

$$R_S(z) = \Gamma_o \left[-\frac{(\hbar cK)^2}{z^2} \cos^2\theta - \frac{i\hbar\omega_o}{\sqrt{z^2-(\hbar cK)^2}} \left[1 - \left(\frac{\hbar cK}{z} \cos\theta\right)^2 \right] \right]$$

with $\Gamma_o = 2\pi f\alpha\hbar\omega_o$, $\alpha = e^2/\hbar c$, f is an oscillator factor, $\hbar\omega_o$ is
the excitonic transition dipole, θ the angle of the exciton wave
vector with the material elementary dipole of transition supposed
to lie in the lattice plane; z is the energy variable of the
Green function.
The eigen energies of the coupled system should satisfy the
relation :

$$z^2 - (\hbar\omega_o)^2 = 2\hbar\omega_o R_S(z) \tag{2}$$

They are found to be of two kinds according to their position with
respect to the photon continuum.
i) a radiatively stable branch of states, beneath the bottom of
the continuum ($z < \hbar cK$), which we called polariton states. For
$K \gg \omega_o/c$, this state is mainly an exciton, for $K \ll \omega_o/c$ it is
a photon like state, splitted off from the continuum bottom, and,
in this way, "trapped" in the lattice.
(ii) the continuum of photon states, with, when $\hbar cK < \hbar\omega_o$, a
resonance nearly at the energy $\hbar\omega_o$ of the unperturbed exciton.
This can be interpreted as a radiatively unstable branch of
excitons (as their energy is greater than $\hbar cK$, their decay
towards photons $\vec{K} + \vec{q}$ is allowed). These features are summarized
in figure 4, where, for each K-sub-surface, the projection of the
exciton state on the eigenstates is presented as a function of the
energy. The δ-peaks represent the discrete polariton states, while
the continuous part is the "contamination" of the photon continuum
by the exciton (or density of coupled-states). The Fourier transform of this distribution of states gives the probability of still
finding the exciton during its desintegration. At large times,
only the δ-peak contributes to this probability, exhibiting the
weight of the exciton in the polariton state (see figure 4).
The radiative widths and shifts of both branches of states are
given as functions of K on figures 5 and 6. The order of the
radiative decay time of an exciton is : $(f\alpha\omega_o)^{-1} \sim 10^{-12}$ s (for
anthracene parameters). This time, $(c/\omega_o a)$ shorter than that of
an isolated molecule (with a, a characteristic molecular
dimension) shows the cooperative character of the surface emission,
sometimes compared to superradiant behavior, with only one
excitation in the initial state. This very short time allows
the fluorescence channel to be in efficient competition with
other relaxation mechanisms in condensed media (for instance,
surface phonons or traps and relaxation to the bulk). The study

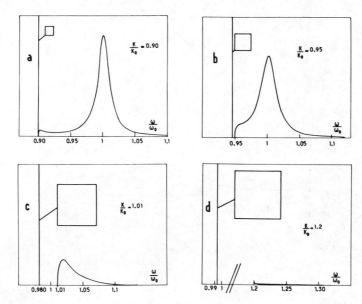

Fig.4 : Weight of the exciton state as a function of the energy
of the eigenstates of the total system, for different
values of the wave-vector. Contribution of the polariton-
· δ-peak is represented by the square area.

of a two-dimensional lattice provides a model of 3-dimensional
crystal : by piling up N such planes, we get a good simulation
of anthracene crystals, where interplane electron exchange is
negligible. Indeed, it is possible to derive the bulk properties
from those, previously obtained, of the surface, so getting more
accurate results than by assuming continuity approximations for
the medium. Another advantage of this approach is that it easily
accounts for intraplane relaxations by phonons or impurities,
which are dominant phenomena above the transition energy of the
real crystal. We are chiefly concerned, here, with the optical
properties of a surface layer coupled by radiation to a quasi-
resonant bulk (polyacene crystal). These properties can be drawn
from both the surface exciton shift operator (1) and from the
bulk reflectivity amplitude $r_B(\omega)$, given by a suitable model (for
instance the above mentioned model, including relaxation process,
or the semi-classical model, with an imaginary part for the dielectric). Then, if the system does not involve any other interaction between bulk and surface than the dipole-dipole retarded
interaction, we can partition in a simple way the total space of
states into surface and bulk states. The reflectivity amplitude
$r(\omega)$ of the total system becomes :

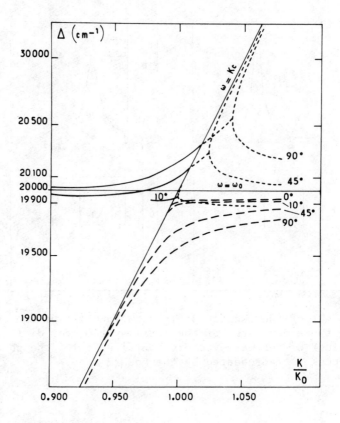

<u>Fig.5</u> : Dispersion curve of the two branches of states :
          - the upper branch with ω > cK is radiatively unstable
          - the lower branch (ω < cK) is stable.

$$r(\omega) = \frac{r_B(\omega)(\omega-\omega_o-\Delta-R_S(\omega)) + R_S(\omega)}{\omega-\omega_o-\Delta-R_S(\omega)(1+r_B(\omega))} \qquad (3)$$

h$(\omega_o+\Delta)$ is the surface transition energy, $r_B(\omega)$ is the bulk
reflectivity amplitude.
The emission from a surface excited state is proportional to :

$$\xi(\omega) = \left| \frac{1 + r_B(\omega)}{\omega-\omega_o-\Delta-R_S(\omega)(1+r_B(\omega))} \right|^2 \qquad (4)$$

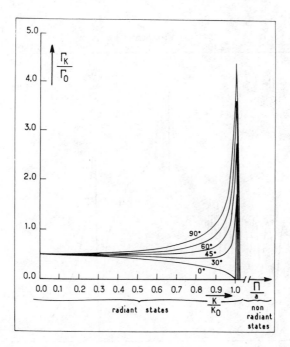

Fig.6 : Radiative width of the upper excitonic branch of the preceding diagram, as a function of $\frac{cK}{\omega}$, for different values of θ, in units of Γ_o ($K = 0$, $\theta^o = 0$). The eigen energy $h(\omega_o + \Delta_K - \frac{i}{2} \Gamma_K)$ loses its meaning in the region $\frac{cK}{\omega_o} \simeq 1$, where the resonance in the continuum (see Fig.4) is no longer lorentzian.

The widths of the structures represented by (3) and (4) are of the same order Γ_S :

$$\Gamma_S = 2 |R_S| (1 + Re(r_B(\omega_o + \Delta))) \qquad (5)$$

thus showing a renormalization of the surface radiative characteristics by the bulk.

The results agree fairly well with our experimental observations (see figure 1), to the extent that the reflection amplitude of the bulk is experimentally known. Near the bulk resonance energy, the reflection is "quasi-metallic" so that $r_B \simeq -1$ and the radiative decay rate (5) is severaly decreased for small Δ. This is the situation of the second and third surface layers, the radiative decays of which are slowed dawn by the bulk reflection, and thus may be quenched by non-radiative transitions. On the other

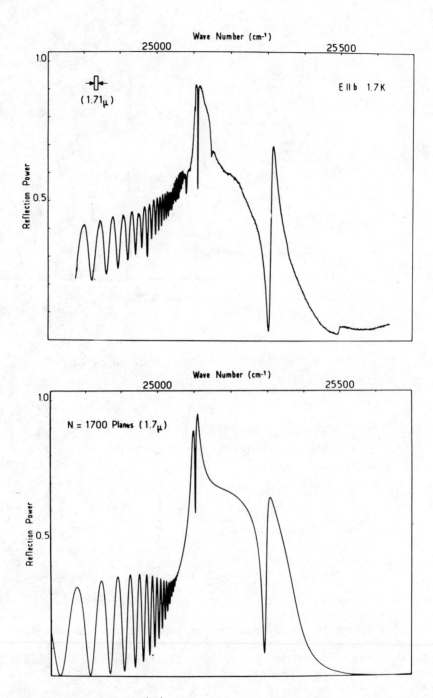

<u>Fig.7</u> : Experimental $\vec{E}\|\vec{b}$ reflection spectrum (upper curve) and
theoretical simulation for the same experimental conditions.

hand, the first surface layer, with $\Delta \sim 200$ cm^{-1}, lies in a region where Re $r_B(\omega_o + \Delta) \sim -0,5$: therefore, the radiative width, of order 10 cm^{-1} is thought to be the dominant broadening at least for the high-energy wing of its emission line. Last, the evolution of the surface structure under gas coating is in qualitative agreement with our considerations and with the variations of the reflectivity in the explored region (ω_o, $\omega_o + 200$ cm^{-1}). These theoretical calculations of the surface radiative couplings, to both free photons and bulk polaritons, are a pre-requisite for the evaluation, from the experimental data, of the non-radiative channels and of their dependence on parameters such as temperatures, gas coating and traps concentration.

REFERENCES

1. M.R. Philpott and J.M. Turlet, Journal of Chemical Physics 64, 3852 (1976).

2. J.M. Turlet, J. Bernard and Ph. Kottis, Chemical Physics Letters 59, 506 (1978).

3. M. Orrit, C. Aslangul and Ph. Kottis, submitted to Physical Review B.

4. See also references quoted in these three papers.

AUTHOR INDEX

SUBJECT INDEX